NUMERICAL METHODS FOR ENGINEERING APPLICATION

NUMERICAL METHODS FOR ENGINEERING APPLICATION

Second Edition

JOEL H. FERZIGER

A Wiley-Interscience Publication

JOHN WILEY & SONS, INC.

New York / Chichester / Weinheim / Brisbane / Singapore / Toronto

This book is printed on acid-free paper.

Published simultaneously in Canada.

Library of Congress Cataloging-in-Publication Data:
Ferziger, Joel H.
Numerical methods for engineering application / Joel H. Ferziger. –2nd ed.
p. cm.
"A Wiley-Interscience publication."
Includes bibliographical references and index.
ISBN 0-471-11621-1 (cloth : alk. paper)
1. Engineering mathematics. 2. Numerical analysis. I. Title.
TA335.F47 1998
515′.35–dc21 97-35864

Printed in the United States of America.

10 9 8 7 6 5 4 3 2 1

CONTENTS

PREFACE TO THE SECOND EDITION

It has been nearly 15 years since the first edition of this book appeared; that is far longer than an author should wait to publish a second edition. A lot has happened in computational engineering in that period. From the point of view of readers of this book some of the most important developments are the following:

- The development and widespread distribution of personal computers and workstations. Small computers of today have capabilities that rival (in both speed and memory) the largest computers of 15 or 20 years ago.
- Although networks existed 15 years ago, no one would have foreseen the possibility of the kind of communication that is now available to essentially everyone via the Internet. Indeed, the programs referenced in this text are available to the reader over the Internet.
- The availability of software packages. Software for solving an enormous range of problems has become available at modest prices to everyone. For the readers of this book this includes the availability of powerful general-purpose mathematics packages such as Mathematica, Maple, and Matlab. There are also specialized packages available for particular areas that are too numerous to mention here. The availability of plotting packages and the text formatting capability of LaTeX make the job of authorship much easier.
- New algorithms. The methods that are used to solve differential equations have improved considerably over the past several years. In particular, adaptive methods have found widespread application as have multigrid

and conjugate gradient methods. These were essentially in their infancy 15 years ago.

These developments have had a direct impact on this book in that it was written in LATEX and the figures were supplied in Postscript. The availability of modern text editors eased the problem of finding and removing errors, of which there were too many in the first edition. Naturally, I took advantage of these to (hopefully) make the writing clearer in a number of places. It is too much to hope that the number of errors has been reduced to zero, but I hope that none of them will prove disastrous.

Although these developments have meant that more engineers now use "off-the-shelf" software packages, it is still important to understand what is in them and how to use them properly. It is the author's experience that one can obtain this kind of understanding only by actually writing at least some simple programs and "playing" with the methods; one does not really understand a method until one has programmed it. I have met many people who use commercial packages and complain about their experience with them. It often turns out that they are either asking a program to do something it was not intended to do or that they are trying to solve a problem that does not have a solution (an ill-posed problem). A little background in the fundamentals of how the method works, what it is good for, and how it should be used would have helped these people enormously. I hope that this book provides some of that background.

The first edition contained only four chapters, the last of which was almost half of the total book. In an effort to make the chapters less unwieldy, the present edition contains eight chapters, some of which are still very long. The chapters in the present work are related to those of the original book in the following way:

- In the second edition, a new first chapter giving a few results and a short review of linear algebra has been added. This is intended strictly as a review and not as a textbook on the subject. The reader should have some background in linear algebra before tackling other areas of numerical analysis, and this chapter is intended to remind the reader that he or she should know this material in order to follow the remainder of the book.
- The original Chapters 1 and 2 have become Chapters 2 and 3 of the current version. They are only slightly changed from the old versions; in Chapter 3, a section on Monte Carlo integration has been added, and the discussion of adaptive quadrature has been expanded to set the stage for adaptive methods covered later. I debated whether to add material on the important subject of surface fitting (which is central to computer-aided design and manufacturing) but decided that justice could not be done to it in the space available.
- Chapter 3 of the first edition (on ordinary differential equations) has been broken into two chapters on initial and boundary value problems (Chap-

ters 4 and 5 of the present work). The first of these is changed a bit from the earlier version, mainly by the addition of more methods that have proven useful in recent years. On the other hand, Chapter 5 is a considerable expansion of the material on boundary value problems in the original Chapter 3 and many more examples are given.

- Chapter 4 of the original book (on partial differential equations) has been split into three chapters on parabolic, elliptic, and hyperbolic equations, Chapters 6, 7, and 8, respectively. The first of these is relatively unchanged. In Chapter 7, the reader will find new material about conjugate gradient, multigrid, and adaptive methods, which are some of the powerful techniques that have taken over the field in recent years. Chapter 8 has also been expanded to give an introduction to some of the methods that are currently being used to solve complex problems of this type.

The original edition had embedded in it print-outs of the codes used to generate the examples. This was the only reasonable way to transfer them to the reader in those days. The Internet has rendered that unnecessary, so the codes have been removed from the text and are available to the reader over the Internet (see Appendix A). Note that these codes are meant to illustrate the methods and are far from the most sophisticated implementations of the methods. I do hope, however, that they are transparent enough that the reader can see how the translation from the description on the printed page to the code was done. To help in this, I have added pseudo-programs in a few places (probably not enough).

Finally, no author can ever do the full job alone. I need to thank my colleagues Profs. Gene Golub and Joseph Oliger, who taught me a great deal about numerical methods; my co-author of another book, Prof. Milovan Perić, from whom I learned a lot and from whom I have borrowed some material; and my recent students who helped with proofreading and suggestions.

JOEL H. FERZIGER

Stanford, CA
June 1997

PREFACE TO THE FIRST EDITION

Many of the traditional fields of engineering have undergone enormous changes in the past 30 years. In place of the handbook correlation methods that were once commonly used, engineers find themselves going ever more frequently to methods based on fundamental principles and the equations derived from them. A major reason for this is that the rate of technological development has increased so rapidly that it is simply impossible to produce the data that traditional methods would require.

One of the key technological changes is the development of the electronic digital computer whose beginnings are only about 30 years old. The best machines of that day could hardly compete with the cheap and readily available hand calculator of today. We have come a long way and there is no reason to believe that the end is yet in sight. Over the past 30 years the capability of computers has increased by something like two orders of magnitude per decade and, while the cost of machines has gone up, it has not increased at anything like the rate at which their performance has. As a result, the effective cost of computation has decreased by more than an order of magnitude per decade. While we can hardly expect these trends to last forever, there is still no indication of a deviation from the curve of cost versus time and the trend can be expected to continue for at least another decade.

The total cost of doing computing is the sum of the hardware (machine) costs and the software (human) costs. In many applications, hardware costs have become a small part of the total and this is causing a kind of minirevolution in the computer business. In many applications of interest to engineers, however, hardware cost remains dominant. This means that the cost of doing computing will probably continue to decrease relative to the cost of other engineering functions. Consequently, we can anticipate a continuation of the trend toward the computerization of many traditional engineering tasks. This trend covers all areas of engineering, but not all to the same degree.

Not all engineers will use computers heavily in their work, but a large enough number will to render courses in the application of computers to the solution of engineering problems a necessary part of the modern engineering curriculum. Such courses are offered by mathematics and computer science departments and are appropriate for many engineering students. But there is a wide variety of students in engineering, each of whom has a different and legitimate set of needs. It is difficult to meet all of these needs with a single set of courses and there is a tendency toward the teaching of some mathematics and computer science courses by engineering departments.

The material in this book comes from a sequence of courses that the author has offered in the Mechanical Engineering Department at Stanford University and has proven to be of interest to students in related departments. As presently structured, there are four courses in the sequence: the first covers linear algebra, including its numerical aspects and its application to the solution of nonlinear problems; the second covers analytical methods for the solution of partial differential equations; the third deals with numerical analysis, especially those aspects related to the solution of ordinary and partial differential equations; and the fourth covers the application of numerical methods to fluid mechanics. Other courses in numerical methods, including some in the finite element method, are offered by other departments.

This book is based on the third and part of the fourth in the sequence of courses described above, principally the numerical solution of ordinary and partial differential equations. (Notes for the others are available from the author.) It is intended for people who will be using the methods rather than those who will be developing new methods. In practical terms, this means that the emphasis is on intuitive understanding of what makes a good numerical method and what distinguishes it from poorer methods, and not on the analysis of methods. In particular, the analysis of the stability and accuracy of particular methods, which rightly plays a central role in numerical analysis, is given less attention in this work than in standard numerical analysis texts. Correct results will be presented here, but the long analyses necessary to derive them are sometimes referenced rather than given. This is not to say that analysis is not important; it is. The audience to which this book is addressed, however, is one that will generally accept the result without seeing the proof, as long as it can be assured that the proof does exist.

The arrangement of the subjects in this work is a fairly standard one. Interpolation is treated first because it is the cornerstone on which other methods are based. Quadrature, a subject that finds considerable use by engineers, is treated next. This is followed by a chapter on numerical methods for ordinary differential equations. Initial value problems are the focus of the first half of this chapter and boundary value problems are dealt with in the second half. The book closes with a chapter on partial differential equations.

The approach is one in which each subject is built on the subject(s) preceding it. This seems to be a near optimum approach, as it allows the student to see the interconnections between methods that sometimes appear disjoint. It is also simpler to grasp the entire subject when the framework is visible.

The point of view taken is that only "good" methods (defined as those that are efficient and well used) should be presented. Better a few pearls than a beach of sand. For this reason, this book does not contain a compendium of all known methods. It is devoted to a relatively few methods chosen either for the pedagogical reason that they are easy to understand and analyze or because they are commonly used. There are, of course, other good methods that could have been included and were not. The methods described are as good as any.

The author would like to end this preface by giving thanks to the people without whom this work could not have been completed. I learned much of what I know about numerical analysis (especially in how to think about it) from my good friend Dr. Samuel Schechter; Professors Joseph Oliger and Gene Golub were also important contributors in this regard. Help in assembling the examples was very cheerfully provided by Sr. Juan Bardina and is greatly appreciated; and a number of his students and colleagues, too numerous to mention by name, also played an important role in suggesting changes and correcting errors in the text. Ms. Ann Ibaraki did service beyond the call of her job in typing and assembling the original version. Later versions were typed by Mss. Ruth Korb, Romen Rey, and Charlean Hampton. Finally, thanks are due to my colleague, friend, and department head Professor Bill Reynolds for his encouragement.

JOEL H. FERZIGER

Stanford University
February 1981

NUMERICAL METHODS FOR ENGINEERING APPLICATION

CHAPTER 1

SHORT REVIEW OF LINEAR ALGEBRA

1.1. INTRODUCTION AND NOTATION

Although this book is primarily concerned with the solution of differential equations (ordinary and partial), any method for solving differential equations on a computer requires reduction of the problem to one that can be solved in a finite number of operations. In practice, this means that the problem must be approximated by a finite set of algebraic equations. Thus linear algebra is basic to the computer solution of any kind of problem. For this reason, a short review of linear algebra is in order. This chapter is not intended as a tutorial in linear algebra for those not already familiar with the subject; those who have no experience with linear algebra should consult a text on the topic such as the one by Strang (1988). The numerical aspects of linear algebra can be found in the recent work of Trefethen and Bau (1996).

Only those parts of linear algebra having direct bearing on the methods covered in the later chapters of this book will be discussed here; in practice, this means that this chapter is concerned only with numerical methods for solving systems of linear equations. Much of the material that is usually considered to be the heart of linear algebra, namely the theory of vector spaces, is not dealt with at all.

We begin with a short overview of the notation used for systems of linear equations. For small systems (those with few variables) one may use a notation that assigns a different name (e.g., x, y, z) to each variable and to each coefficient (a, b, c, etc.). That notation becomes unwieldy for more than a few equations so it is advantageous to use index notation for both the variables and the coefficients:

$$
\begin{aligned}
a_{11}x_1 + a_{12}x_2 + \cdots + a_{1n}x_n &= b_1 \\
a_{21}x_1 + a_{22}x_2 + \cdots + a_{2n}x_n &= b_2 \\
\vdots \quad + \quad \vdots \quad + \ddots + \quad \vdots \quad &= \vdots \\
a_{n1}x_1 + a_{n2}x_2 + \cdots + a_{nn}x_n &= b_n
\end{aligned}
\tag{1.1}
$$

This set of equations can be written in the more compact form:

$$
\sum_{j=1}^{n} a_{ij}x_j = b_i \qquad i = 1,2,\ldots,n \tag{1.2}
$$

The ordering of the equations and variables may be arbitrary, but, in many cases, there is a logical reason for numbering them in a particular way.

It is even more convenient to regard the elements of the solution $(x_1, x_2, \ldots, x_n)$ and those of the forcing terms $(b_1, b_2, \ldots, b_n)$ as elements of n-component vectors:

$$
\mathbf{x} = \begin{pmatrix} x_1 \\ x_2 \\ \vdots \\ x_n \end{pmatrix} \qquad \mathbf{b} = \begin{pmatrix} b_1 \\ b_2 \\ \vdots \\ b_n \end{pmatrix} \tag{1.3}
$$

and consider these vectors as entities in their own right. The coefficients, each of which has two indices, may be regarded as the elements of a two-dimensional array or matrix:

$$
A = \begin{pmatrix} a_{11} & a_{12} & \cdots & a_{1n} \\ a_{21} & a_{22} & \cdots & a_{2n} \\ \vdots & \vdots & \cdots & \vdots \\ a_{n1} & a_{n2} & \cdots & a_{nn} \end{pmatrix} \tag{1.4}
$$

The system of equations can then be written in the compact form:

$$
A\mathbf{x} = \mathbf{b} \tag{1.5}
$$

Comparison of Eqs. (1.2) and (1.5) gives the definition of the product of a matrix and a vector. The form (1.5) allows the system of equations to be given geometric interpretations that are interesting and generate insight into the nature of the solutions; for these, the reader is again referred to texts on linear algebra.

The solution of systems of equations of the type just described is the main issue of this chapter. We shall be interested only in systems in which the numbers of unknowns and equations are equal. There are interesting geometric interpretations of the equations, but we shall not go into that here.

1.2. GAUSS ELIMINATION AND *LU* DECOMPOSITION

In the majority of cases of interest, the system of equations (1.5) has a unique solution. This solution may be formally written:

$$\mathbf{x} = A^{-1}\mathbf{b} \tag{1.6}$$

where A^{-1} is called the matrix inverse of A (division is undefined for matrices). In practice, it is rarely economical to compute the inverse of a matrix. When A has an inverse, it is said to be nonsingular, and Eq. (1.6) represents the unique solution for any right-hand side $\mathbf{b}$.

When most of the elements of the matrix are nonzero, the matrix is said to be full. The most direct method of finding the solution to a full system of equations is Gauss elimination or one of the variations on it. We now present the simplest version of Gauss elimination; it is nothing more than a systematic procedure for the elimination of variables from the equations. A FORTRAN realization of this procedure, called GAUSS, together with an example of its use is available on the Internet site mentioned in Appendix A.

The method begins by eliminating x_1 from the second equation. This is accomplished by multiplying the first equation by a_{21}/a_{11} and subtracting it from the second equation. In a similar way, x_1 can be eliminated from equations $3, 4, \ldots, n$; for a generic equation, say the jth, the first equation multiplied is a_{j1}/a_{11} and subtracted from the jth. When this process has been applied to Eqs. $2, 3, \ldots, n$, none of these equations contain x_1 any longer, that is, they are a system of $n-1$ equations for the $n-1$ variables, $x_2, x_3, \ldots, x_n$. This smaller system can be treated in exactly the way the original one was, reducing it to a system of $n-2$ equations. The process is repeated until, finally, the last equation contains only the variable x_n. The matrix of the resulting system has the form:

$$U = \begin{pmatrix} a_{11} & a_{12} & a_{13} & \cdots & a_{1n} \\ 0 & a_{22} & a_{23} & \cdots & a_{2n} \\ 0 & 0 & a_{33} & \cdots & a_{3n} \\ \vdots & \vdots & \vdots & \ddots & \vdots \\ 0 & 0 & 0 & \cdots & a_{nn} \end{pmatrix} \tag{1.7}$$

where the name U signifies that the matrix is upper triangular, that is, it is a matrix with no nonzero elements below the main diagonal. None of the elements of this matrix, other than those in the first row, are identical to the original ones. In a computer, the elements of the modified matrix may be stored in the same locations as the original elements as the latter are no longer needed.

After the forward elimination process is complete, the last equation is easily solved for $x_n = b_n/a_{nn}$. Having x_n, it is not difficult to solve the next to last equation for x_{n-1} and, by repetition of this process, we can solve for all of the variables. We give the formula from which x_i is computed:

$$x_i = \frac{b_i - \sum_{j=i+1}^{n} a_{ij} x_j}{a_{ii}} \tag{1.8}$$

It may be shown that the entire process requires approximately $n^3/3$ each of additions (or subtractions) and multiplications. The largest part of this cost is associated with the forward elimination portion of the method; the backward substitution requires only about $n^2/2$ operations of each type. Since the method requires division by the diagonal elements of the matrix, a_{ii} (the pivots), it will fail if any of these elements should happen to become zero during the process. For a nonsingular matrix, this problem can be overcome by exchanging the equation in which the zero pivot occurs with any equation below it. Fortunately, the systems of equations we shall encounter in later chapters are almost always nonsingular.

For general systems of algebraic equations, that is, ones not connected to differential equations, the Gauss elimination method can produce large numerical errors. These errors arise because some of the diagonal elements (the pivots) can become very small during the elimination process and, when they are used as divisors, round-off errors are amplified. This problem can be cured by interchanging the equation containing the small pivot with one that has a larger element in the same column, that is, by the use of pivoting. Fortunately, pivoting is rarely needed for the equations that will be encountered in this book and we shall say no more about it.

A variation on Gauss elimination is the Gauss–Jordan method. The forward elimination part of this method is identical to the Gauss elimination method described above; the difference lies in what happens afterward. To begin the second part of the Gauss–Jordan method, instead of solving the last equation, we divide it by a_{nn} (which amounts to the same thing). Then we eliminate the element $a_{n,n-1}$ by multiplying the last equation by it and subtracting the result from equation $n-1$. The same process is then used to eliminate $a_{n-2,n}, a_{n-3,n}, \ldots, a_{1,n}$. The first $n-1$ equations then have the triangular form that the set of n equations had before the backward procedure was begun. The same process is then applied to this smaller equation set.

At the end of the procedure, the matrix has been replaced by the identity matrix (a matrix with ones on its principal diagonal and zeros everywhere else) and the original forcing vector **b** has been replaced by the solution vector **x**.

Another important variant of Gauss elimination is *LU* decomposition. The Gauss elimination process described above may also be used to demonstrate that any matrix that does not produce zero pivots during the process may be factored into the product of a lower and an upper triangular matrix:

$$A = LU \tag{1.9}$$

Indeed, the upper triangular matrix is exactly the one that results from Gauss elimination. The lower triangular matrix L is also easily constructed. Its diagonal elements, l_{ii}, are all unity while the off-diagonal elements are the factors used in the Gauss elimination process. As a result, the factors L and U can be found by applying Gauss elimination to the matrix (the vector **b** is not required at this point).

Having the factorization, one can write the system of equations (1.5) as

$$LU\mathbf{x} = \mathbf{b} \tag{1.10}$$

which may be broken into two systems of equations:

$$L\mathbf{y} = \mathbf{b} \tag{1.11}$$

$$U\mathbf{x} = \mathbf{y} \tag{1.12}$$

The matrix of each of these systems is triangular and may be solved by the method used in the back substitution portion of Gauss elimination or the Gauss–Jordan process. The only significant difference is that the solution of the first system must start with the first equation and proceed in the downward direction.

The cost of solving a system of equations by *LU* decomposition is identical to the cost of Gauss elimination. The advantage of the *LU* method is that, once the factorization has been performed, the solution of a system of equations costs only n^2 operations. More importantly, the solution of many systems containing the same matrix need not be carried out simultaneously. As we shall see later, there are times when this can be a great advantage.

We close this section with a short discussion of an important problem associated with the numerical solution of large systems of linear algebraic equations—ill-conditioning. When a system containing an ill-conditioned matrix is solved, the result is very sensitive to small changes in the elements of the matrix or the forcing vector; the two types of modifications of the system are in fact equivalent. Errors resulting from round-off (required by the finite size with which a computer represents a number) may also be interpreted as changes in the coefficients or forcing vector, and so they, too, can make the results unreliable. Consequently, the results may depend on the computer used, the compiler, the programming, and other variables that would not normally be expected to have an effect. Ill-conditioning is difficult to deal with. The best cures are to try to rewrite the equations in a way that reduces the ill-conditioning of the matrix (preconditioning) or to use higher numerical precision (e.g., double-precision arithmetic) in computing the solution.

The error in the solution may be bounded by introducing a condition number $C(A)$:

$$\frac{\|\delta x\|}{\|x\|} \leq C(A)\frac{\|\delta b\|}{\|b\|} \tag{1.13}$$

where $\|\cdot\|$ represents the norm (length) of a vector. As noted above, errors in the elements of the matrix or round-off treated as contributions to δb. It may be shown that an estimate of the condition number is

$$C(A) = \|A\| \cdot \|A^{-1}\| \tag{1.14}$$

which shows that an ill-conditioned matrix is one that is nearly singular. In practice, it may be difficult to distinguish between a matrix that is singular and one that is ill-conditioned.

Program 1.1: Gauss Elimination It is sometimes difficult for readers to understand how one proceeds from a description of a method to an actual program. As an aid to doing so, we shall include "pseudo-programs" (set of statements that can be translated into actual programs) for some of the key algorithms in this book. These pseudo-programs are written in a MATLAB-like script and should be relatively easy to understand; a percent sign indicates a comment. Loops are started with `for` and finish with `end` statements. Indenting is used for clarity. Actual programs in FORTRAN77 can be found on the Internet site described in Appendix A.

We start with the Gauss elimination algorithm without pivoting.

```
\% forward elimination
for i = 1:n
  for j = i+1:n
  for k = i+1:n
    a(j,k) = a(j,k) - a(j,i) * a(i,k) / a(i,i)
  end
    b(j) = b(j) - a(j,i) * b(i) / a(i,i)
  end
end
\% back substitution
x(n) = b(n)/a(n,n)
```

```
for i = 1:n-1
  t = b(n-i)
  for j = 1:i
    t = t - a(n-i,n-i+j) * x(n-i+j)
  end
  x(n-i) = t / a(n-i,n-i)
end
```

Note that the elimination part of the algorithm contains three nested loops; this is why the number of operations scales as n^3. The back substitution portion contains only two nest loops and its cost scales as n^2.

Example 1.1: Solution of a System of Equations Throughout this book we shall present examples to illustrate the methods. Let us apply the Gauss elimination algorithm to the solution of the system of equations:

$$x_1 + 2x_2 - x_3 = 2$$
$$2x_1 - x_2 + x_3 = 1$$
$$x_1 + x_2 - 2x_3 = 2$$

This data was input to the code that called the subroutine GAUSS and the following output was obtained:

$$1.0000 \quad 0.0000 \quad -1.0000$$

1.3. TRIDIAGONAL AND OTHER BANDED SYSTEMS

The systems of algebraic equations obtained by discretizing differential equations are rarely full. Indeed, they are often not only far from full but have a simple structure. Among the simplest such systems are the ones arising from the boundary value problems associated with second-order ordinary differential equations. They often lead to systems of equations with tridiagonal matrices, ones with no nonzero elements located more than one element away from the main diagonal. Such a matrix has the form:

$$A = \begin{pmatrix} a_{11} & a_{12} & 0 & 0 & \cdots & 0 \\ a_{21} & a_{22} & a_{23} & 0 & \cdots & 0 \\ 0 & a_{32} & a_{33} & a_{34} & \cdots & 0 \\ \vdots & \vdots & \ddots & \ddots & \vdots & \vdots \\ \vdots & \vdots & \vdots & \ddots & \ddots & \vdots \\ 0 & 0 & 0 & 0 & a_{n-1,n} & a_{nn} \end{pmatrix} \tag{1.15}$$

The existence of only three diagonals with nonzero elements allows a simpler notation to be used. We can write A as

$$A = \begin{pmatrix} b_1 & c_1 & 0 & 0 & \cdots & 0 & 0 \\ a_2 & b_2 & c_2 & 0 & \cdots & 0 & 0 \\ 0 & a_3 & b_3 & c_3 & \cdots & 0 & 0 \\ \vdots & \vdots & \ddots & \ddots & \ddots & \vdots & \vdots \\ \vdots & \vdots & \vdots & \ddots & \ddots & \ddots & \vdots \\ 0 & 0 & 0 & 0 & \cdots & b_{n-1} & c_{n-1} \\ 0 & 0 & 0 & 0 & \cdots & a_n & b_n \end{pmatrix} \tag{1.16}$$

This change in notation allows the matrix to be stored as three arrays of length n rather than a single array of size n^2, an enormous advantage if n is large. The name of the forcing vector must be changed when this notation is used, but that is a trivial matter.

To solve tridiagonal systems, we again use Gauss elimination, but the cost can be reduced considerably by taking advantage of the structure of the matrix. In the forward elimination part of the Gauss elimination, only the coefficient a_{i+1} (which multiplies x_i) needs to be eliminated from equation $(i + 1)$; the only coefficient that is changed in this equation is b_{i+1}; the element of the forcing vector on the right hand side (r_{i+1}) of the equation is changed as well. Furthermore, there are no nonzero elements in the ith column below the $(i + 1)$st row, that is, variable x_i does not appear in any equation beyond the $(i + 1)$st. These modifications mean that the cost of solving a tridiagonal system is proportional to n rather than to n^3. A code, TRDIAG, for solving tridiagonal systems is found on the Internet site given in Appendix A; a pseudo-code version is given below. Finally, we note that the tridiagonal (or Thomas) algorithm is so efficient that there is no reason to use *LU* decomposition for tridiagonal systems.

A somewhat more difficult case is that of the pentadiagonal matrix, one which has two nonzero diagonals on either side of the main diagonal. Like the tridiagonal system, it is solved by a simplified version of Gauss elimination. In this case, two elements have to be modified in the row below the pivot element and one element of the second row below the pivot is also modified. The modifications cause the cost of solution to be about three times that of solving a tridiagonal system.

When periodic boundary conditions are applied to a problem, the tridiagonal matrix is replaced by a periodic tridiagonal matrix. This type of matrix is identical to the tridiagonal one, except that there are extra elements in the top-right and lower-left corners. The extra effort required to deal with these elements increases the solution cost to double that of the tridiagonal system.

Program 1.2: Tridiagonal Systems As mentioned above, the algorithm for solving tridiagonal systems is simpler than full Gauss elimination. As pseudo-code version of it follows:

```
\% forward elimination
for i = 1:n
  b(i+1) = b(i+1) - a(i+1) * c(i) / b(i)
  r(1+1) = r(i+1) - a(i+1) * r(i) / b(i)
end
\% back substitution
x(n) = b(n)/a(n,n)
for i = 1:n-1
  x(n-i) = (r(n-i) - c(n-i) * x(n-i+1) / b(n-i)
end
```

Note that both the forward elimination and back substitution are accomplished with a single loop, indicating that the number of operations is proportional to n. The version of the algorithm just given modifies the diagonal elements of the matrix. With a small modification, it is possible to leave all of the elements of the matrix unchanged; this is useful when a number of systems with the same matrix need to be solved in succession, a situation that arises quite commonly in the solution of partial differential equations.

Example 1.2: Solution of Tridiagonal System Now let us solve the system of equations:

$$-2x_1 + x_2 = -1$$

$$x_1 - 2x_2 + x_3 = 0$$

$$x_2 - 2x_3 = -1$$

using the tridiagonal subroutine. The result is

1.0000 1.0000 1.0000

1.4. BLOCK SYSTEMS

Still another important special case, one that arises from both systems of ordinary differential equations and from some solution methods for partial differential equations is the block tridiagonal system. The matrix of such a system may be represented:

$$L = \begin{pmatrix} B_1 & C_1 & 0 & 0 & \cdots & 0 & 0 \\ A_2 & B_2 & C_2 & 0 & \cdots & 0 & 0 \\ 0 & A_3 & B_3 & C_3 & \cdots & 0 & 0 \\ \vdots & \vdots & \ddots & \ddots & \ddots & \vdots & \vdots \\ \vdots & \vdots & \vdots & \ddots & \ddots & \ddots & \vdots \\ 0 & 0 & 0 & 0 & \cdots & B_{n-1} & C_{n-1} \\ 0 & 0 & 0 & 0 & \cdots & A_n & B_n \end{pmatrix} \tag{1.17}$$

where A_i, B_i, and C_i are themselves $m \times m$ matrices, where m is usually small. In a similar way, the solution vector $\mathbf{x}$ and the forcing vector $\mathbf{b}$ become vectors whose elements are themselves vectors:

$$\mathbf{x} = \begin{pmatrix} \mathbf{x}_1 \\ \mathbf{x}_2 \\ \vdots \\ \mathbf{x}_n \end{pmatrix} \qquad \mathbf{b} = \begin{pmatrix} \mathbf{b}_1 \\ \mathbf{b}_2 \\ \vdots \\ \mathbf{b}_n \end{pmatrix} \tag{1.18}$$

The solution method mimics the one used for ordinary tridiagonal systems. The difference is that where we divided by a pivot element, b_i, we must now multiply by the inverse of the pivot matrix, B_i. This is the formal description of the method. In practice, inverse matrices are never computed. Instead, we begin by treating the first block of equations, that is, the first m equations as a separate system and use the Gauss–Jordan method to replace the matrix B_1 by the identity matrix. In the process, the matrix C_1 is replaced by $B_1^{-1}C_1$, and the first forcing block vector is replaced by $B_1^{-1}\mathbf{b}_1$.

The first block is then multiplied by A_2 and subtracted from the second block; this replaces A_2 by a zero matrix. Again, this is the formal procedure. In practice, a specialized version of the Gauss elimination procedure, one that takes advantage of the fact that B_1 has been reduced to the identity matrix is used. After A_2 has been eliminated, the last $n - 1$ blocks are a block tridiagonal system of equations to which the procedure just outlined may be applied anew.

At the end of the forward elimination, the system is upper block bidiagonal with the diagonal blocks being identity matrices. The last block is easily solved for $\mathbf{x}_n$ and, having $\mathbf{x}_n$, we substitute it into the $(n - 1)$st block of equations and solve the latter for $\mathbf{x}_{n-1}$. By continuing in this way, all of the elements of the solution may be found.

The cost of solving a block tridiagonal system is proportional to m^3n, where m is the size of an individual block and n is the number of blocks in the system. It is clear that, if the blocks are large, this method can be rather expensive. A realization of the algorithm just described is found in the BLKTRI code found at the Internet site given in Appendix A.

1.5. EIGENVALUES

In many applications one has to deal with matrices that depend on a parameter. The simplest possibility is that the matrix is square and the dependence on the parameter (say λ) is linear. The canonical example of such a matrix is $A - \lambda I$, but the more general case $A - \lambda B$ is also of interest. The latter can be dealt with in much the same way as the former, so we follow the traditional approach of considering the canonical case.

In general, the dependence of the matrix $A - \lambda I$ on the parameter λ is smooth and, for most values of λ, the matrix is nonsingular. There are, however, values of λ for which the matrix becomes singular. In this case, we must have

$$\det(A - \lambda I) = 0 \tag{1.19}$$

This equation is called the characteristic equation of the matrix. Its left-hand side can be shown to be a polynomial of degree n in λ so the equation has exactly n roots; these values of λ are called the *eigenvalues* of the matrix A. For the more general matrix, $A - \lambda B$, they are called generalized eigenvalues.

One may show, by recursively using the development in minors of a determinant, that the determinant in (1.19) is a polynomial of degree n in the parameter λ, where n is the size of the matrix. According to the fundamental theorem of algebra, this polynomial has exactly n roots that may be real or complex (which must occur as conjugate pairs if the matrix is real). Corresponding to each distinct eigenvalue, there is at least one solution of the system:

$$(A - \lambda I)\mathbf{x} = 0 \tag{1.20}$$

or

$$A\mathbf{x} = \lambda \mathbf{x} \tag{1.21}$$

The solution to this system is called an *eigenvector* of the matrix A.

If the number of eigenvectors is equal to the size of the matrix (this is guaranteed if the eigenvalues are all different or if the matrix is symmetric and may be so in other cases), one can construct a square matrix whose columns are the eigenvectors:

$$S = (\mathbf{x}_1, \mathbf{x}_2, \ldots, \mathbf{x}_n) \tag{1.22}$$

and it is not difficult to show that

$$S^{-1}AS = \Lambda \tag{1.23}$$

where Λ is a diagonal matrix whose elements are the eigenvalues of A (in the order corresponding to the ordering of the eigenvectors in S). This is known as a diagonalizing transformation.

An important special case is that of symmetric matrices, that is, matrices for which

$$a_{ij} = a_{ji} \tag{1.24}$$

or

$$A = A^T \tag{1.25}$$

where the superscript T denotes the transposed matrix, the one obtained from A by interchanging the rows and columns. For these, the eigenvalues are guaranteed to be real and the eigenvectors can be chosen to be orthogonal. (Eigenvectors belonging to different eigenvalues are definitely orthogonal; those belonging to multiple eigenvalues can be orthogonalized.) Furthermore, the normalization of eigenvectors is arbitrary (any multiple of an eigenvector is also an eigenvector).

The matrix S that diagonalizes a symmetric matrix therefore has the property that its columns are mutually orthogonal and they may also be normalized. For such a matrix, it may be shown that the rows are also orthonormal and, more importantly, that its inverse is simply its transpose. Matrices with these properties are called orthogonal and are usually represented by the symbol Q. We thus have

$$Q^{-1} = Q^T \tag{1.26}$$

and transformations of the type:

$$A' = Q^T A Q = Q^{-1} A Q \tag{1.27}$$

are called orthogonal transformations. Putting these results together, we see that a symmetric matrix can be diagonalized by an orthogonal transformation with the matrix being the special one whose columns are the normalized eigenvectors of the matrix.

The determination of the eigenvalues and eigenvectors of symmetric matrices is usually accomplished by making a sequence of orthogonal transformations (which is itself an orthogonal transformation); these methods are iterative; each iteration is designed to bring the matrix closer to a diagonal state. For nonsymmetric matrices, the procedure is similar but a bit more more complicated and more expensive.

The details of how eigenvalues and eigenvectors are computed will not be described here but may be found in the standard texts on linear algebra. Computer routines based on these procedures are contained in linear algebra packages such as LINPACK or LAPACK that are found on many machines.

They are also available in commercial mathematical packages such as Mathematica and Matlab.

Finally, we mention that ill-conditioning, which was discussed above in connection with the solution of linear systems of equations, is related to the eigenvalues of the matrix. An alternative definition of the condition number is

$$C(A) = \lambda_{\max}/\lambda_{\min} \tag{1.28}$$

where $\lambda_{\max}$ and $\lambda_{\min}$ are the largest and smallest magnitudes of the eigenvalues of the matrix, respectively. Note that a singular matrix has at least one zero eigenvalue and possesses no inverse; it therefore has an infinite condition number according to either definition. A system of differential equations that gives rise to an ill-conditioned system of algebraic equations is said to be *stiff*.

CHAPTER 2

INTERPOLATION

Interpolation is the process of "reading between the lines" of a table or of fitting a smooth curve to a limited set of data. We take it up here for a number of reasons, only one of which is that interpolation is frequently used to estimate quantities from tabulated values. A more important application is to numerical differentiation and integration, as we shall see in the following chapters.

There are two principal kinds of interpolation, depending on the type of data provided and the kind of result desired. In standard interpolation, a set of data points is given and a curve that passes smoothly through them is required. In the other type of interpolation, the data may have uncertainty associated with them, and the task is to find a smooth curve that passes sufficiently close to the data points; the method of least squares is the tool most commonly used for this purpose. In standard interpolation, the equation defining the approximating curve must have as many parameters as there are data points; in least-squares fitting, the number of parameters is typically much smaller than the number of data points. We will treat only the standard case in this chapter. Least-squares fitting is equally important, especially for experimentalists but is not included here principally because it does not relate directly to the topics covered in later chapters.

The central problem of interpolation may be stated as follows: Given a set of data points (x_i, y_i), $i = 1, 2, \ldots n$, find a smooth function $f(x)$ that passes through each of them. The points at which the data are given are usually called *knots*. In this kind of interpolation, the following properties are required of the interpolation function:

- From the requirement that the curve fit the data, we must have

$$f(x_i) = y_i \qquad i = 1,2,\ldots,n \tag{2.1}$$

- The function should be easy to evaluate.
- It should be easy to integrate and differentiate.
- It should be linear in the adjustable parameters (to simplify the problem of finding them).

The choice of interpolating function depends on how smoothness is defined and on the nature of the data to be fitted. Many kinds of functions have been used as interpolants; the most common ones are polynomials because they satisfy the second and third criteria better than any other class of function. There are many types of polynomial interpolation; we begin with the simplest type—Lagrange interpolation. The approach presented is an intuitive one rather than a formal one; as noted above, the main purpose is to set the stage for the methods to be presented in the following chapters.

2.1. LAGRANGE INTERPOLATION

2.1.1. Theory

Lagrange interpolation passes a polynomial of lowest possible degree through the n given data points. Since n parameters are needed to fit the data, the lowest order polynomial capable of performing the task is of degree $n-1$:

$$f(x) = a_{n-1}x^{n-1} + a_{n-2}x^{n-2} + \cdots + a_1 x + a_0 \tag{2.2}$$

The straightforward approach to finding the coefficients of this polynomial is to substitute Eq. (2.2) into Eq. (2.1) to obtain

$$a_{n-1}x_i^{n-1} + a_{n-2}x_i^{n-2} + \cdots + a_1 x_i + a_0 = y_i \qquad i = 1,2,\ldots,n \tag{2.3}$$

which can be regarded as a set of n linear algebraic equations for the n unknown coefficients, $a_0, a_1, \ldots, a_{n-1}$, since x_i and y_i are given. This set of equations can be solved by the Gauss elimination or *LU* decomposition methods given in Chapter 1, but this is not a good approach for two reasons. The primary reason is that the equations become very ill-conditioned for n larger than about 5, making accurate solution of them very difficult; a short discussion of ill-conditioning was given in Chapter 1. A second reason is that a closed-form expression for the interpolation function will be preferred for some of the applications that follow.

There is another approach that produces a closed-form solution for the interpolation polynomial. Since Eqs. (2.3) are a set of linear algebraic equations for the coefficients $a_0, a_1, \ldots, a_{n-1}$, these coefficients must be linear combinations of the y_i. The most general expression that is both linear in each of the y_i and a polynomial of degree $n-1$ in x is

$$f(x) = \sum_{k=1}^{n} y_k L_k(x) \tag{2.4}$$

where the $L_k(x)$ are themselves polynomials of degree $n-1$. Furthermore, from linear algebra, we know that the polynomials $L_k(x)$ depend on the abscissas x_i and not on the y_i. (If this were not true, the right-hand side of Eq. (2.4) would not be linear in y_k.

The problem is then to find the $L_i(x)$. It is important to note that they do not depend on the data, y_i. Consequently, we may choose the y_i to simplify the problem. A useful choice is to let one y_i, say the one with $i = j$, be unity while all of the others are zero. Then substitution of Eq. (2.4) into Eq. (2.1) gives

$$L_j(x_i) = \delta_{ij} \qquad i = 1, 2, \ldots n \tag{2.5}$$

where δ_{ij} is the Kronecker symbol:

$$\delta_{ij} = \begin{cases} 1 & i = j \\ 0 & i \neq j \end{cases} \tag{2.6}$$

In Eq. (2.5), j is arbitrary, so this equation must hold for every value of j. Thus $L_j(x)$ is a polynomial of degree $n-1$ that is zero when $x = x_1, x_2, \ldots x_{j-1}, x_{j+1}, \ldots$, or x_n and unity when $x = x_j$. Now, a fundamental result of algebra shows that any polynomial of degree n can be factored into a constant multiple of a product of n factors $(x - x_l)$, where the x_l are the zeros of the polynomial. Since $L_j(x)$ is a polynomial of degree $n-1$ whose zeros are all known and are given above, it must have the form

$$L_j(x) = C_j(x - x_1)\cdots(x - x_{j-1})(x - x_{j+1})\cdots(x - x_n) \tag{2.7}$$

where C_j is a constant whose value is determined by requiring that $L_j(x_j) = 1$; with a simple calculation, we find

$$C_j = \frac{1}{(x_j - x_1)(x_j - x_2)\cdots(x_j - x_{j-1})(x_j - x_{j+1})\cdots(x_j - x_n)} \tag{2.8}$$

So L_j is uniquely determined:

$$L_j(x) = \frac{(x - x_1)\cdots(x - x_{j-1})(x - x_{j+1})\cdots(x - x_n)}{(x_j - x_1)\cdots(x_j - x_{j-1})(x_j - x_{j+1})\cdots(x_j - x_n)} \tag{2.9}$$

For later applications it is convenient to introduce the polynomial of degree n:

$$F(x) = (x - x_1)(x - x_2) \cdots (x - x_{n-1})(x - x_n) \tag{2.10}$$

In terms of this polynomial,

$$L_j(x) = C_j F(x)/(x - x_j) \tag{2.11}$$

which, with Eq. (2.4), provides the desired interpolation.

2.1.2. Error Analysis

Any numerical method produces an approximation to the quantity we wish to compute; rarely is the exact value needed, but there is almost always a minimum acceptable tolerance. To assure that it is attained, it is important to be able to determine the error in the numerical approximation. As Lagrange interpolation is the first numerical method to be described in this work and because it underlies much of what will be done later, we shall present the error analysis for it in some detail.

To determine the error in Lagrange interpolation, suppose that the given data are the exact values of some smooth function $y(x)$ at the points $x_1, x_2, \ldots, x_n$. If $f(x)$ is the Lagrange interpolation polynomial, then, $f(x) - y(x)$ is a function that is zero at each of the n data points. Furthermore, $F(x)$, which is defined by Eq. (2.10), is a polynomial that is zero at each of these points. Now consider the function

$$g(x) = y(x) - f(x) - AF(x) \tag{2.12}$$

where A is a constant that will be determined later; whatever the value of A, the function $g(x)$ is zero at each of the data points. We may choose A so that $g(x) = 0$ at some additional point x_0 in the interval, $x_1 < x_0 < x_n$. Then $g(x)$ has at least $n + 1$ zeros at $x_0, x_1, \ldots, x_n$, all lying in the interpolation interval. Since g is smooth and bounded, it must have a minimum or maximum between each pair of zeros. Therefore, within the interpolation interval, $g'(x)$ has at least n zeros, $g''(x)$ has at least $n - 1$ zeros, and, by continuation, $g^{(n)}(x)$ has at least one zero in the interval. Let ξ be this zero, so that $g^{(n)}(\xi) = 0$. Since f is a polynomial of degree $n - 1$, $f^{(n)} = 0$. Also, by differentiating Eq. (2.10), we find $F^{(n)} = n!$ so that

$$g^{(n)}(\xi) = y^{(n)}(\xi) - An! = 0 \tag{2.13}$$

Thus, solving for A, we have

$$A = \frac{y^{(n)}(\xi)}{n!} \tag{2.14}$$

and substituting into Eq. (2.12), we find

$$y(x) = f(x) - \frac{y^{(n)}(\xi)}{n!} F(x) \tag{2.15}$$

where $x_1 < \xi < x_n$. The last term is the desired error estimate for Lagrange interpolation.

This estimate demonstrates some important properties of Lagrange interpolation. As might be expected, the larger $y^{(n)}(\xi)$, that is, the less smooth the function, the greater the error in the interpolation. Also, we see from the definition of $F(x)$ that the error is greater for wider spacing between the data points and near the endpoints, x_1 and x_n, than near the center of the range. Finally, if we use Eq. (2.4) for extrapolation, the errors can become very large. All of this accords with our intuition.

Later in this section, we shall give a number of examples that illustrate the properties of Lagrange interpolation. The examples are chosen to illustrate some important properties of the method; the results will later be compared with the results produced by other methods. In the examples, simple functions will be used to allow easy computation of the actual error. Before doing so, we give an easier method of computing the Lagrange interpolation of a function.

2.1.3. Divided Differences

In practice, the computation of the interpolated values is rarely done by evaluating the polynomial using the explicit formulas given above, although this method can be used for low-order interpolation. The method of divided differences, originally developed by Newton, is more efficient for this purpose. We now present it without development. The reader should verify, at least for a low-order case, that the results it produces are identical to what the polynomial gives.

Divided differences are computed as follows. Given the values of the independent and dependent variables $x_i, y_i = f(x_i)$; $i = 1,2,\ldots,n$, we first compute the first divided differences:

$$y[i-1,i] = \frac{y_i - y_{i-1}}{x_i - x_{i-1}} \qquad i = 2,3,\ldots,n \tag{2.16}$$

which, as we shall see later, are also approximations to the first derivative of y. Then we compute the second divided differences:

$$y[i-2,i-1,i] = \frac{y[i-1,i] - y[i-2,i-1]}{x_i - x_{i-2}} \qquad i = 3,4,\ldots,n \tag{2.17}$$

The process is continued using the formula:

$$y[i-j,\ldots,i-1,i] = \frac{y[i-j+1,\ldots,i] - y[i-j,\ldots,i-1]}{x_i - x_{i-j}}$$

$$i = j+1, j+2, \ldots, n \qquad (2.18)$$

until the nth divided difference $y[1,2,\ldots,n]$ has been obtained. It is impossible to continue further. The results can be presented as a table but there is no need for it here.

The interpolant is then the polynomial:

$$\begin{aligned} y(x) = {} & y[1] + (x-x_1)y[1,2] + (x-x_1)(x-x_2)y[1,2,3] \\ & + \cdots + (x-x_1)\cdots(x-x_{n-1})y[1,2,\ldots,n] \end{aligned} \qquad (2.19)$$

where $y[1] = y_1$.

The divided difference method, which is more efficient than the method given earlier, was used to generate the results presented in the examples below. The FORTRAN program, called DIVDIFF, which was used to generate the results for one of the functions used in the examples, can be found on the Internet site described in Appendix A. Results for other examples were generated by simply changing the function. The algorithm consists of a part that computes the divided differences, noting that, according to Eq. (2.19), only those with first index equal to 1 need to be retained; then the polynomial is evaluated for a particular value of x, noting that each polynomial is the preceding one multiplied by $x - x_{k-1}$; finally the sum in Eq. (2.19) is computed.

We now proceed to the examples.

2.1.4. Examples

We give a number of examples designed to illustrate the most important properties of Lagrange (or, indeed, any other type of) interpolation, beginning with a smooth function and then going on to successively more difficult functions.

Example 2.1: Lagrange Interpolation of a Smooth Function For the first example we interpolate $y = e^x$ on the interval $0 < x < 1$. This is a smooth function that should be easily fit by any interpolation method. If we use only three points, the calculation can be done by hand. With $x_1 = 0$, $x_2 = 0.5$, and $x_3 = 1$, we have

$$\begin{aligned} f(x) \approx {} & f(x_1)\frac{(x-x_2)(x-x_3)}{(x_1-x_2)(x_1-x_3)} + f(x_2)\frac{(x-x_1)(x-x_3)}{(x_2-x_1)(x_2-x_3)} \\ & + f(x_3)\frac{(x-x_1)(x-x_2)}{(x_3-x_1)(x_3-x_2)} \end{aligned} \qquad (2.20)$$

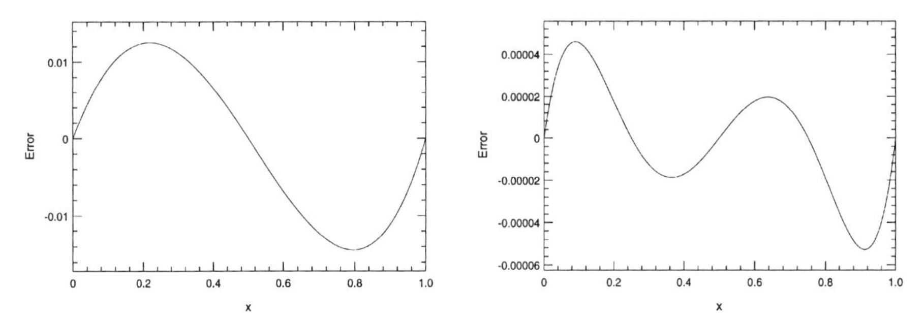

FIGURE 2.1. Error in Lagrange interpolation of e^x using three and five points.

Evaluating this at $x = 0.25$, we have

$$f(0.25) \approx (1)\frac{(-0.25)(-0.75)}{(-0.5)(-1)} + (1.648721)\frac{(0.25)(-0.75)}{(1)(-0.5)}$$
$$+ (2.718282)\frac{(0.25)(-0.25)}{(1)(0.5)}$$
$$= 0.375000 + 1.236541 - 0.339785 = 1.271756$$

The exact value of the function is $e^{0.25} = 1.2840254$ so that the error is 0.0123 or just about 1%. This must be considered rather good given the coarseness of the interpolation.

For this function, the interpolation is sufficiently accurate that graphical presentation of the interpolant and the exact function would not show the differences very clearly. Instead, the error (defined as difference between the exact function and the interpolant) over the full range $0 < x < 1$ is shown in the left side of Figure 2.1. As expected, the error changes sign at the data points (at which it must be zero) and therefore oscillates. It is relatively evenly distributed over the range and is largest about halfway between the data points. These observations are fairly typical.

Before considering a case with a larger number of points, let us see how Lagrange interpolation works as an extrapolator. At $x = -0.25$, we find $f(x) = 0.83345$ compared to an exact value of $e^{-0.25} = 0.77880$. The error is 0.055, more than four times what it was at $x = 0.25$ on an absolute basis and much worse in terms of relative error. Thus, as expected, we see that *extrapolation is much more prone to error than interpolation.*

Interpolating e^x using five-point Lagrange interpolation produces the results shown on the right side of Figure 2.1. As might be expected, the errors are much smaller. As before, they oscillate, and the largest errors are found near the ends of the interval. This accords with expectation.

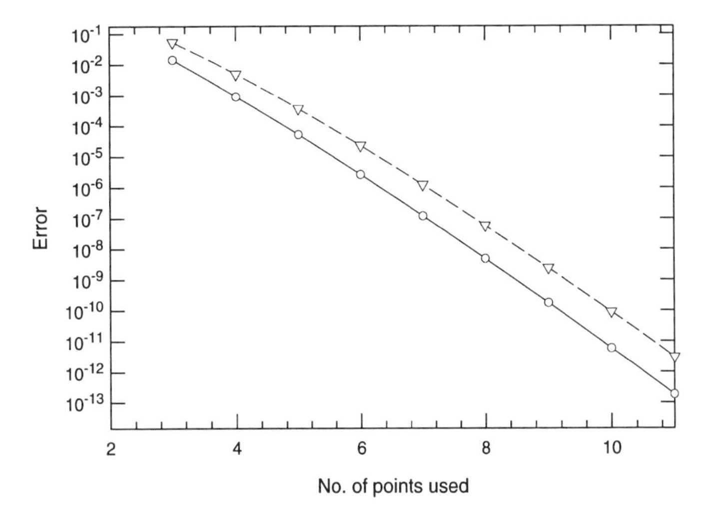

FIGURE 2.2. Maximum error in Lagrange interpolation of e^x versus number of points used. The dashed line is the estimate obtained from Eq. (2.15).

It is interesting to compare the error estimate [Eq. (2.15)] with the actual error. Two difficulties arise in making such a comparison. The first is that we don't know the value of ξ that makes Eq. (2.15) exact. The safest approach is to choose the value that gives the largest error estimate; in this case, this means that we must choose $\xi = 1$. A more serious problem is that the nth derivative of the function required is often difficult (or even impossible) to calculate. In this example, the difficulty disappears; due to the simplicity of the function, its derivatives are easily computed. The maximum error in Lagrange interpolation of e^x on $0 \leq x \leq 1$ as a function of the number of data points and the estimate found from Eq. (2.15) are shown in Figure 2.2; we see that the formula overestimates the actual error; in fact, the estimate is nearly double the actual error. This is not too bad; error estimates of this kind are often much worse than this. In computing the actual errors, double precision arithmetic was used. Had single precision arithmetic been used, the error would level out in the range 10^{-7} to 10^{-6}. The machines used in these calculations (and all others in this book) were personal computers with Intel chips. When run with FORTRAN, the relative precision of these machines is about 10^{-7}, meaning that a number is normally represented with a precision of about seven significant figures and an error smaller than this cannot be produced. This accuracy is typical of a wide range of machines. Using more points increases the number of numerical operations, increasing the round-off error. It is important to emphasize that one should never use a method of accuracy greater than that of the computer used; the cost will increase with no gain in accuracy.

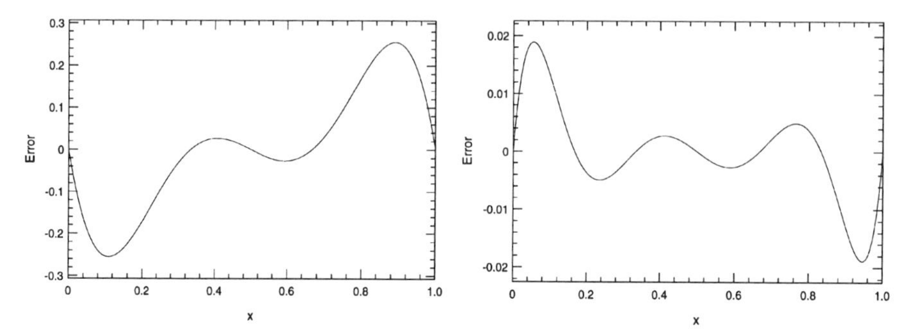

FIGURE 2.3. Error in Lagrange interpolation of $\sin 2\pi x$ using four and seven points.

The combination of higher precision arithmetic and an accurate algorithm is the best way to achieve high accuracy. However, increasing the precision also increases the cost of a calculation; this is rarely important in something as simple as interpolation but can become significant in more complex problems. It was used here to demonstrate the behavior of the method, but high precision computation is needed only occasionally in engineering work; its use should be reserved for those cases in which it is really necessary. Double precision can be often used to determine whether a bad result is due to round-off accumulation, the design of the algorithm, or the programming.

Example 2.2: Lagrange Interpolation of an Oscillating Function As a second case, we take the function $\sin 2\pi x$ on the interval $0 < x < 1$. We wish to determine how many points are required to describe a full period of this function accurately. If only two or three points are used, the values on which the interpolation is based would all be zero, and the interpolation would be meaningless so we must use more points.

Figure 2.3 shows the error obtained using four and seven points, respectively. We see that the maximum error is much larger than for the exponential function; it also decreases much less rapidly as the number of points increases.

The behavior of the maximum error as a function of the number of data points in Figure 2.4. We see that, as expected, the maximum error is much larger than for the exponential function and that it decreases more slowly with increasing number of points (note the different scales in Figures 2.2 and 2.4). The decrease of the maximum error is not as smooth as it is for the exponential function; the reason is that the choice of the data points on the sine curve is very important. This is what one expects intuitively—wiggly functions are much more difficult to fit with polynomials than smooth ones. In particular, to reduce the error to 10% requires at least 6 points and, for 1% accuracy, a minimum of 10 points is needed.

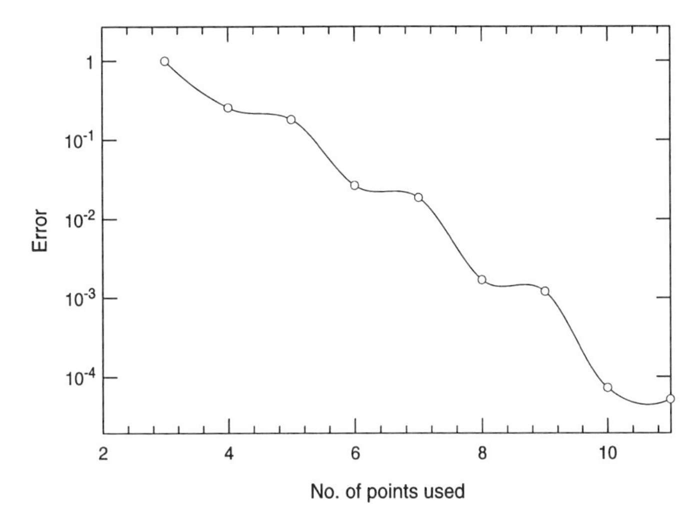

FIGURE 2.4. Maximum error in Lagrange interpolation of $\sin 2\pi x$ versus number of points used.

As comparison of these two examples shows, a method can be quite good in one case just as poor in another. One of the keys to good numerical work is choosing a method appropriate to the problem.

Example 2.3: Lagrange Interpolation of Difficult Functions This example presents another type of function that poses difficulty for interpolation procedures: ones with concentrated curvature and/or large slopes. As our example we take the so-called superellipse:

$$y(x) = (1 - x^m)^{1/m} \tag{2.21}$$

For $m = 2$ the curve is a circle; for large values of m, it is almost a square. Superellipses are sometimes used to fair between a circular cross section and a square one. These curves have infinite slope at $x = 1$, which is impossible for a polynomial to reproduce. The results for a circle ($m = 2$) are shown in Figure 2.5. The error is again much larger than it was for the exponential function and is concentrated near $x = 1$, as is to be expected from the nature of the function. Also, the maximum error does not fall off very rapidly as the number of points is increased as can be seen in Figure 2.6. The reason is that as more data points near $x = 1$ are included, the polynomial is asked to deal with ever larger slopes in this vicinity, something it does not do very well.

The case of $m = 4$ is shown in Figures 2.7 and 2.8. The results are similar to those obtained for $m = 2$, but much more exaggerated. In particular, the error is even more concentrated where the curvature is large and it falls off more slowly with increasing number of points.

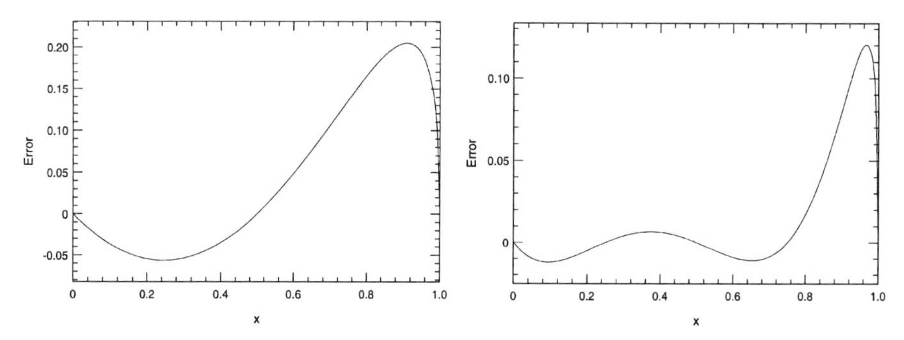

FIGURE 2.5. Error in Lagrange interpolation of $(1 - x^2)^{1/2}$ using three and five points.

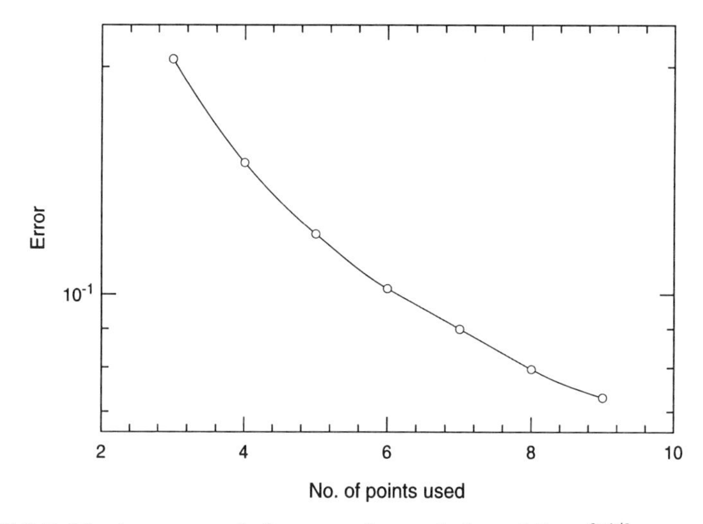

FIGURE 2.6. Maximum error in Lagrange interpolation of $(1 - x^2)^{1/2}$ versus number of points used.

It might be anticipated that since the error is concentrated near $x = 1$, improvement could be obtained by using unequal spacing of the mesh points with the points more closely spaced in the neighborhood of $x = 1$. This is the case, but the improvement is not dramatic. A later example will demonstrate this more clearly.

2.1.5. Piecewise Polynomial Interpolation

A simple method of approximating functions is piecewise interpolation. Instead of fitting a single high-order polynomial to all of the data, we fit low-order polynomials to portions or subsets of the data. This method provides

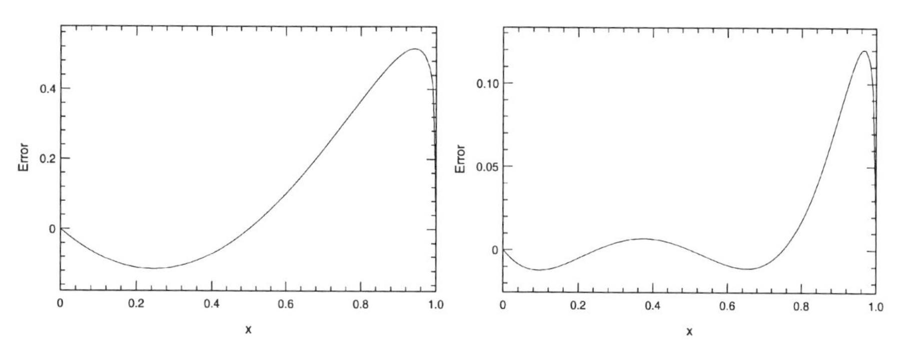

FIGURE 2.7. Error in Lagrange interpolation of $(1-x^4)^{1/4}$ using three and five points.

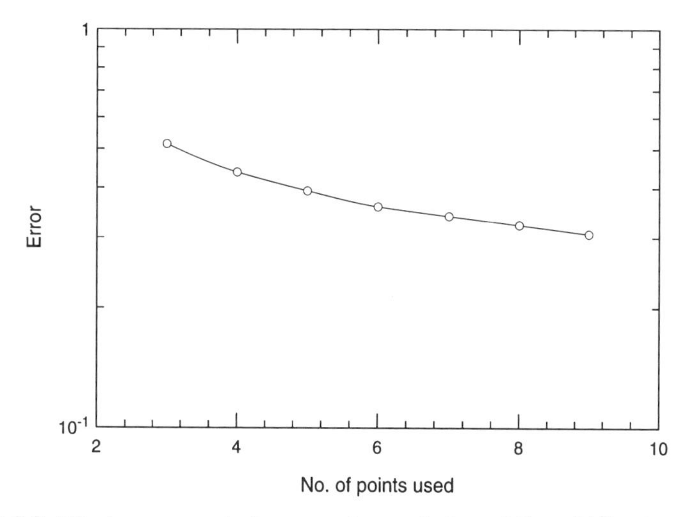

FIGURE 2.8. Maximum error in Lagrange interpolation of $(1-x^4)^{1/4}$ using 10 points.

considerable flexibility since it is easy to use narrow spacing of the data points in regions where the function has large curvature or high slope. The results can be quite satisfactory.

The simplest example of piecewise Lagrange interpolation is the familiar piecewise linear (the well-known "connect-the-dots") method. As we shall see later, piecewise Lagrange interpolation (linear and higher order) forms the basis for a number of important numerical methods.

Example 2.4: Piecewise Linear Interpolation Let us apply piecewise linear interpolation to the first two functions (the exponential and the sine) of the above examples; 10 equally spaced intervals (11 data points or knots) are used to interpolate these functions for $0 \leq x \leq 1$. Figure 2.9 shows the error in the interpolation of the exponential. To understand this result, we note that, by def-

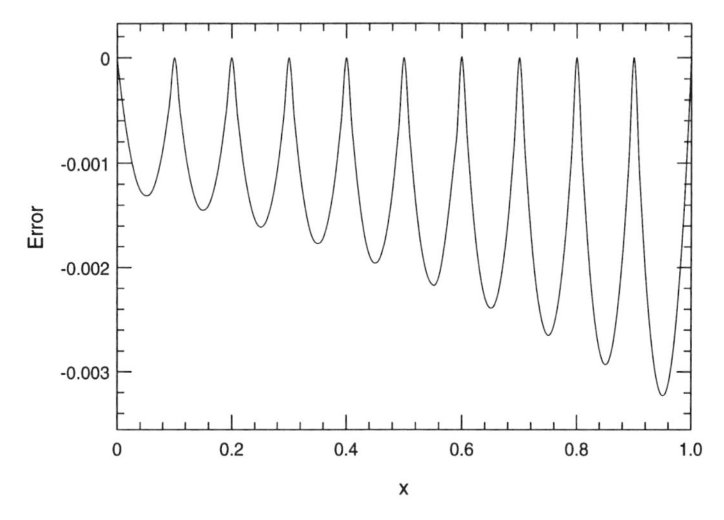

FIGURE 2.9. Error in piecewise linear interpolation of e^x using 11 points.

inition, the interpolant is exact at each of the knots, $0, 0.1, 0.2, \ldots, 1$. Between them, it is not difficult to show that the error has an approximately parabolic shape and the maximum error midway between knots $i-1$ and i is approximately $(x_i - x_{i-1})^2 f''(x_{i-1})/8$. Note that, as the number of knots increases, the error decreases more slowly than the error of the Lagrange method. On the other hand, the amount of computation required to evaluate the interpolant is much smaller for this method. With 10 points, the maximum error of the piecewise method is about 0.5% or about three orders of magnitude greater than the error in tenth-order Lagrange interpolation.

As expected, the error in interpolating the sine function, shown in Figure 2.10, is much larger than for the exponential. The ratio of the maximum error to that for tenth-order Lagrange interpolation is about two orders of magnitude.

Although piecewise interpolation gives larger errors than Lagrange interpolation for a given number of data points, it has a number of significant advantages. As noted, less computation is needed. More, importantly, if more accuracy is required, it is easy to insert new data points between the original ones; it is not necessary to recompute any of the values outside the affected region. This is something that will be used to advantage in constructing adaptive methods in later chapters. One disadvantage of piecewise linear interpolation is that the derivative of the interpolant is discontinuous at the knots; this can be cured by using piecewise quadratic (or higher order) interpolation.

2.1.6. Summary

These examples show a number of important characteristics of polynomial interpolation:

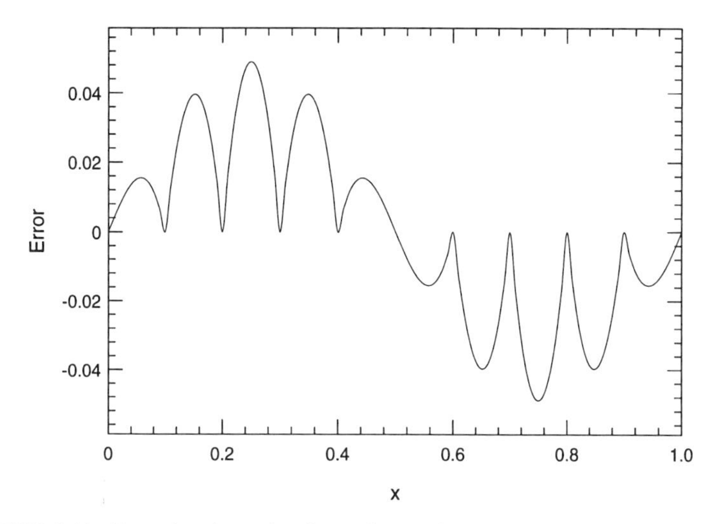

FIGURE 2.10. Error in piecewise linear interpolation of $\sin 2\pi x$ using 11 points.

- Lagrange interpolants tend to oscillate about the exact function.
- Smooth functions are interpolated more accurately than ones that oscillate or have concentrated curvature.
- The decrease in the error as the number of points and order of the polynomial increase depends strongly on the nature of the function as well as the interpolation method.
- Extrapolation yields much larger errors than interpolation.

Lagrange interpolation with more than three or four points is only rarely used due to its cost and inflexibility. Piecewise Lagrange interpolation offers some improvement but suffers from having discontinuous derivatives at the joints between the segments, which may cause trouble if the result is to be differentiated. The methods described in the following sections attempt to overcome these difficulties.

Low-order Lagrange interpolation is often used by itself. High-order methods are not used much, but they do form the basis for other methods; this is the principal reason why we have discussed them here.

2.2. HERMITE INTERPOLATION

There are occasions when smoothness greater than that provided by Lagrange interpolation is required. Additional smoothness can be obtained in a number of ways. A simple one is to force the interpolant to fit not only the function y_i at each point, but the derivative y_i' as well. This approach is called Hermite interpolation. Since derivative data are usually not available, this method is not often used as an interpolation technique. It does, however, provide the basis for important numerical techniques (Gauss quadrature, in particular) and is presented here for this reason.

The problem may be stated as follows. Given y_i and y_i' at n data points $x_1, x_2, \ldots, x_n$, find a polynomial of degree $2n-1$ that fits all the data. The method of constructing the polynomial is similar to the one used for Lagrange interpolation, so it will be described only briefly. The polynomial must be linear in all the data (y_i and y_i'), so the interpolant has the form:

$$f(x) = \sum_{k=1}^{n} U_k(x) y_k + \sum_{k=1}^{n} V_k(x) y_k' \tag{2.22}$$

where $U_k(x)$ and $V_k(x)$ are polynomials of degree $2n-1$ with the following properties:

$$\begin{aligned} U_k(x_j) &= \delta_{jk} \qquad & U_k'(x_j) &= 0 \\ V_k(x_j) &= 0 \qquad & V_k'(x_j) &= \delta_{jk} \end{aligned} \tag{2.23}$$

The required polynomials can be constructed by using the properties of the Lagrange polynomial and the results may be expressed in terms of $L_k(x)$. The results are:

$$\begin{aligned} U_k(x) &= [1 - 2L_k'(x_k)(x - x_k)]L_k^2(x) \\ V_k(x) &= (x - x_k)L_k^2(x) \end{aligned} \tag{2.24}$$

Finally, the error can be estimated by

$$y(x) = f(x) + \frac{y^{(2n)}(\xi)}{(2n)!} F^2(x) \tag{2.25}$$

where ξ again is some point in the interval $x_1 < \xi < x_n$.

The properties of Hermite interpolation are generally similar to those of Lagrange interpolation. Hermite interpolation produces smoother and more accurate results for a given number of points but it also suffers from some of the same difficulties, that is, it tends to produce oscillating errors, it may perform poorly for oscillating functions or ones with large derivatives, and it does not extrapolate well.

2.3. SPLINES

2.3.1. Definition and Development

Some of the disadvantages of Lagrange and Hermite interpolation are that they can be difficult to compute (especially when the number of data points is large), they are rather inflexible (adding points requires recomputing everything), and the piecewise versions have discontinuous derivatives at the knots. They are useful for deriving numerical methods, but as curve fits they are generally less than satisfactory.

For curve-fitting purposes, we would like to create a smooth curve similar to one a draftsperson might produce. An old drafting method uses a spline—a thin, flexible strip of material (usually wood) that can be bent to fit the given data points. The spline is made to fit the data by holding the drafting board vertically and placing weights on the spline. From mechanics, we know that the shape of the spline is a solution of the differential equation:

$$EI\frac{d^4y}{dx^4} = F(x) \tag{2.26}$$

where E and I are Young's modulus of the material and the moment of inertia of a cross section (these are not important for our purposes) and F is the applied force. Since the force is applied only at discrete points, $F = 0$ between the weights so the curve is cubic between each pair of weights.

Since the weights are idealized so that they act at single points and their integrated effect is finite, the force is actually undefined at the knots. Technically, it must be regarded as a multiple of a Dirac delta function, a function defined so that:

$$\delta(x) = 0 \qquad x \neq 0 \tag{2.27}$$

but

$$\int \delta(x)\,dx = 1 \tag{2.28}$$

where the interval of integration can be arbitrary so long as it includes $x = 0$. Furthermore, if $f(x)$ is any smooth function,

$$\int f(x)\,\delta(x)\,dx = f(0) \tag{2.29}$$

and, by a change of variable it is easy to see that

$$\int f(x)\,\delta(x-a)\,dx = f(a) \tag{2.30}$$

Now if, near the node at x_n, $F(x)$ in Eq. (2.26) is $F\delta(x - x_0)$, then integration of that equation from a point just to the left of the node to one just to its right shows that the third derivative of y jumps by an amount F at the node. When the third derivative is integrated, we will find that the second derivative has a cusp—a point at which the function is continuous but the derivative is not. The first derivative and the function itself are also continuous at these nodes.

Thus we define a cubic spline interpolating curve by the following criteria:

- The curve is piecewise cubic, that is, the coefficients are different on each interval (x_i, x_{i+1}).
- The curve passes through the given data points, (x_i, y_i), $i = 1, 2, \ldots, n$.
- The first and second derivatives are continuous at the node points x_i. The endpoints require special treatment, which we will discuss below.

We begin by noting that the criteria given above require that $f''(x)$ be piecewise linear and continuous. Thus if the values of the second derivative are known at the node points, we can use the linear Lagrange formula with $n = 1$ to compute $f''(x)$:

$$f_i''(x) = f''(x_i)\frac{x_{i+1} - x}{x_{i+1} - x_i} + f''(x_{i+1})\frac{x - x_i}{x_{i+1} - x_i} \tag{2.31}$$

The subscript on f'' on the left-hand side indicates that this formula holds only for $x_i \leq x \leq x_{i+1}$. Equation (2.31) may be integrated twice to give a cubic containing two constants of integration. These constants may be evaluated by requiring the function to fit the data, that is, we enforce $f_i(x_i) = y_i$ and $f_i(x_{i+1}) = y_{i+1}$. After some straightforward but tedious calculation, we find

$$\begin{aligned} f_i(x) = {} & f''(x_i)\frac{(x_{i+1} - x)^3}{6\Delta_i} + f''(x_{i+1})\frac{(x - x_i)^3}{6\Delta_i} \\ & + \left[\frac{y_i}{\Delta_i} - \frac{\Delta_i}{6}f''(x_i)\right](x_{i+1} - x) + \left[\frac{y_{i+1}}{\Delta_i} - \frac{\Delta_i}{6}f''(x_{i+1})\right](x - x_i) \end{aligned} \tag{2.32}$$

where $\Delta_i = x_{i+1} - x_i$. If the $f''(x_i)$ can be determined, everything on the right-hand side of Eq. (6.72) is known and the interpolation function can be evaluated.

The criteria that define the spline that were given above are already satisfied with the exception of continuity of the first derivative. We must therefore use this condition to determine the $f''(x_i)$. Continuity of f' is enforced by carrying out the following operations:

- Differentiate $f_i(x)$ and set $x = x_{i+1}$; that is, evaluate the derivative of f at the right end of interval i.
- Differentiate $f_{i+1}(x)$ and set $x = x_{i+1}$; that is, evaluate the derivative of f at the left end of interval $i + 1$.
- Equate the results of the two computations.

The result of these operations is the set of equations:

$$\frac{\Delta_{i-1}}{6} f''(x_{i-1}) + \frac{(\Delta_{i-1} + \Delta_i)}{6} f''(x_i) + \frac{\Delta_i}{3} f''(x_{i+1}) = \frac{y_{i+1} - y_i}{\Delta_i} - \frac{y_i - y_{i-1}}{\Delta_{i-1}} \tag{2.33}$$

This is a set of linear algebraic equations for the as-yet unknown second derivatives, $f''(x_i)$. Since the set of equations (2.33) is tridiagonal, that is, the ith equation contains only $f''(x_{i-1})$, $f''(x_i)$, and $f''(x_{i+1})$, it may be solved by the method given in Chapter 1; the cost of solution is proportional to the number of points used and so is quite reasonable.

The remaining problem is that, for $i = 1$ or $i = n$, continuity of f' is not applicable, so Eqs. (2.33) represent only $n - 2$ equations for the n unknown second derivatives, and two further equations are needed. These must be based on assumptions or approximations applied at the endpoints. There are many choices, all of which have been used in applications. Some of the better end conditions are:

1. **Periodic Spline.** This assumption is useful when the data are part of a periodic curve. Periodicity requires that the following conditions be applied:

$$f''(x_1) = f''(x_{n-1}) \qquad f''(x_2) = f''(x_n) \tag{2.34}$$

   This modifies the tridiagonal nature of the system of equations, and a special method must be used to solve the system of equations (2.33).

2. **Parabolic Run-out.** We set

$$f''(x_1) = f''(x_2) \qquad f''(x_n) = f''(x_{n-1}) \tag{2.35}$$

   which makes f'' constant on both end intervals; f is therefore quadratic on the end intervals.

3. **Free End.** If the end of the draftsperson's spline is free, there is no moment at the end and the appropriate condition is

$$f''(x_1) = f''(x_n) = 0 \tag{2.36}$$

   This gives the so-called natural spline—the curve that minimizes the total curvature of the spline. Physically, it corresponds to pinning the ends; these are called simple supports in beam theory.

4. **Cantilever End.** A combination of conditions 2 and 3 gives the following condition:

$$f''(x_1) = \lambda f''(x_2) \qquad f''(x_n) = \lambda f''(x_{n-1}) \qquad 0 \leq \lambda \leq 1 \tag{2.37}$$

which represents a cantilevered beam. This leaves the problem of picking λ; $\lambda = 0$ corresponds to condition 3 and $\lambda = 1$ to condition 2. The usual choices lie between these two values.

5. **Mixed End Condition.** Still another choice is given by Forsythe, Malcolm, and Moler (1977), who suggest passing a Lagrange cubic through the first four data points and then matching the third derivative of this cubic to that of the spline at the endpoint. This makes the end condition:

$$-\Delta_1 f''(x_1) + \Delta_1 f''(x_2) = \alpha_1 y_1 + \alpha_2 y_2 + \alpha_3 y_3 + \alpha_4 y_4 \qquad (2.38)$$

where the α_i are the coefficients of the difference formula. For a uniform mesh, $\alpha_1 = -\alpha_4 = (\frac{1}{6}\Delta)$ and $\alpha_2 = -\alpha_3 = (-\frac{1}{2}\Delta)$. A similar treatment is used at the other end.

For equally spaced intervals the system of equations (2.33) becomes

$$\tfrac{1}{6}[f''(x_{i-1}) + 4f''(x_i) + f''(x_{i+1})] = \frac{y_{i+1} - 2y_i + y_{i-1}}{\Delta^2} \qquad (2.39)$$

One can show that for this case the error is given by

$$f(x) - y(x) = \frac{\Delta^4}{96} y^{\text{iv}}_{\text{max}} \qquad (2.40)$$

which is similar to the error in cubic Lagrange interpolation. The advantages and disadvantages of the spline are best illustrated by examples.

2.3.2. Examples and Programs

Example 2.5: A Simple Test This example uses a method of investigating numerical methods that will be used a number of times in this book. By considering the simplest nontrivial case, it is possible to gain insight into the behavior of a method. For spline interpolation, the simplest case that illustrates the important features has four data points. This gives one segment in which the function is cubic and two end intervals. For further simplicity, we assume that the data points are equally spaced so that Eq. (2.39) can be used. For $i = 2$, we have

$$f''(x_1) + 4f''(x_2) + f''(x_3) = \frac{6}{\Delta^2}(y_1 - 2y_2 + y_3) \qquad (2.41)$$

If we abbreviate $f''(x_i) = f_i''$ and use the end condition (2.37), this becomes

$$(4 + \lambda)f_2'' + f_3'' = \frac{6}{\Delta^2}(y_1 - 2y_2 + y_3) \qquad (2.42)$$

Similarly, for $i = 3$,

$$f_2'' + (4 + \lambda) f_3'' = \frac{6}{\Delta^2}(y_2 - 2y_3 + y_4) \tag{2.43}$$

These equations are readily solved for f_2'' and f_3''. To make things still simpler, we will compute the function at the midpoint $x = (x_2 + x_3)/2$ for which Eq. (6.72) gives

$$f\left(\frac{x_2 + x_3}{2}\right) = \frac{1}{2}(y_2 + y_3) - \frac{\Delta^2}{16}(f_2'' + f_3'') \tag{2.44}$$

The first term on the right-hand side is the result produced by linear interpolation between the two central data points. The last term is a correction that can be evaluated from Eqs. (2.42) and (2.43); the result is

$$f\left(\frac{x_2 + x_3}{2}\right) = \left[\frac{1}{2} + \frac{3}{8(5 + \lambda)}\right](y_2 + y_3) - \frac{3}{8(5 + \lambda)}(y_1 + y_4) \tag{2.45}$$

The interpolated value does not depend strongly on the value of λ for $0 < \lambda < 1$ (as we would hope).

This result can be applied to the functions of the examples in the preceding section. For $y = e^x$, we find that Eq. (2.45) with $\lambda = 0$ underestimates $e^{0.5}$ by about 5×10^{-3} and with $\lambda = 1$ the interpolated value is low by 5×10^{-4}. Clearly, both results are quite good, again showing that the exponential is easily interpolated.

Interpolation of $y = \sin 2\pi x$ using the spline gives $y(0.5) = 0$, the exact value, but this is due to the symmetry of the function and not the accuracy of the method. It is therefore more interesting to consider $\sin \pi x$, for which $y(0.5) = 1$. Then $\lambda = 0$ gives $y(0.5) = 0.9959$ and $\lambda = 1$ gives $y(0.5) = 0.9743$, which indicates that the pinned end works best for this function. This is the case for a wide variety of functions.

Finally, we apply the result to the superellipse $(1 - x^4)^{1/4}$. We find $f(0.5) = 1.0306$ for $\lambda = 0$ and $f(0.5) = 1.0424$ for $\lambda = 1$. Both of these are rather far above the exact value 0.9840 and are in fact above the maximum value of 1.0000 that the function takes on the interval $0 \leq x \leq 1$. This, too, is typical. The extreme concentration of curvature of this function puts a large bending moment on the end of the spline, which is transmitted throughout the curve as "porpoising" or undesirable oscillation.

These observations are supported further by the next example. First we describe the programs used to generate these examples.

Program 2.1: Spline Interpolation Since splines are frequently used as an interpolation tool, FORTRAN subroutines for cubic spline fitting are found on the Internet site described in Appendix A. There are two subroutines; both are found in the SPLINE file together with the driver routine used to derive some of the results presented below. The first subroutine (SPLINE) computes the second derivatives at the knots and the second (SPEVAL) evaluates the spline for a given value of x.

We now give a simple description of the algorithm.

```
\% Assemble the right hand side
for i=2:n-1
  compute the right hand side of Eq (2.30)
end
compute the right hand sides of the first and last
equations (the ends)
\% Solve the system
call subroutine to solve the tridiagonal system of
equations
\% Given m values of x, evaluate the spline
for j=1:m
  evaluate spline for x(j)
end
```

Example 2.6: Spline Interpolation with Many Points We now look at some of results of cubic spline interpolation using several points. Since many of the results are similar to those for Lagrange interpolation, fewer results will be presented.

Figure 2.11 shows the results for the exponential using the cantilever end condition. In this case both spline and Lagrange interpolation are quite accurate but the latter is superior. Other spline end conditions might produce better results, but since experience shows that the best end conditions depend on the function, it is not a worthwhile exercise to do the computations. The results for the sine function are similar.

For the circle, the spline, like Lagrange interpolation, produces the largest error where the slope is large, near $x = 1$. The maximum errors are shown in Figure 2.12. The Lagrange error is slightly smaller, but neither method is very good.

Because the spline is nearly as accurate as Lagrange interpolation for smooth functions and may be better than Lagrange interpolation when the function oscillates, and is more flexible, it is usually the preferred method of fitting a function, especially when the result is to be graphed and when the number of data points is large. In this case, the cost of computing the Lagrange polynomial is quite large and the tendency to oscillate about the correct result becomes more pronounced. The spline suffers from neither problem and is generally well behaved when the number of points is large.

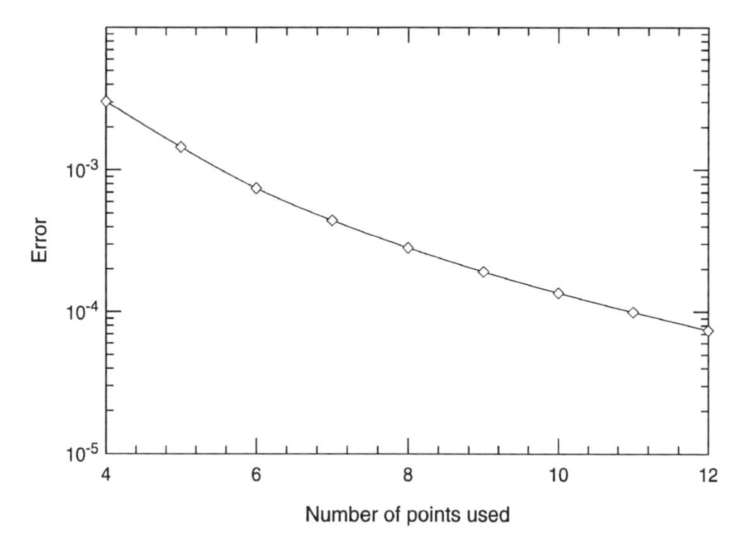

FIGURE 2.11. Maximum error in cubic spline interpolation of e^x versus number of points.

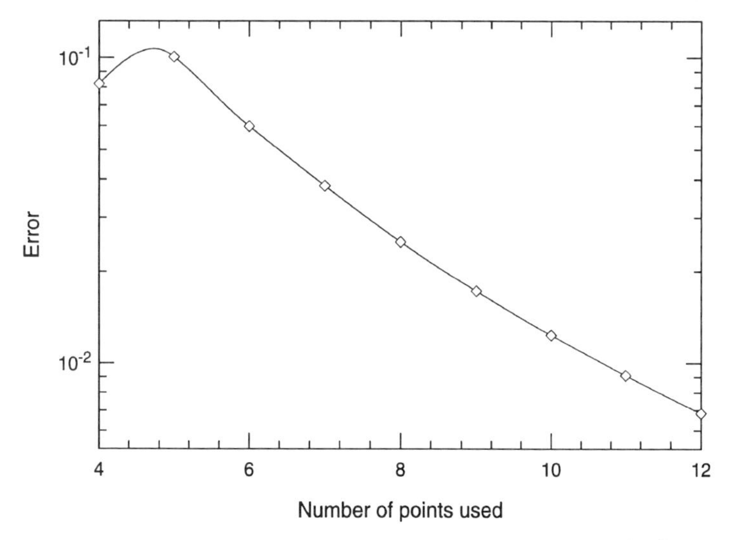

FIGURE 2.12. Maximum error in cubic spline interpolation of $(1 - x^2)^{1/2}$ versus number of points.

2.3.3. B-Splines

The cubic spline can be written in another form. Just as the Lagrange polynomial can be written as a sum of products of the data values y_i and polynomials that are independent of the function values [cf. Eq. (2.4)], the cubic spline can be written in the form:

$$f(x) = \sum_{i=1}^{n} y_i f_i(x) \tag{2.46}$$

This is again due to the linear dependence of the spline on the data. The functions $f_i(x)$ in Eq. (2.46) are called basis splines or B-splines for short; $f_i(x)$ is itself a cubic spline function that takes the value unity when $x = x_i$, is zero at all of the other knots, and obeys the specified end conditions.

The B-spline polynomials can be precomputed for a given set of knots. If the data is changed, it is not necessary to recompute the derivatives at the knots.

By giving up some of the continuity of the function or derivatives at the knots, it is possible to create B-splines that are nonzero only over a limited portion of the domain. These functions are much easier to evaluate. It is also much easier to insert new knots when this approach is used and this can be a great advantage. We shall say a little more about these methods at the end of this chapter.

2.3.4. Final Remarks

Finally, the concept of a spline—a curve that is made of pieces that are individually very smooth and are connected by compatibility conditions at the nodes—can be and has been extended to many functions other than cubic polynomials. A linear spline is nothing more than the piecewise linear function discussed earlier. Quadratic splines are also used; they require only one end condition and the solution of a bidiagonal set of equations can be solved by the back substitution part of the tridiagonal solution algorithm given in the preceding chapter. Exponential splines have also been used; they are computationally more difficult (if a constant in the exponent is allowed to be one of the parameters, the problem is nonlinear).

Still another kind of spline is based on the Bézier curve, a piecewise cubic function that passes through the data points. It differs from the standard cubic spline by having the shape of each section controlled by two points not on the curve itself. This provides additional flexibility as well as other advantages; the Bézier curve can be computed by a recursive method similar to the one used in divided differences, and it is not difficult to compute the result in a rotated coordinate system—an enormous advantage for graphics. For more details about splines and their properties, the reader is referred to the book by de Boor (1978).

The next two sections address two remaining problems—dealing with functions with concentrated curvature and dealing with functions with infinite slope.

2.4. TENSION SPLINES

In most cases the cubic spline gives a smooth fit to data. In some severe cases, however, the interpolation curve wiggles too much. We can find a cure by considering the mechanical analogy used to introduce the spline. If the thin strip of wood yields a curve that oscillates too much, this can be remedied by putting it in tension, that is, we pull on the ends of the thin spline in the direction tangential to the curve. If no tension is exerted, we get the ordinary spline described in the preceding section but, if the tension is extremely large, a linear spline (the piecewise linear function described above) results.

From mechanics, one can show that the effect of adding tension to the spline causes Eq. (2.31) to be replaced by

$$f_i'' - \sigma^2 f_i = [f''(x_i) - \sigma^2 y_i]\frac{x_{i+1} - x}{x_{i+1} - x_i} + [f''(x_{i+1}) - \sigma^2 y_{i+1}]\frac{x - x_i}{x_{i+1} - x_i} \tag{2.47}$$

On integrating this equation twice and requiring that the curve pass through the given node points, we obtain

$$\begin{aligned} f(x) = {} & \frac{f''(x_i)}{\sigma^2}\frac{\sinh\sigma(x_{i+1} - x)}{\sinh\sigma(x_{i+1} - x_i)} + \left[y_i - \frac{f''(x_i)}{\sigma^2}\right]\frac{x_{i+1} - x}{x_{i+1} - x_i} \\ & + \frac{f''(x_{i+1})}{\sigma^2}\frac{\sinh\sigma(x - x_i)}{\sinh\sigma(x_{i+1} - x_i)} + \left[y_{i+1} - \frac{f''(x_{i+1})}{\sigma^2}\right]\frac{x - x_i}{x_{i+1} - x_i} \\ & x_i < x < x_{i+1} \end{aligned} \tag{2.48}$$

where sinh denotes the hyperbolic sine function: $\sinh x = [\exp(x) - \exp(-x)]/2$.

Then on applying continuity of the first derivative, we find the equations that need to be solved for the second derivatives at the knots, $f''(x_i)$, to be

$$\begin{aligned} & \left[\frac{1}{\Delta_{i-1}} - \frac{\sigma}{\sinh\sigma\Delta_{i-1}}\right]\frac{f''(x_{i-1})}{\sigma^2} \\ & + \left[\sigma\frac{\cosh\sigma\Delta_{i-1}}{\sinh\sigma\Delta_{i-1}} - \frac{1}{\Delta_{i-1}} - \sigma\frac{\cosh\sigma\Delta_i}{\sinh\sigma\Delta_i} - \frac{1}{\Delta_i}\right]\frac{f''(x_i)}{\sigma^2} \\ & + \left[\frac{1}{\Delta_i} - \frac{\sigma}{\sinh\sigma\Delta_i}\right]\frac{f''(x_{i+1})}{\sigma^2} = \frac{y_{i+1} - y_i}{\Delta_i} - \frac{y_i - y_{i-1}}{\Delta_{i-1}} \end{aligned} \tag{2.49}$$

Again we have a tridiagonal system of equations for the second derivatives. The coefficients are more difficult to evaluate, as is the resulting spline, but excellent results have been obtained with the use of tension splines. Generally, if the natural spline is not satisfactory, one should try the tension spline with the tension parameter set equal to unity as a first guess. If the results still wiggle too much, the tension can be increased.

Endpoint conditions for the tension spline are modeled after those for the natural spline. Any of the conditions given in the preceding section may be used, but one should replace $f''(x_i)$ by $f''(x_i) - \sigma^2 y_i$.

Example 2.7: Tension and Nonuniform Splines There is no reason why the points or knots on which the data are given need to be uniformly spaced (as they have been in all the examples to this point). It makes sense that points ought to be more concentrated in the region the difficulty lies.

A particularly simple nonuniform grid is the so-called compound interest grid, which is useful when the extra resolution is needed near one end of the domain. This grid is defined by

$$\Delta_{i+1} = \alpha \Delta_i \tag{2.50}$$

where $\Delta_i = x_i - x_{i-1}$ and α is the parameter that controls the grid spacing. For $\alpha > 1$ the points are more closely spaced near the left boundary; the resolution is higher near the right boundary when $\alpha < 1$. This kind of grid is useful when the difficulty is concentrated at one end of the domain. Using the well-known formula for the sum of a geometric series, one can show that, if the domain length is L, and n intervals are used,

$$\Delta_1 = \frac{\alpha - 1}{\alpha^n - 1} L \tag{2.51}$$

We shall now use this grid to investigate some properties of splines applied to the superellipse function $(1 - x^4)^{1/4}$. Results obtained using 10 intervals are shown in Figure 2.13 along with the exact values of the function. On a uniform grid, the standard cubic spline produces severe oscillations, leading to the large errors that we have already noted. The tension spline is not much better on this grid; these results are not shown. On a nonuniform compound interest grid with $\alpha = 0.8$, which is close to the best value for the number of points and the function used in this example, the cubic spline yields results that are almost as bad as on the uniform grid. The problem is that, in the attempt to reproduce the large slope and curvature of the function, the spline produces large overshoots. When strong tension is added ($\sigma = 10$) on a nonuniform grid with $\alpha = 0.6$), the result is fairly smooth; in fact, the maximum error is about 0.03.

2.5. PARAMETRIC AND MULTIDIMENSIONAL INTERPOLATION—COMPUTER GRAPHICS

As the examples have shown, none of the methods discussed so far is capable of dealing with infinite slopes. Although, in simple cases, the problem can be

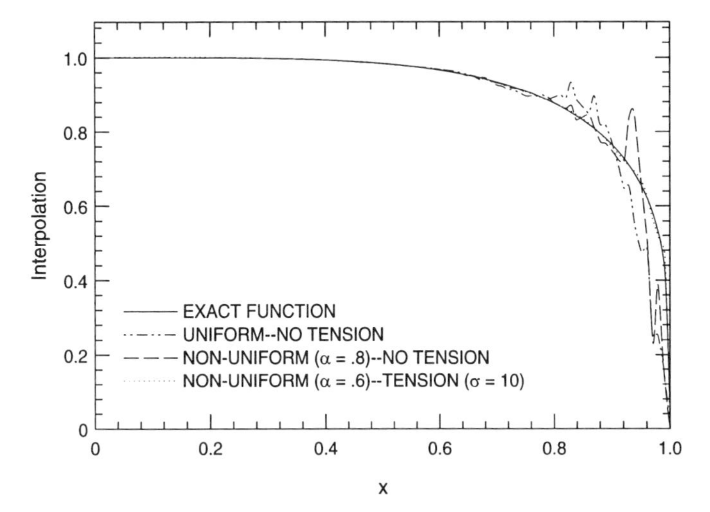

FIGURE 2.13. Cubic spline and tension spline interpolation of $(1 - x^4)^{1/4}$ on uniform and nonuniform grids with 11 points (10 intervals).

solved by interchanging the roles of the independent and dependent variables, this method does not work for curves that have closed loops. This problem commonly arises when a function of two independent variables, $f(x, y)$, is to be represented in terms of a contour plot showing lines with $f = \text{const.}$ Construction of the contour lines requires two-dimensional interpolation to locate the contours and parametric interpolation to draw them smoothly. These topics are described briefly in this section. We begin with the latter.

2.5.1. Parametric Interpolation

Suppose we wish to draw a smooth curve through a set of data that contains at least one closed loop. One issue is that the function is multivalued; for each x there are at least two values of y and vice versa. It is difficult to create an interpolation method that does not jump from one branch of the curve to another. One way out of the difficulty is to restate the problem so that we no longer need to deal with the infinite slope. This can be accomplished with parametric interpolation.

Parametric interpolation regards both x and y as functions of a third variable, which we will call s. The choice of the parameter s is somewhat arbitrary; the only requirement is it increase monotonically along the curve. A natural choice is the arc length along the curve or some approximation to it. The actual arc length is hard to compute, but the straight-line distance between successive data points can supply the values at the nodes. The first point is assigned $s = s_1 = 0$; at the second point, $s = s_2 = [(x_2 - x_1)^2 + (y_2 - y_1)^2]^{1/2}$, at the third, $s = s_3 = s_2 + [(x_3 - x_2)^2 + (y_3 - y_2)^2]^{1/2}$, and so on. The parametric functions $x(s)$ and $y(s)$ are created by the usual spline method (or any other interpolation method). For a closed curve, periodic end conditions are appropriate; for other cases, one of the other end conditions described earlier can be applied. The desired interpolation then is obtained by computing both x and y, the coordinates of a point on the curve, for a given value of s. If the value of y is needed for a particular x, we first find the value of s corresponding to the given x and then compute $y(s)$. Finding s requires solving a cubic equation; the extra computation involved is the chief drawback of the method. Nevertheless, the method is important enough to be in common use.

Example 2.8: Parametric Fit to a Circle To see how the method works, suppose we are given a number of points on a circle of radius R. In this case parametric interpolation is almost equivalent to spline fitting $x(s) = \cos(s/R)$ and $y(s) = \sin(s/R)$. (The representation would be exact if s were the exact arc length $R\theta$.) We already know how the spline deals with these functions, so there is no need to give detailed numerical results here. If enough points are used, the results are quite good.

2.5.2. Multidimensional Interpolation

The simplest methods for multidimensional interpolation are extensions of one-dimensional methods. For example, if a function of x and y is given on a regular array of points, that is, the nodes of a rectangular grid, one can do the following. On each line $x = x_i$ on which the data are given, compute the second derivatives of the spline fit in y by the method of the preceding section. These second derivatives may be considered functions of x and may be interpolated by spline fitting in x. To evaluate the interpolated value at a given (x, y), we first determine the box in which it is located, then compute the second derivatives $f''(x, y_i)$ and $f''(x, y_{i+1})$. Using these second derivatives, it is not difficult to evaluate $f(x, y)$.

Many other methods exist for fitting data in two or more dimensions. These include methods for data given on irregular sets of points, but they will not be presented here. For further details, see Chui and co-workers (1987).

2.5.3. Graphics and Design

The method for multidimensional interpolation just given has some disadvantages. Computer graphics and computer-aided design are major areas that require fitting of surfaces of objects with simple and easy to use methods that are also smooth and accurate. In both of these applications it is important to be able to translate, rotate, or stretch the object; this requires a method that is more coordinate-free than the one described above. This is a major area that requires more attention than it can be given here; indeed, it deserves a separate text and the one by Farin (1993) is highly recommended. We shall give just a few of the major ideas here.

The most common approach to surface fitting is to use B-splines based on Bézier curves. The advantage of the Bézier method is that, given two data points and the two control points that define the cubic between two knots, the curve can be computed by recursive linear interpolation. If the surface is to be translated, rotated, or stretched, this can be accomplished by applying the appropriate transformation to the points from which the curve is constructed. The method also generalizes easily to surfaces in three dimensions, and this is the reason why it is used so often in computer graphics and computer-aided design.

PROBLEMS

1. For the case in which there only three evenly spaced data points, show that the divided difference and Lagrange polynomial approaches give the same parabolic fit.

2. Another set of functions that are frequently used as test cases are $y = x^n$ on $0 < x < 1$. The lower order of these functions are very smooth, but the higher order ones are sharply curved. Use linear and quadratic Lagrange interpolation on these functions for $n = 3$, 6, and 10, respectively. Find the maximum error and the point at which it occurs. Compare the result obtained with the error estimate given in the text.

3. Apply Hermite interpolation using data at two points ($x = 0$ and $x = 1$) to compute the value of $e^{0.5}$.

4. In Chapter 4 we will use interpolation formulas to derive approximations to derivatives. This is done by differentiating the interpolation formula and evaluating it at the point at which the derivative is required.

   a. Given the values of $f(x_1)$ and $f(x_2)$, use linear Lagrange interpolation to find $f'(x_1)$ and $f'(x_2)$.

   b. Given $f(x_1)$, $f(x_2)$, and $f(x_3)$, use quadratic Lagrange interpolation to find f' and f'' at each of the three points.

5. Determine how Lagrange interpolation performs on the sinusoidal function with a larger number of points, say 10, 15, and 20.

6. A quadratic spline is a continuous curve that has a continuous first derivative, but not necessarily a continuous second derivative. Derive a set of equations for the first derivative of the quadratic spline at the node points. What are appropriate end condition(s) for this interpolation?

7. Using four points, apply spline interpolation with the cantilever end condition to the function $f(x) = x^6$ on $0 < x < 1$. Which value of λ in gives the best result for $x = 0.5$? Repeat the calculation for $x = 0.95$.

8. Compute the spline interpolation of $\sin 2\pi x$ with 10, 15, and 20 points, using the program given (or your own). Plot the error versus x.

9. Apply parametric interpolation to the ellipse $x^2 + (y^2/2) = 1$. Use 8, 12, 16, and 20 points that are approximately evenly spaced in the arc length coordinate s. Fit only the part of the curve in the first quadrant and obtain the rest by suitable reflections. Use 2, 3, 4, and 5 points in the quadrant.

CHAPTER 3

INTEGRATION

Engineers frequently compute integrals; the process of doing so is called *quadrature*. Integration methods can also be used, as we will see in the following chapter, to derive methods for solving ordinary differential equations. The method used to compute an integral will depend, naturally enough, on the nature of the integral; thus, we need to classify the types of integrals. The simplest case is an integral between finite limits of a function that is smooth and bounded and can be evaluated at any point in the range of integration. In more difficult integrals, one or more of these conditions is violated. If the function becomes infinite at some point on the range of integration or at least one of the limits is infinite, the integral is said to be *singular*. Special methods are required for the evaluation of such integrals; some of them will be discussed at the end of this chapter. In another interesting case, the values of the function are given at a fixed set of points and may be uncertain. This kind of problem usually arises in conjunction with experimental data and is best treated by fitting a least-squares curve to the data and integrating the resulting function. Since least-squares curve fitting was not treated in the preceding chapter, this case will not be treated here.

We look first at the nonsingular case. All quadrature formulas considered in this chapter approximate the integral by a weighted sum of the values of the integrand at particular points on the interval of integration, that is, by

$$\int_a^b f(x)\,dx = \sum_{i=1}^{n} w_i f(x_i) \tag{3.1}$$

where $a \leq x_1 < x_2 \cdots < x_n \leq b$. The numbers w_i are called the weights and the points x_i at which the function is evaluated are called the abscissas; both the abscissas and weights depend on the method and on the number of intervals, n. A quadrature formula consists simply of a prescription of the weights and abscissas. In the simplest approach, the abscissas are evenly spaced and the weights are chosen to give the best approximation to the integral in some sense. This method has the advantage that it can be systematically improved by successively halving the interval size (see below). A more sophisticated approach allows the abscissas as well as the weights to be chosen so as to maximize the accuracy for a given number of data points. In either case, the parameters depend on the criterion used to measure the accuracy of the method.

Any quadrature formula should yield the exact integral when the number of points used becomes large. An obvious approach to obtaining an accurate result is to estimate the integral using a quadrature formula with, say, n points, repeat the procedure using the same method with twice as many points, and compare the two results. If they agree within the desired tolerance, we can probably accept the last result as accurate. If they do not agree, we can increase the number of points again and continue until convergence is achieved. It is easiest to carry out this type of procedure when the abscissas are evenly spaced because the results already computed can be reused. For functions that are not smooth, the best approach is to divide the range of integration into a number of subranges; each subrange can then be treated with equally spaced abscissas and special attention can be paid to those regions where the function is either very large or rapidly varying.

For quadrature formulas with equally spaced abscissas, the error in the approximation is usually proportional to an integral power of the interval between successive abscissas. With $n + 1$ evenly spaced points, including the endpoints a and b, the interval is $h = (b - a)/n$. If, as h is decreased (n increased), the error goes to zero as h^m, we say that the method is an mth-order approximation and the order of the method is an index of its quality. An alternative measure of quality is the highest order of polynomial that a quadrature formula integrate exactly. The choice of the measure of accuracy depends on the type of method used. Naturally, we apply the one that is easiest to work with for each method.

The first quadrature formulas to be considered in this chapter are based on the use of Lagrange interpolation on an equally spaced set of points. We shall derive some of the formulas and their properties and discuss how to choose a formula of this class. When high accuracy is desired, a trick that not only decreases the size of the interval but simultaneously increases the accuracy of the approximation may prove valuable. The concept behind this method is introduced in Section 3.2 and is applied to the acceleration of the convergence of quadrature formulas in Section 3.3.

Another approach is to allow the computer to estimate the error in the integral from the results themselves. This can be done for each interval; one can then improve the accuracy only in those intervals in which the accuracy is not yet satisfactory. These are called *adaptive methods* and are discussed in Section 3.4.

In some cases the evaluation of the integrand is costly, and it is important to use a quadrature method that yields maximum accuracy for a given number of evaluations of the function; a different measure of quality is then desirable and the method of choice becomes Gauss quadrature. This method and some of its relatives are discussed in Section 3.5.

A method based on the use of random numbers (the Monte Carlo method) is described in Section 3.6. Finally, Section 3.7 considers methods for dealing with singular integrals, particularly integrals whose integrands become infinite somewhere in the range of integration.

3.1. NEWTON–COTES FORMULAS

One of the easiest ways to obtain useful quadrature formulas is to use Lagrange interpolation on an equally spaced mesh and integrate the result. In this manner one obtains the Newton–Cotes formulas; they come in two varieties. The closed formulas use the endpoints of the range of integration as abscissas; the open ones do not.

We start by considering the closed formulas. To derive them, we divide the range of integration into n equal intervals. Including the endpoints, there are $n+1$ data points; the abscissas corresponding to these points are

$$x_j = a + jh \qquad j = 0, 1, 2, \ldots, n \tag{3.2}$$

where $h = (b-a)/n$. We can pass a Lagrange interpolation polynomial of degree n through the points $(x_j, f(x_j))$, $j = 0, 1, 2, \ldots, n$. From the results of Chapter 2, we know that the result can be expressed in the form:

$$P(x) = \sum_{k=0}^{n} L_k(x) f(x_k) \tag{3.3}$$

This polynomial can be integrated to give the desired result

$$\int_a^b f(x)\,dx \approx \int_a^b P(x)\,dx = \sum_{k=0}^{n} f(x_k) \int_a^b L_k(x)\,dx = (b-a) \sum_{k=0}^{n} C_k^n f(x_k) \tag{3.4}$$

Since the Lagrange polynomials $L_k(x)$ depend only on the location of the data points and not on the values of the function, the C_k^n are pure numbers that are defined by

$$C_k^n = (b-a)^{-1} \int_a^b L_k(x)\,dx \tag{3.5}$$

TABLE 3.1. Closed Newton–Cotes Coefficients

n	N	NC_0^n	NC_1^n	NC_2^n	NC_3^n	NC_4^n	NC_5^n	NC_6^n	Error
1	2	1	1						$8.3 \times 10^{-2} \Delta^3 f''$
2	6	1	4	1					$3.5 \times 10^{-4} \Delta^5 f^{iv}$
3	8	1	3	3	1				$1.6 \times 10^{-4} \Delta^5 f^{iv}$
4	90	7	32	12	32	7			$5.2 \times 10^{-7} \Delta^7 f^{vi}$
5	288	19	75	50	50	75	19		$3.6 \times 10^{-7} \Delta^7 f^{vi}$
6	840	41	216	27	272	27	216	41	$6.4 \times 10^{-10} \Delta^9 f^{viii}$

They are called Cotes numbers; their values for $n \leq 6$ are tabulated in Table 3.1.

A few properties of the Cotes numbers are easily discovered. Since the Lagrange polynomial obviously fits a constant exactly, the function $f(x) = 1$ must be integrated exactly by the Newton–Cotes formulas of any order. Substituting this into Eq. (3.4), we find

$$\sum_{k=0}^{n} C_k^n = 1 \tag{3.6}$$

Furthermore, we can reverse the direction of integration; that is, we could integrate from b to a rather than from a to b. Consequently, Cotes numbers must possess the symmetry property:

$$C_k^n = C_{n-k}^n \tag{3.7}$$

All that remains to be done is to compute the Cotes numbers and to determine the accuracy of these quadrature formulas.

The first two Newton–Cotes formulas are well known and of special interest. For $n = 1$ we obtain the *trapezoid rule*:

$$\int_a^b f(x)\,dx = \tfrac{1}{2}(b-a)[f(a) + f(b)] \tag{3.8}$$

For $n = 2$ the Cotes formula is *Simpson's rule*, the formula obtained by passing a parabola through three points:

$$\int_a^b f(x)\,dx = \tfrac{1}{6}(b-a)[f(a) + 4f(\tfrac{1}{2}(a+b)) + f(b)] \tag{3.9}$$

The higher-order Cotes formulas are less well known, but Cotes numbers for them are listed in Table 3.1.

In practice, one rarely computes a difficult integral by applying a single Cotes formula to the entire interval. For complicated functions, many points would be required, and it is difficult to compute the coefficients of the Newton–Cotes formulas of high order. A better approach is to break the range of integration into subintervals, which may or may not be equal in size, and apply the quadrature formula separately to each interval. If the interval (a,b) is broken into n equal intervals each of width $h = (b - a)/n$, and the trapezoid rule is applied to each subinterval, we obtain the quadrature formula:

$$\int_a^b f(x)\,dx = h\left\{\sum_{k=0}^{n} f(x_k) - \tfrac{1}{2}[f(x_0) + f(x_n)]\right\}$$
$$x_k = a + \frac{k}{n}(b - a) \tag{3.10}$$

Simpson's rule may be applied in the same manner, but the number of subintervals must be even, since the formula is applied to pairs of intervals. One finds

$$\int_a^b f(x)\,dx = \tfrac{1}{3}h[f(x_0) + 4f(x_1) + 2f(x_2) + \cdots + 2f(x_{n-2}) + 4f(x_{n-1}) + f(x_n)] \tag{3.11}$$

We also need to inquire about the accuracy of these formulas. A rough estimate of the error can be obtained from the error estimate for Lagrange interpolation. The problem with this approach is that the value of ξ in Eq. (2.15) depends on x, so it is difficult to obtain an accurate estimate of the error by this approach. In fact, accurate estimation of the error is rather difficult so the derivation of the formula is not given here; the reader interested in following it is referred to one of the more advanced numerical analysis texts; see, for example, Isaacson and Keller (1966). The results are given in the last column of Table 3.1. Note that the error estimates given in the table are for a single interval. If one of the Newton–Cotes formulas is applied to many subintervals, as in Eqs. (3.10) and (3.11), the formula should be applied to each subinterval and the results summed.

Table 3.1 shows that each even-numbered approximation is a significant improvement over the odd one preceding it, but the odd approximations are not much better than their even predecessors. For this reason, the odd approximations (other than the trapezoid rule) are almost never used.

Another question that is frequently asked is: Which approximation is best for a particular application? A reasonable rule of thumb is that one should use the approximation for which the numerical coefficient given in the last column of Table 3.1 is closest to the desired relative error. Alternative approaches that avoid much of this difficulty are the Romberg method presented in Section 3.3 and the adaptive method presented in Section 3.4.

The formulas given above are the *closed* Newton–Cotes formulas because the values of the function at the endpoints are used in the approximation. There are also *open* Newton–Cotes formulas, which do not use the endpoints

TABLE 3.2. Piecewise Trapezoid Integration of e^x

Intervals n	Error ϵ	ϵ/h^2
1	0.141	0.141
2	0.036	0.144
3	0.016	0.144
4	0.0089	0.143
5	0.0057	0.143

as abscissas. These formulas are not as accurate as the closed formulas and are rarely used. The first of these formulas is an exception worth noting, however, as it will be used a number of times in later chapters:

$$\int_a^b f(x)\,dx = (b-a)f(\tfrac{1}{2}(a+b)) \tag{3.12}$$

This is known as the *midpoint rule* and may be obtained by assuming that the function is constant over the interval of integration.

We now do a number of examples that display some of the important properties of Newton–Cotes integration.

Example 3.1: Piecewise Trapezoid Integration The function e^x, used in the interpolation examples in Chapter 2, is very smooth so its integral should be easy to compute. The exact result is

$$I = \int_0^1 e^x\,dx = e - 1 = 1.718281828$$

First we apply the trapezoid rule with various numbers of subintervals. Since the error should scale like the square of the interval size, we also compute ϵ/h^2 where ϵ is the error and h the interval size. With $h = 1$, we find

$$I \approx \tfrac{1}{2}(e^0 + e^1) = 1.859$$

which is larger than the exact result by 0.141 or about 8%. The errors for larger numbers of subintervals are given in Table 3.2. For a function as smooth as the exponential, the error is proportional to h^2 even for relatively large interval sizes, as is shown by the last column of the table.

Example 3.2: Variable-Order Newton–Cotes Integration Computing the same integral using a single Newton–Cotes formula applied to the entire interval gives the results shown in Table 3.3. These errors follow the estimates given in

TABLE 3.3. Newton–Cotes Integration of e^x

n	ϵ
1	0.141
2	5.79×10^{-4}
3	2.58×10^{-4}
4	8.6×10^{-7}
5	2.8×10^{-7}

Table 3.1 very closely. Since the number of function evaluations in Simpson's rule is the same as that for two-interval trapezoid integration, a comparison of the results for $n = 2$ in Tables 3.2 and 3.3 show that, for the same amount of effort, Simpson's rule produces better results than the trapezoid rule. In the same manner, comparison of corresponding entries for larger n shows that when high accuracy is desired, using higher-order formulas is more efficient than using more intervals. This result applies to many numerical methods, but the advantage is offset somewhat by the inconvenience of having to precompute many numerical coefficients.

Example 3.3: Trapezoid Integration of Other Functions The full-wave sine function is not a good choice for an example of integration. The exact result

$$\int_0^1 \sin 2\pi x \, dx = 0$$

will be produced by many quadrature formulas as a consequence of the symmetry of the function, not the quality of the method. We consider, therefore, the integral

$$I = \int_0^1 (1 - x^2)^{1/2} \, dx = \tfrac{1}{4}\pi = 0.785398$$

which contains another of the functions used in the interpolation examples in Chapter 1. The results are shown in Table 3.4. The error decreases less rapidly than for the exponential. In fact, the error does not become proportional to h^2 even for the smallest values of h because the derivatives of the integrand are unbounded near the endpoint $x = 1$. This shows that caution and judgment are required when applying numerical methods.

For the same reason, Newton–Cotes integration of this function converges to the correct result very slowly. The results are given in Table 3.5. As noted, the decrease of the error is much slower than it was for the exponential, but greater improvement is obtained in going from an odd to an even approximation than vice versa.

An advantage of piecewise integration with uniform intervals is that, when the interval is subdivided, the values used to compute the previous estimate of

TABLE 3.4. Trapezoid Integration of $(1 - x^2)^{1/2}$

n	Integral	ϵ	ϵ/h^2
1	0.5000	0.2854	0.285
2	0.6830	0.1024	0.410
3	0.7294	0.0560	0.504
4	0.7489	0.0365	0.584
5	0.7593	0.0261	0.653

TABLE 3.5. Newton–Cotes Integration of $(1 - x^2)^{1/2}$

n	Integral	ϵ
1	0.5000	0.2854
2	0.7440	0.0414
3	0.7581	0.0273
4	0.7727	0.0127
5	0.7754	0.0100
6	0.7792	0.0062

the integral are also needed to produce the refined estimate; only the values at the new abscissas need to be freshly computed. Thus, for the complete method, the number of times the integrand is evaluated is exactly the number required to compute the integral using the finest interval. Since the computation of the values of the integrand is normally the most expensive part of a quadrature method, this is an important advantage for this approach.

We now consider a method for improving the accuracy of numerical calculations in general and quadrature in particular.

3.2. RICHARDSON EXTRAPOLATION AND ERROR ESTIMATION

An ideal integration subroutine is one that allows the user to provide the function to be integrated (a routine for evaluating it), the range of integration, and the desired accuracy; the code should then produce the result within the requested accuracy without further intervention on the part of the user. Ideally, this should be accomplished at minimal cost. In principle, any of the Newton–Cotes quadrature formulas can produce any desired accuracy by repeatedly halving the interval until the result converges. For difficult functions, this can be costly so better approaches have been developed. Romberg integration is one such procedure. It is based on the concept of Richardson extrapolation (also called deferred approach to the limit), a useful numerical technique in its own right and the subject of this section; its application to integration, which produces the Romberg method, is given in the following section. Other applications of the method will be found in later chapters.

As with many good methods, the concept behind it is simple. Suppose we wish to compute some quantity g, not necessarily an integral, which cannot be evaluated exactly. Suppose further that we have some method of approximating it. Most approximations depend on a parameter, say h, which can be made small; we denote the approximate value of g obtained with this value of the parameter as $g(h)$. The quadrature formulas of the preceding section (with g being the integral and h the size of the subinterval) are approximations of this kind. It often happens that the approximation can be represented by a Taylor series in the small parameter h, so that

$$g(h) = g + c_1 h + c_2 h^2 + \cdots \tag{3.13}$$

where $c_1, c_2, \ldots$ are constants. Some terms of this series may be missing. For example, the series may contain only the even terms; that is, it is actually a Taylor series in h^2. All of the terms except the first (which is the exact value) represent errors that we would prefer to eliminate.

Suppose that we have calculated $g(h)$ for some particular value of h. In the preceding section it was suggested that the computation could be improved by repeating the calculation with h replaced by $h/2$. Then we would obtain

$$g(\tfrac{1}{2}h) = g + \tfrac{1}{2}c_1 h + \tfrac{1}{4}c_2 h^2 + \cdots \tag{3.14}$$

In making this calculation we have roughly doubled the number of numerical operations (which is a good measure of cost) and halved the error.

The idea behind Richardson extrapolation is that by combining $g(h)$ and $g(h/2)$ we can obtain a better approximation. More precisely, we subtract Eq. (3.13) from twice Eq. (3.14) to get

$$g_1(h) = 2g(\tfrac{1}{2}h) - g(h) = g + c_2' h^2 + c_3' h^3 + \cdots \tag{3.15}$$

where $c_2', c_3', \ldots$ are fractions of $c_2, c_3, \ldots$. The leading error term is now proportional to h^2, and, if h is small, $g_1(h)$ is a considerably more accurate approximation to the exact value than either of the two values from which it was derived. Furthermore, the procedure can be continued. Having g_1 we can compute

$$g_2(h) = \tfrac{1}{3}[4g_1(\tfrac{1}{2}h) - g_1(h)] = g + c_3'' h^3 + \cdots \tag{3.16}$$

which is still more accurate. Continuing in this way, at the nth stage of this process, we compute

$$g_n(h) = \frac{2^n g_{n-1}(h/2) - g_{n-1}(h)}{2^n - 1} = g + O(h^{n+1}) \tag{3.17}$$

If it is known a priori that the approximation $g(h)$ contains only even powers, one can use just the even n operations in the sequence. (No harm will come from using the odd operations other than a waste of computational effort.) In this process, it is not necessary to halve the interval each time. Other fractions can be used, but halving makes maximum use of the previous results and is much easier if the process is to be continued an indefinite number of times.

The same idea can be used to estimate the error in an approximation. By simply subtracting Eq. (3.13) from Eq. (3.14), we obtain

$$g(\tfrac{1}{2}h) - g(h) = -\tfrac{1}{2}c_1 h - \tfrac{3}{4}c_2 h^2 + \cdots \tag{3.18}$$

On the other hand, by rearranging Eq. (3.14) we find

$$g - g(\tfrac{1}{2}h) \approx -\tfrac{1}{2}c_1 h - \tfrac{1}{4}c_2 h^2 + \cdots \tag{3.19}$$

If h is small enough, the terms proportional to h^2 in these equations (and all higher-order terms) are much smaller than the ones proportional to h and can be neglected. Thus we see that

$$\epsilon(\tfrac{1}{2}h) \approx g - g(\tfrac{1}{2}h) \tag{3.20}$$

is a reasonable estimate of the error in the $g(h/2)$ for small h. When h is not small, this estimate will not be very accurate, but both the estimate and the actual error will be quite large. This is often all the information that one needs; it tells the user that the computed value contains a significant error and that a more accurate calculation is probably necessary.

Example 3.4: Richardson Estimation of 2π A simple and interesting illustration of these ideas is the calculation of 2π by computing the perimeter of regular polygons inscribed in a unit circle. A little trigonometry suffices to show that the perimeter of an n-sided inscribed polygon is

$$P_0^n = 2n\sin\frac{\pi}{n} = 2\pi - \frac{\pi^3}{3n^2} + O\left(\frac{1}{n^4}\right) \tag{3.21}$$

where the expansion is valid for large n; $1/n$ plays the role of the small parameter is this case. From Eq. (3.21), we see that only even-order terms occur. One might object that the value of π is needed to calculate $\sin\pi/n$, but this is not so because the trigonometric formula for the sine of the half angle allows P_0^n to be computed by taking square roots.

The results of applying Richardson extrapolation to this problem are shown in Table 3.6. The first column gives the values computed from Eq. (3.21) and the succeeding columns give the extrapolated values. Since the error in

TABLE 3.6. Richardson Calculation of 2π (Exact = 6.283185308)

n	P_n^0	P_n^1	P_n^2	P_n^3	P_n^4
1	0.0000				
2	4.0000	5.3333			
4	5.6569	6.2091	6.2675		
8	6.1229	6.2783	6.2829	6.2831496	
16	6.2429	6.2828	6.28318	6.2831852	6.2831853
32	6.2731				
64	6.2807				
128	6.2826				
256	6.28303				
512	6.28315				

Eq. (3.21) contains only even powers of n^{-2}, we need to use just the even-numbered steps of the Richardson extrapolation procedure:

$$P_n^k = \frac{4^k P_{k-1}^n - P_{k-1}^{n/2}}{4^k - 1} \tag{3.22}$$

In Table 3.6, values are presented with either four-digit accuracy or the number of correct digits, whichever is higher; in a few cases more digits are shown for comparison purposes. It is amazing that we obtain eight-place accuracy using Richardson extrapolation with $n = 16$, whereas the original formula with $n = 512$ yields only five-place accuracy. This calculation does not cost much, but in a more complicated procedure the savings might be a factor of 32. Richardson extrapolation has many applications in numerical analysis, one of which is given in the next section.

3.3. ROMBERG INTEGRATION

Romberg integration is nothing more than a combination of the methods presented in the preceding two sections. The idea is to take a relatively inaccurate quadrature method and improve it through the use of Richardson extrapolation. We can start with a method as simple as the trapezoid rule.

The method begins by estimating the integral using the trapezoid rule with a single interval; needless to say, the result is not likely to be accurate. The integral is then recalculated with two intervals, four intervals, eight intervals, and so on; call the results $I_0^1, I_0^2, I_0^4, \ldots$. In the calculation of I_0^n, half of the needed function values have been computed earlier (for the calculation of $I_0^{n/2}$) and can be stored so that they need not be recalculated. Assuming that the major cost is that of evaluating the function, the cost of the complete computation is just a little more than that of the most accurate trapezoid rule evaluation of the integral.

With considerable difficulty (the interested reader is referred to the more advanced numerical analysis texts) one can show that the error of the trapezoid rule can be expressed as a series in h^2, the square of the grid size. Knowing this, we can obtain improved values of the integral using Richardson extrapolation using only the even-order steps (as in the example of computing π above). In the first extrapolation step, we calculate

$$I_1^n = \frac{4I_0^n - I_0^{n/2}}{3} \tag{3.23}$$

The results obtained from this formula are precisely what we would get by applying Simpson's rule with n intervals. The error in I_1^n is proportional to h^4, that is, it is fourth-order accurate (this is known to be true for Simpson's rule).

Since the error in I_1^n is proportional to h^4, we may apply Richardson extrapolation to it using the formula

$$I_2^n = \frac{16I_1^n - I_1^{n/2}}{15} \tag{3.24}$$

and this result is sixth-order accurate, that is, the error is proportional to h^6; however, it is not the sixth-order Newton–Cotes formula. The process may be continued using the general formula

$$I_k^n = \frac{4^k I_{k-1}^n - I_{k-1}^{n/2}}{4^k - 1} \tag{3.25}$$

until the desired accuracy is obtained. The stopping criterion is discussed in the following paragraph.

The order in which the estimates of the integral are actually computed is $I_0^1, I_0^2, I_1^2, I_0^4, I_1^4, I_2^4, I_0^8, \ldots$, as this makes the most efficient use of the computed function values.

When the maximum extrapolation that can be done with a given number of points is reached, that is, when $I_n^{2^n}$ has been calculated, it is compared with $I_{n-1}^{2^n}$, the next most accurate result. If the agreement is satisfactory, the calculation is terminated. If not, the trapezoid integral with 2^{n+1} points is computed and extrapolated; the entire process is repeated until the desired accuracy is obtained. A schematic of the method is given in Figure 3.1.

In principle, arbitrary accuracy can be achieved. In fact, the accuracy is limited by the numerical accuracy of the computer and of the function evaluation algorithm so the highest accuracy that can be expected is about one order of magnitude less than the numerical accuracy of either the machine or the function evaluation algorithm.

Program 3.1: Romberg Integration The following pseudo-program gives the essence of how the Romberg method described above is programmed.

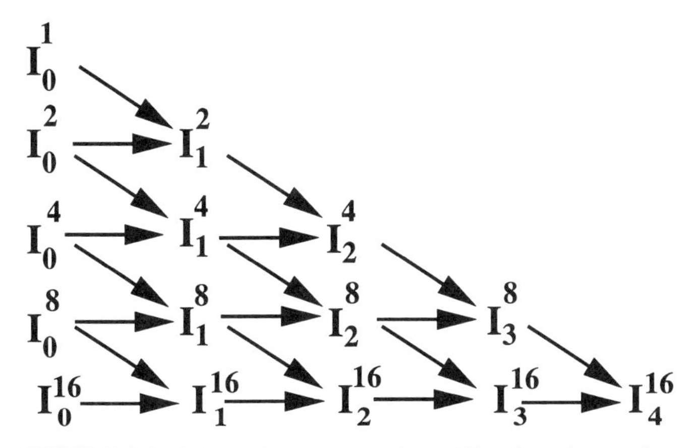

FIGURE 3.1. Schematic representation of Romberg integration.

```
\% Romberg integration
\% It is assumed that a procedure for evaluating the
function f(x) is attached

h = b - a
epsilon = 1.e-4     \% Desired accuracy
delta = 1.          \% Actual error

do while delta > epsilon
  level = 1

\% Trapezoid integration
  int(0,level) = -.5 * (f(a) + f(b))
  x = a
  do while x <= b
    int(0,level) = i(0,level) + h * f(x)
    x = x + h
  end

\% Extrapolation
  for k = 1:level
    i(k,level) = (4^k * i(k-1,level) - i(k-1,level-1)) /
    (4^k - 1)
  end

\% Convergence test
  delta = int(level,level) - int(level-1,level)

end
```

TABLE 3.7. Romberg Integration of e^x (Exact = 1.7182818)

n	I_n^0	I_n^1	I_n^2	I_n^3	I_n^4
1	1.859141				
2	1.753931	1.718861			
4	1.727222	1.718319	1.718283		
8	1.720519	1.718284	1.718282	1.718282	
16	1.718841	1.718282	1.718282	1.718282	1.718282
			Error		
1	1.4×10^{-1}				
2	3.6×10^{-2}	5.8×10^{-4}			
4	8.9×10^{-3}	3.7×10^{-5}	8.6×10^{-7}		
8	2.2×10^{-3}	2.3×10^{-6}	1.4×10^{-8}	3.4×10^{-10}	
16	5.6×10^{-4}	1.5×10^{-7}	2.2×10^{-10}	1.3×10^{-12}	3.3×10^{-14}

Example 3.5: Romberg Integration of Various Functions The function e^x is easy to integrate and the efficiency of Romberg integration is shown in Table 3.7 which gives both the result and, because the error is so small that it cannot be detected in the results, the error as well. It is clear that the Romberg method is very efficient for the computation of the integrals of smooth functions. Extremely high accuracy is obtained at very low computational cost.

As would be expected from the previous results, the function $(1 - x^2)^{1/2}$ is difficult to integrate with the Romberg method; however, as it has a singularity at $x = 1$, we put off dealing with it until Section 3.6. There is another type of function that is difficult to integrate and, since these integrals occur more often in practice, we will consider one of them. These are functions in which the major contribution to the integral comes from a relatively small portion of the range of integration. A good example of such an integral is

$$I = \int_0^1 e^{-a^2x^2}\, dx = \left(\frac{\sqrt{\pi}}{2a}\right) \operatorname{erf}(a) \tag{3.26}$$

where erf(a) is the error function and is tabulated in a number of sources, for example, Abramowitz and Stegun (1970). The function in the integrand is called the Gaussian. For small a, the integrand is very smooth; almost any method will do well, and the Romberg method converges in very much the same way as it does for the integral of e^x. For large a, the integrand is highly peaked near $x = 0$ and the function becomes much more difficult to integrate.

We perform this integration with the aid of a computer program found on the Internet site described in Appendix A. The driver program handles the input–output functions while the major work is done by subroutine ROMBRG. A routine for evaluating the function must be supplied separately.

TABLE 3.8. Romberg Integration of e^{-x^2}

Evaluations	Error
3	3×10^{-4}
5	9×10^{-6}
9	1×10^{-7}
17	1.4×10^{-10}

TABLE 3.9. Romberg Integration of e^{-100x^2}

Evaluations	Error
3	5.8×10^{-1}
5	4.5×10^{-2}
9	1.1×10^{-2}
17	1.2×10^{-2}
33	9.5×10^{-4}
65	8.2×10^{-6}
129	4.4×10^{-7}
257	3.5×10^{-7}

For this function, evaluation is relatively simple; in other cases, the function evaluation may be quite elaborate; for example, each evaluation might require solving a differential equation. Usually, the evaluation of the integrand is the most time-consuming part of computing an integral so the number of evaluations to obtain the required accuracy is a measure of the quality of a method. The subroutine assumes that only one parameter (a) is needed to define the function but this is easily changed.

The results for $a = 1$ are shown in Table 3.8; as anticipated, the method performs very well in this case. A much more severe test of the method is provided by the case with $a = 10$ for which the integrand is strongly peaked near $x = 0$. The results are shown in Table 3.9. A large number of evaluations is required. The accuracy obtained with 129 points is the best we can do with single-precision arithmetic because round-off error is beginning to become important. This is indicated by the small improvement in going from 129 points to 257 points.

3.4. ADAPTIVE QUADRATURE

Although the Romberg method is capable of any desired accuracy (within the limits of the computer used), it is not the most efficient method. As mentioned above, function evaluations usually represent the bulk of the cost of a quadrature, so it is desirable to minimize their number. This and the following sections give methods designed to do this.

One reason why the Romberg method requires so many function evaluations for functions with sharp peaks is that the abscissas are evenly spaced throughout the interval of integration. Many of the evaluations are in regions that make only a small contribution to the integral. Intuition tells us that it would be better to use fewer points in regions where the function is either small or slowly varying and more points where the function is large and/or varies rapidly.

We can reduce the number of function evaluations by using the following procedure. First, the interval of integration is broken into a number of subintervals. Then each subintegral is computed, and the error in the result is estimated using the Richardson procedure described earlier. Those subintervals on which the accuracy is not within the desired tolerance can be further subdivided. No further calculation is done on intervals where the accuracy is sufficient; this is where the major savings come from. Methods of this kind are called *adaptive quadrature* methods; many of them have been proposed. As they bear a close resemblance to each other we will describe only one such method here.

Suppose we wish to compute the integral

$$I = \int_a^b f(x)\,dx \tag{3.27}$$

with accuracy ϵ. We start by computing the integral with the trapezoid rule using a single interval for the entire domain. The panel size is $h = b - a$ and we call the result I^h; it is a poor estimate of the integral except in rare cases. We next break the interval (a,b) into two subintervals of size $h/2$ and use the trapezoid rule to estimate each subintegral; call these results $I_1^{h/2}$ and $I_2^{h/2}$. Knowing that the error in the trapezoid approximation is proportional to h^2 allows us to estimate it by means of the Richardson technique described earlier. The error in the sum of the two subintegrals, that is, $I_1^{h/2} + I_2^{h/2}$, is approximately one fourth the error in the original approximation, I^h; thus we can write

$$\begin{aligned} I^h &\approx I_{\text{exact}} + ch^2 \\ I^{h/2} = I_1^{h/2} + I_2^{h/2} &\approx I_{\text{exact}} + (\tfrac{1}{4}c)h^2 \end{aligned} \tag{3.28}$$

Then $(I^{h/2} - I^h)/3$ is an estimate of the error in $I^{h/2}$ and is easily computed from information already available. If this estimated error is less than the required error limit ϵ, we accept the result. If the error is too large, as is normal in the early stages of the procedure, we apply the method used for the full interval (dividing it and estimating the integral in each half) to each of the two subintervals, accepting either result only if the estimated error is $\epsilon/2$ or less. The process is continued until the error is within the desired accuracy. Perhaps the method is best illustrated by an example. First, we give the algorithm in the form of a pseudo-program.

Program 3.2: Adaptive Quadrature It is essential to note that the data pattern in all programs based on adaptive methods (including this one) is irregular; one cannot predict in advance how many data points will be needed or what data they represent. The adaptive quadrature code requires arrays that contain the abscissas, the function value at each abscissa, and the partial integrals (they are needed for error estimation). The unpredictability and irregularity of the data pattern makes it difficult to set up a data structure in advance. For this reason, all data-containing arrays in the program are one dimensional; they are lists containing all the values of a quantity used in the computation. In the program used in the example below, some data are stored in more than one location in order to simplify the logical structure of the code. To locate a particular datum in each array, a pointer is needed; a pointer is an index that tells the program where to find the datum. Since the numbers of intervals (partial integrals) and data points differ, two sets of pointers are used in the program; this is not necessary but it simplifies the program. The use of pointers is natural in the C language and the most recent implementation of FORTRAN (FORTRAN90), which make explicit provision for them; however, they can be implemented in FORTRAN77. (The code was written before FORTRAN90 was available; having it would have made the job much easier.) Because the adaptive quadrature program is logically more complex than most of the programs in this book, it will be described in words as well as a pseudo-program.

When a new level of grid refinement is required, the code checks whether each partial integral at the preceding level is acceptable or not; a logical variable array that is set when the error estimate is computed carries this information. If the result is accurate enough, the code continues to search until it finds a region where the integral is not accurate enough. When such an interval is found, the code creates new abscissas and the corresponding function values (those already computed are copied), computes the new integrals, and compares them with the values at the preceding level to determine whether they are of acceptable accuracy. When all the intervals at the preceding level have been exhausted, all the integrals required on the new level have been computed and the code proceeds to the next level. The computation stops when all of the values at a level are acceptable.

```
\% initialize parameters
  level = 1
  indrun = 2 \% index of last point
  intrun = 1 \% index of last interval
  indlast(0) = 0 \% index of last point at level 0
```

```
  intlast(0) = 0 \% index of last interval at level 0
  res = 0. \% resulting integral
  delx = b - a \% interval size
  x(1) = a \% first abscissa
  x(2) = b \% second abscissa

\% compute integral on lowest level
  compute f(1), f(2), rint(1) \% partial integral
  iacc(1) = .false. \% rint(1) not acceptable
  istop = .false.

c main loop
  dowhile (.not.istop)

c set up indices on old and new levels
    delx = .5 * delx; error = .5 * error  values at new
    level
    indlast(level) = indrun  index of last point at
    level
    intlast(level) = intrun  index of last interval at
    level
    i1 = intlast(level - 1) + 1  first interval on old
    level
    i3 = indlast(level - 1) + 1  first point on new
    level
    i2 = intrun  last interval at old level
    indrun = indrun + 1
    level = level + 1
\% main loop
    istop = .true.  will set to false if bad result is
    found
    do i=i1,i2
      if(iacc(i)) then \% result on previous level
      acceptable
        i3 = i3 + 1
      else \% result on previous level not acceptable
        if (i.Eqi1.or.x(indrun).ne.x(i3)) then  first or
        last point
          x(indrun) = x(i3); f(indrun) = f(i3)
        endif
\% set grid, compute function, integral, error on next
level
        x(indrun+1) = x(indrun) + delx
        x(indrun+2) = x(i3+1)
        compute f(indrun+1), f(indrun+2)
```

```
        compute rint(intrun+1), rint(intrun+2)
        indrun = indrun + 2; intrun = intrun + 2
        i3 = i3 + 1
        errnew = (rint(intrun-1) + rint(intrun) -
        rint(i)) / 3.
c check error, set flag, and update result
        if (abs(errnew).gt.error) then \% error not
        acceptable
          iacc(intrun) = .false.
          iacc(intrun-1) = .false.
        else \% error acceptable
          iacc(intrun) = .true.
          iacc(intrun-1) = .true.
          res = res + rint(intrun-1) + rint(intrun) \%
          update integral
        endif
\% another level is necessary
        istop = .false.
      endif
    enddo
  enddo

  return
  end
```

Example 3.6: Integration of e^{-100x^2} with Adaptive Quadrature The function e^{-100x^2} used in Example 3.5 provides a good illustration of what adaptive quadrature can do. We will use trapezoid integration as the basic method and demand relatively low accuracy, $\epsilon = 0.01$; this is done so that all of the partial results can be shown in a reasonable amount of space. For purposes of this example let $T(a,b)$ denote the trapezoid estimate of

$$\int_a^b e^{-100x^2}\,dx$$

and let $I(a,b)$ denote the exact integral.

We begin by computing the result at the coarsest level using a single interval:

$$T(0,1) = 0.50000$$

Next we compute the result on the second level at which there are two intervals:

$$T(0,0.5) = 0.25000$$

$$T(0.5,1) = 0.00000$$

The sum of these gives 0.25000 as the new estimate of $I(0,1)$. The estimated error in this value is $(0.25-0.50)/3 \approx -0.083$, which is larger in magnitude than the requested error, so the process must be continued.

To limit the length of the description, we shall consider just one of the subintervals at level 3. The interval (0,0.5) is broken into two and we compute

$$T(0,0.25) = 0.12524$$

$$T(0.25,0.5) = 0.00024$$

which gives a new estimate $I(0,0.5) = 0.12548$ and an error estimate of approximately 0.04. This is more than $\frac{1}{2}(0.01)$ so we must continue the calculation. For the first subinterval at the third level (0,0.25), we compute

$$T(0,0.125) = 0.07560$$

$$T(0.125,0.25) = 0.01322$$

obtaining the new estimate $I(0,0.25) = 0.08882$ and $\epsilon \approx 0.012 > \frac{1}{4}(0.01)$ so the process must be continued still further.

The numbers generated in the entire process of computing this integral are given in Table 3.10. In this table, the first line at each level gives the parameters for that level: the interval size and acceptable error. The second line gives the abscissas used at that level; the third line gives the values of the function at those abscissas; those that were evaluated at a previous level are shown but not recalculated. The trapezoid estimate of the integral on each interval is given on the fourth line and the sums of pairs of integrals (which will be subtracted from the corresponding integrals on the preceding level to obtain error estimates) are found on the fifth line. Finally, the last line gives the error estimate; if it is less than the tolerance for that level, the result is accepted.

Several points about adaptive quadrature are illustrated by this example. The method does concentrate the abscissas in the region responsible for the largest contribution to the integral, just as it was designed to do. The total number of abscissas used, which is also the number of function evaluations, is only 17. All but two of these are in the first half of the range and all but four are in the first quarter of the range. The relative error in the estimates of some of the partial integrals, for example, $T(0.25,0.375)$, are quite large, but they are accepted because their contributions to the total integral are small and therefore their contributions to the total error are also small.

In fact, for a case like this, in which the major contribution to the integral comes from a small part of the range, the criterion for acceptance of a result is

TABLE 3.10. Adaptive Quadrature of e^{-100x^2} [a]

Level	1	Interval	1	Tolerance	0.01
Abscissa	0	1			
Function	1	3.7×10^{-44}			
Integral		0.5			
Level	2	Interval	0.5	Tolerance	0.005
Abscissa	0	0.5	1		
Function	1	1.4×10^{-11}	4×10^{-44}		
Integral		0.25	3×10^{-12}		
Sum			0.25		
Error			0.08333		
Level	3	Interval	0.25	Tolerance	0.0025
Abscissa	0	0.25	0.5	0.75	1
Function	1	0.00193	1×10^{-11}	3.72×10^{-25}	3.7×10^{-44}
Integral		0.125241	0.00024	1.74×10^{-12}	4.7×10^{-26}
Sum			0.12548		1.7×10^{-12}
Error			0.04151		5.8×10^{-13}
Level	4	Interval	0.125	Tolerance	0.00125
Abscissa	0	0.125	0.25	0.375	0.5
Function	1	0.209611	0.00193	7.81×10^{-7}	1.4×10^{-11}
Integral		0.075601	0.01322	0.0001207	4.9×10^{-8}
Sum			0.08882		**0.000121**
Error			0.01214		0.00004

(*Continued*)

TABLE 3.10. Adaptive Quadrature of e^{-100x^2} [a] *(Continued)*

Level	5	Interval	0.0625	Tolerance	0.000625					
Abscissa	0	0.0625	0.125	0.1875	0.25					
Function	1	0.676634	0.20961	0.0297292	0.00193					
Integral		0.052395	0.0277	0.0074794	0.000989					
Sum			0.08009		0.008469					
Error			−0.0015		0.001584					
Level	6	Interval	0.03125	Tolerance	0.000313					
Abscissa	0	0.03125	0.0625	0.09375	0.125	0.15625	0.1875	0.21875	0.25	
Function	1	0.906961	0.67663	0.4152368	0.209611	0.087038	0.029729	0.008353	0.00193	
Integral	0.029796	0.02474	0.0170605	0.009763	0.004635	0.001824	0.000595	0.000161		
Sum			0.05454		**0.026824**		0.00646		**0.000756**	
Error			−0.0007		0.00029		0.00034		0.000078	
Level	7	Interval	0.01563	Tolerance	0.000156					
Abscissa	0	0.015625	0.03125	0.046875	0.0625	0.125	0.140625	0.15625	0.171875	0.1875
Function	1	0.975882	0.90696	0.8027383	0.676634	0.209611	0.138409	0.087038	0.052125	0.029729
Integral		0.015437	0.01471	0.013357	0.011558		0.002719	0.001761	0.001087	0.000639
Sum			**0.03015**		**0.024915**			**0.00448**		**0.001727**
Error			−0.0001		-5.7×10^{-5}			0.000052		0.000033
Sum of values in boldface			0.088968							
Exact value			0.088623							
Difference			0.000345							

[a] The accepted values, which are summed to give the final result, are shown in boldface type.

quite conservative and the final result is actually considerably more accurate than the requested error. The actual difference between the computed integral and the exact result is 0.000345, which is more than an order of magnitude smaller than the requested accuracy of 0.01.

If Romberg integration were run until it reached a grid as fine as the finest grid used in the adaptive quadrature, 129 evaluations would have been required. By this measure, the adaptive method is almost an order of magnitude more efficient. In reality, Romberg integration increases the accuracy of the method as the grid is refined and would not need to use this fine a grid; 32 points are more than sufficient; see Example 3.5.

Also note that, at the finest level, there is a gap between the subintervals (0,0.0625) and (0.125,0.1875); this is because, although the contribution to the integral from the interval (0.0625,0.125) is larger than that arising from the interval (0.125,0.1875), the error in the latter is larger due to the derivatives of the integrand being larger there. Because it uses error as the criterion for refinement, the adaptive method requires reevaluation of the integral only where it is really necessary.

There is a tradeoff in adaptive quadrature, just as there is for most methods. If a more accurate method such as Simpson's rule were used as the basic method, fewer subdivisions of the integral would have been necessary. On the other hand, the function will be evaluated more times than necessary in some regions and resulting in a waste of computation. General-purpose adaptive quadrature routines designed for high accuracy (relative accuracy of 10^{-4} or less) are usually based on Newton–Cotes methods with $n = 4$ to $n = 8$.

A subroutine for adaptive quadrature (ADAPQUAD) is given on the Internet site described in Appendix A. As with the other methods found there, it is intended to illustrate the method and is not the most efficient code of its type. More efficient routines can be found in numerical analysis packages that are available commercially.

The concept of adaptive methods has become important in numerical methods in recent years. We shall meet it again in subsequent chapters.

3.5. GAUSS QUADRATURE

We now consider a method that is optimum in the sense of providing the maximum accuracy for a prescribed number of function evaluations. The price paid for this property is the difficulty of systematic error reduction.

All of the methods introduced so far use evenly spaced abscissas, at least on subintervals. With a fixed number of abscissas picked a priori, it is very difficult to obtain methods of significantly greater accuracy than ones already introduced. We can obtain higher accuracy for a given number of abscissas, by allowing the abscissas to be adjustable and choosing them to maximize the accuracy of the formula. In Gauss quadrature, the ability to integrate polynomials of the highest possible order exactly is taken as the criterion of goodness. Since the general formula (3.1) contains n weights and n abscissas, all of which are now adjustable, $2n$ parameters are at our disposal. With all of them adjustable, we should be able to integrate polynomials of degree $2n-1$ exactly with an n point formula. The problem is to find the parameters that accomplish this.

We begin by noting that with the transformation

$$\xi = \frac{2x-(a+b)}{b-a} \tag{3.29}$$

any integral between the finite limits (a,b) is transformed into an integral between the limits -1 and $+1$, so it is sufficient to treat integrals with limits $(-1,1)$. As we are trying to integrate polynomials of degree $2n-1$ using n abscissas, a logical starting point is an interpolation formula that fits such a polynomial to n points; the Hermite interpolation formula, Eq. (2.22), is exactly what is needed. Integrating that formula we obtain

$$\int_{-1}^{1} w(x)f(x)\,dx = \sum_{k=1}^{n} w_k f(x_k) + \sum_{k=1}^{n} v_k f'(x_k) \tag{3.30}$$

where $w(x)$ is a weight function (not directly related to the weights, w_k, which we have inserted for later convenience, and w_k and v_k are defined by

$$w_k = \int_{-1}^{1} w(x)U_k(x)\,dx \tag{3.31}$$

$$v_k = \int_{-1}^{1} w(x)V_k(x)\,dx \tag{3.32}$$

and U_k and V_k are the polynomials given by Eq. (2.24). For Eq. (3.30) to be a quadrature formula of the form (3.1), that is, one that does not contain the derivatives $f'(x_k)$, all of the v_k must be zero. This can be assured by selecting the abscissas properly. For the present, it suffices to take $w(x) = 1$. Then, with the explicit expression (2.25) for V_k, we have

$$\int_{-1}^{1} (x-x_k)L_k^2(x)\,dx \qquad k = 1,2,\ldots,n \tag{3.33}$$

as the quantities that must be zero. Using Eq. (2.10), we can write this as

$$\int_{-1}^{1} F(x)L_k(x)\,dx \qquad k = 1,2,\ldots,n \tag{3.34}$$

Recall that $F(x)$ is a polynomial of degree n and the $L_k(x)$ are polynomials of degree $n-1$. Also note that, if the abscissas x_k are all different, the polynomials $L_k(x)$ form a linearly independent set. Since there can be no more than n linearly independent polynomials of degree $n-1$, the $L_k(x)$ are a complete set (i.e., there are no polynomials of degree n that are linearly independent of the $L_k(x)$). Hence Eq. (3.34) can be read as stating that $F(x)$ is a polynomial of degree n that is orthogonal to any polynomial of degree $n-1$. This polynomial is unique (aside from a multiplicative constant) and is well known in applied mathematics; $F(x)$ must be the Legendre polynomial of degree n, $P_n(x)$.

The Legendre polynomials, which have the orthogonality property

$$\int_{-1}^{1} P_n(x)P_m(x)\,dx = N_n\,\delta_{nm} \tag{3.35}$$

where $N_n = 2/2n+1$ is a normalization constant, have been well studied, and their properties may be found in many textbooks on applied mathematics. The analysis just completed shows that the abscissas of the method that we are trying to construct must be the zeros of the Legendre polynomial of degree n; it is possible to show that all of them lie between -1 and $+1$. The weights can be computed by substituting Eq. (2.25) for U_k into the integral in (3.31) defining w_k. The resulting integral can be simplified somewhat, but it can only be evaluated numerically. Some values will be given below.

The result of this calculation is the quadrature formula

$$\int_{-1}^{1} f(x)\,dx = \sum_{k=1}^{n} w_k f(x_k) \tag{3.36}$$

which is known as *Gauss quadrature*. The values of the abscissas x_k and the weights w_k for values of n up to 8 are given in Table 3.11, which has been adapted from Abramowitz and Stegun (1970).

The error in Gauss quadrature with n abscissas can shown to be

$$\epsilon \approx \frac{2^{2n+1}(n!)^4}{(2n+1)(2n!)^3} f^{(2n)}(\xi) \tag{3.37}$$

by means of a laborious calculation. This again shows that a polynomial of degree $2n-1$ is integrated exactly.

As stated earlier, Gauss quadrature gives the best accuracy (in the sense of correctly integrating polynomials of highest possible order) for a given number of function evaluations. Examination of Table 3.11 shows, however, that the abscissas for the various orders are all different. If one wishes to improve the accuracy by going to a higher-order method, the new calculation must be done from scratch. Consequently, if one requires a method that gives a prescribed accuracy regardless of the function being integrated, Romberg or adaptive integration remain the methods of choice.

TABLE 3.11. Abscissas and Weights for Gauss Quadrature

Abscissa ($\pm x_i$)		Weight (w_i)
	$n = 2$	
0.5773503		1.0000000
	$n = 3$	
0.0000000		0.8888889
0.7745967		0.5555556
	$n = 4$	
0.3399810		0.6521452
0.8611363		0.3478548
	$n = 5$	
0.0000000		0.5688889
0.5384693		0.4786287
0.9061798		0.2369269
	$n = 6$	
0.2386192		0.4679139
0.6612094		0.3607616
0.9324695		0.1713245
	$n = 7$	
0.0000000		0.4179592
0.4058452		0.3818301
0.7415312		0.2797054
0.9491079		0.1294850
	$n = 8$	
0.1834346		0.3626838
0.5255324		0.3137066
0.7966665		0.2223810
0.9602899		0.1012285

There are a number of variations on the theme of Gauss quadrature that are quite important. We briefly describe two of the more popular methods of this kind; there are others. The first one deals with integrals with one infinite limit. Such an integral can be transformed so that the range of integration runs from zero to infinity and the quadrature formula takes the form

$$\int_0^\infty e^{-x} f(x)\, dx = \sum_{k=1}^{n} w_k f(x_k) \tag{3.38}$$

The exponential weight function is used for convenience. The abscissas then turn out to be the zeros of the nth polynomial of the set defined by the orthogonality property

$$\int_0^\infty e^{-x} L_m(x) L_n(x)\, dx = \delta_{mn} \tag{3.39}$$

TABLE 3.12. Gauss Quadrature of e^{-x^2}

n	Error
2	-2.3×10^{-4}
4	-3.4×10^{-7}
6	$< 1.0 \times 10^{-8}$

which are known as the Laguerre polynomials. The abscissas and weights for these formula can be found in the work of Abramowitz and Stegun (1965). The method is called Gauss–Laguerre quadrature.

Finally, when both limits are infinite, we have the formula

$$\int_{-\infty}^{\infty} e^{-x^2} f(x)\,dx = \sum_{k=1}^{n} w_k f(x_k) \tag{3.40}$$

In this case the abscissas are the zeros of the nth Hermite polynomial. The Hermite polynomials are defined by the orthogonality property

$$\int_{-\infty}^{\infty} e^{-x^2} H_m(x) H_n(x)\,dx = \delta_{mn} \tag{3.41}$$

The abscissas and weights for this formula can be found in Abramowitz and Stegun (1965). The method is known as Gauss–Hermite quadrature.

Example 3.7: Gauss Quadrature of $e^{-a^2x^2}$ Gauss quadrature is sufficiently accurate that the error in computing

$$\int_0^1 e^x\,dx$$

is less than 10^{-8} with just four points. To provide a better test, we will try to integrate the function of Example 3.6

$$I(a) = \int_0^1 e^{-a^2x^2}\,dx \tag{3.42}$$

With small a, say $a = 1$, four points are sufficient to achieve an accuracy of 10^{-7}. Table 3.12 gives only the results for the even approximations, but for most functions, including this one, the error reduction tends to be orderly.

Comparing the results in Tables 3.8 and 3.12, we see that the Gauss quadrature does produce a smaller error for a given number of points or, conversely, requires fewer evaluations for a given accuracy than Newton–Cotes integration. A comparison of Tables 3.9 and 3.13 shows that, for a more difficult

TABLE 3.13. Gauss Quadrature of e^{-100x^2}

n	Error
2	-8.3×10^{-2}
4	$+1.8 \times 10^{-2}$
6	-2.0×10^{-3}
8	$+1.5 \times 10^{-4}$
10	-1.7×10^{-5}
12	-3.0×10^{-6}
14	-4.6×10^{-7}
16	$+4.9 \times 10^{-8}$
18	-5.0×10^{-9}

function, the advantage of Gauss quadrature is even greater. However, this advantage can be largely eliminated by using Romberg or adaptive quadrature. The major disadvantage of Gauss quadrature, as noted earlier, is the cost of error reduction.

3.6. MONTE CARLO METHODS

The final method that we shall present here is a rather unusual one. All of the methods presented earlier may be called deterministic; their abscissas and weights are chosen according to a recipe that is fixed in advance. In contrast, the Monte Carlo method (which is named for the famous casino) is based on the use of random numbers and is therefore statistical in nature.

The Monte Carlo method is rather simple in concept and easy to describe. The abscissas are chosen randomly. If the range of integration is (a,b), a random number, x, whose value lies between 0 and 1 is selected. It is converted to a random number in the desired range, z, via the variable change:

$$z = a + (b - a)x \tag{3.43}$$

and the function is evaluated at the point z. This process is repeated a large number of times. The values of the function obtained are summed, divided by the number of trials, and multiplied by the length of the domain to produce an estimate of the error.

As in many methods in which the error is due to random events, we expect that the relative error in the estimated integral will be inversely proportional to the square root of the number of trials. This is indeed the case as is shown in the following example.

Example 3.8: Monte Carlo Integration of e^{-100x^2} We shall apply the Monte Carlo method to the estimation of the integral that was introduced in Example

TABLE 3.14. Monte Carlo Quadrature of e^{-100x^2}

Number of Evaluations	Estimated Integral·	Absolute Error	Error × $\sqrt{N}$
10	0.093566	0.00494	0.016
100	0.060332	0.0253	0.253
1000	0.078577	0.0100	0.317
10000	0.089793	0.00170	0.170
100000	0.089846	0.00122	0.386
1000000	0.088726	0.000103	0.103

3.5 [cf. Eq. (3.26)]. As noted earlier, the integrand is a sharply peaked function that is difficult to integrate by any method. We used a FORTRAN code, MCQUAD, found on the Internet site described in Appendix A to compute this integral. Although random number generators are available on a number of compilers and may be called as a library function, a random number generator subroutine was included as part of the code; it was taken from Press et al. (1992). The results obtained for a particular trial are shown in Table 3.14.

The first thing to notice is that the number of evaluations of the function required to obtain an accurate result is much larger than for any of the other methods. This demonstrates that the Monte Carlo approach is not a good all-purpose method; the types of problems for which the method is best suited are given later. We also note that the error multiplied by the square root of the number of evaluations, which we suggested might be constant, is actually far from constant. The reason is simply that the error is itself statistical in nature and fluctuates considerably. The excellent result achieved for $N = 10$ is pure chance; another initial seed (a number that needs to be provided to the random number generator to get it started) gives a completely different result. Indeed, when the calculation was repeated with a different initial seed, the numbers in the last column were completely different. However, the error does decrease in proportion to $N^{-1/2}$ in a statistical sense.

The enormous number of evaluations required by the Monte Carlo method is its great disadvantage. This number can be decreased considerably, especially if something is known about the integrand. For example, one could break the integral into subintegrals and use more evaluations in regions where the integrand is large and fewer where it is small. There are further tricks that are beyond the scope of this book.

It is clear that Monte Carlo methods are not well suited to the computation of integrals in one dimension. Where they prove valuable is in multidimensional integration. The cost of the deterministic tends to scale as c^n, where c is the cost in one dimension and n is the dimension of the integral, while the cost of the Monte Carlo method is almost independent of the dimension of the integral. They are especially valuable for integrals over domains that are irregularly shaped.

3.7. SINGULARITIES

Occasionally we are faced with the problem of computing an integral whose integrand blows up at some point in the range of integration. Naturally, we assume that the singularity is mild enough for the integral to exist. Such integrals can be computed by the methods already described, provided that a little attention is given to dealing with the behavior near the singular point. Some of the better methods for accomplishing this are now given.

3.7.1. Integration by Parts

In many singular integrals it is possible to factor the integrand into the product of two components: one that contains the singularity but is easily integrated analytically, and a second that is not singular and is smooth enough to be differentiable everywhere in the domain of integration, including the singularity. In such a case, integration by parts will often convert the integral into one that is not singular.

This is best illustrated by an example:

$$\int_0^{\pi} \frac{\cos x}{\sqrt{x}}\, dx = 2\sqrt{x}\cos x|_0^{\pi} + 2\int_0^{\pi} \sqrt{x}\sin x\, dx \tag{3.44}$$

The integrated part is easily evaluated and the remaining integral is not singular. Although it is not the case here, it is sometimes desirable to repeat the integration by parts to obtain a still smoother integral.

3.7.2. Singularity Subtraction

A related but slightly different method is to factor the integral into the sum of two parts: one that contains the singularity but is integrable analytically, and a second that is nonsingular but requires numerical quadrature. This method can also be applied to the integral that was treated in the subsection above:

$$\int_0^{\pi} \frac{\cos x}{\sqrt{x}}\, dx = \int_0^{\pi} \frac{dx}{\sqrt{x}} + \int_0^{\pi} \frac{\cos x - 1}{\sqrt{x}}\, dx \tag{3.45}$$

Again the first part is easily evaluated and the second is nonsingular. The choice between these two methods is one of convenience and of deciding which gives the easier remaining integral.

3.8. CONCLUDING REMARKS

Although we have presented only a few quadrature methods, we have concentrated on those that are the most effective and, therefore, most widely used. A look at numerical analysis texts will show that there are many more methods. For the most part, these have no significant advantages over the methods presented here, except in special cases.

The one case in which it may pay to consider a method other than the ones given here is the case of multiple or iterated integrals. The Monte Carlo method is useful for these (especially if the domain is irregularly shaped) and there are a number of other good methods that can be found in the texts.

PROBLEMS

1. Another example of a peaky function is the Lorentz profile $(1 + x^2/a^2)^{-1}$. For large a it is well behaved but for small a it is strongly peaked near the origin. Integrate this function from 0 to 1 using (a) Newton–Cotes methods of various orders and (b) Simpson's rule with variable numbers of points. Plot the error versus the number of function evaluations for $a = 0.1$ and 1.

2. Integrate the function of Problem 1 using Romberg integration. Use the subroutine in the text or one available at your own computer center. Again, plot the error versus the number of function evaluations.

3. Repeat Problem 1 for the superellipse $(1 - x^4)^{1/4}$.

4. Repeat Problem 2 with the superellipse of Problem 3.

5. Occasionally integral equations occur in physics and engineering. Fredholm integral equations take the form:

$$h(x)f(x) + \int_0^1 K(x,y)f(y)\,dy = g(x)$$

where $K(x,y)$, $h(x)$, and $g(x)$ are given functions; of course, the limits could be different. Propose a method of solving this equation numerically. Suppose that you were asked to solve this equation with an accuracy of ϵ. How would you approach this problem?

6. Numerically solve the integral equation

$$g(\theta) - \frac{1-\epsilon}{4\epsilon}\int_{\theta_0}^{2\pi-\theta_0} g(\theta')\sin\frac{|\theta-\theta'|}{2}\,d\theta'$$
$$= \frac{\sigma T^4}{2}\left[1 + \cos\frac{\theta+\theta_0}{2} + \cos\frac{\theta-\theta_0}{2}\right]$$

Note that the exact solution is

$$g(\theta) = \frac{\epsilon\sigma T^4 \sin\frac{1}{2}\theta_0 \cos\frac{1}{2}\sqrt{\epsilon}(\pi-\theta)}{\sin\frac{1}{2}\theta_0 \cos\frac{1}{2}\sqrt{\epsilon}(\pi-\theta_0) + \cos\frac{1}{2}\theta_0 \sin\frac{1}{2}\sqrt{\epsilon}(\pi-\theta_0)}$$

(This problem arises in radiative heat transfer in a cylinder with a slot.)

7. Numerically compute

$$\int_0^{0.999} \tan(\tfrac{1}{2}\pi)x\,dx$$

Be careful near the upper limit!

CHAPTER 4

ORDINARY DIFFERENTIAL EQUATIONS: I. INITIAL VALUE PROBLEMS

One of the major scientific and engineering applications of numerical methods is to the solution of differential equations, both ordinary and partial. As solvers for partial differential equations are based largely on those for ordinary differential equations (ODEs), the latter are considered first. We begin by classifying the types of problems with which we need to deal. The properties listed in in Table 4.1 provide a useful classification scheme for problems involving ordinary differential equations. We shall begin by discussing issues connected with them.

We first note that nearly any equation of order higher than first can be reduced to a system of first-order equations. Thus the nth-order ordinary differential equation:

$$y^{(n)} = f(x, y, y', \ldots, y^{(n-1)}) \tag{4.1}$$

can be converted to a system of first-order equations by defining

$$\begin{aligned} y_1 &= y \\ y_j &= y^{(j-1)} \qquad j = 2, \ldots, n \end{aligned} \tag{4.2}$$

so that Eq. (4.1) becomes

$$\begin{aligned} y_j' &= y_{j+1} \qquad j = 1, 2, \ldots, n-1 \\ y_n' &= f(x, y, y_1, y_2, \ldots, y^n) \end{aligned} \tag{4.3}$$

TABLE 4.1. Classification of Problems in Ordinary Differential Equations

First order	Higher order
Single equation	System of equations
Linear	Nonlinear
Homogeneous	Inhomogeneous
Initial value	Boundary value

Thus the classifications based on the first two lines of Table 4.1 are essentially equivalent. Furthermore, except for some types of boundary value problems, the methods used for systems of equations are essentially identical to those used for single equations, and it is sufficient, at least initially, to consider the case of a single first-order equation. Special issues that arise for systems of equations will be discussed later.

The distinction between linear and nonlinear equations is not very important as long as the equations may be written in the form of Eq. (4.1)—in other words, as long as it is possible to isolate the highest derivative. A system that is linear in the highest order derivative is called quasi linear; it is fortunate that most engineering problems are of this type. For the most part, the same methods may be applied to both linear and nonlinear equations. The major difference is that, for some methods, nonlinear problems may require the solution of nonlinear algebraic equations at each step and that can increase the cost of solution considerably. Attention will need to be paid to this issue. However, it is essential to understand methods for dealing with linear equations before tackling nonlinear ones, so we shall consider linear equations first.

The distinction between homogeneous and inhomogeneous problems is not important except for one special case, that of linear homogeneous boundary value problems, which may become eigenvalue problems. These are discussed after boundary value problems are considered in the next chapter.

The distinction between initial and boundary value problems is very important. Solution of initial value problems is much more straightforward. There is a definite starting point, say x_0, at which the data are given. Starting at x_0, one can try to find the solution a short distance h away. When this has been accomplished, the results at the point $x_0 + h$ can be considered new initial data, and the solution can be computed at $x_0 + 2h, x_0 + 3h, \dots, x_0 + nh$, that is, the solution can be computed for as long as one desires. In this way, one can *march* the solution in the direction of increasing independent variable. For boundary value problems this is no longer true; there is data at some x removed from where the calculation is started that must be matched. This makes the problem more difficult and different methods are needed. However, as some of the methods for solving boundary value problems are based on ones for initial value problems, it is sensible to begin by considering initial value problems.

Based on this discussion of the problems we need to resolve, the plan in this chapter will be to first study methods of estimating derivatives numerically and then to apply them to the development of schemes for solving single first-order ordinary differential equations. The properties of these methods are examined with special emphasis on the features required for efficient solution. Then we examine systems of equations and look at the difficult problem of stiffness. Methods for solving boundary value problems are considered in the following chapter. Many of the results obtained in this and the next chapter will be applied to the solution of partial differential equations in the chapters to follow.

4.1. NUMERICAL DIFFERENTIATION

Before proceeding to the solution of differential equations, we look at how one might estimate derivatives numerically. The essential problem is that, since a computer can only deal with a finite (even if large) number of discrete values of a function, we need to estimate the derivative of a function from its values at a limited number of points. We assume that the data are the exact values of a smooth function at the data points and, further, that the derivatives are needed only at the data points.

There are at least three approaches to this problem. In the first method we use interpolation to fit a smooth curve through the data points and differentiate the resulting curve to get the desired result. A second approach uses Taylor series to produce substantially the same results with somewhat more effort, but with the added benefit that information about the error is obtained. Finally, one can derive schemes for solving ordinary differential equations directly from numerical quadrature methods.

4.1.1. Interpolation

The first approach uses Lagrange interpolation to provide estimates of derivatives. In the applications to follow, expressions for both the first and second derivatives are needed, so we will develop approximations for both.

Let us systematically apply Lagrange interpolation to the estimation of derivatives. The simplest scheme of this type is linear Lagrange interpolation, which approximates the function by

$$f(x) \approx \frac{x_i - x}{x_i - x_{i-1}} f(x_{i-1}) + \frac{x - x_{i-1}}{x_i - x_{i-1}} f(x_i) \tag{4.4}$$

on the interval $x_{i-1} \leq x \leq x_i$. Differentiating this function, we find that its derivative is constant on this interval. In particular, we get the same value at both endpoints of the interval:

$$f'(x_{i-1}) \approx D_+ f \approx \frac{f_i - f_{i-1}}{h_i} \tag{4.5}$$

and

$$f'(x_i) \approx D_- f \approx \frac{f_i - f_{i-1}}{h_i} \tag{4.6}$$

These are known as the *forward* and *backward difference* formulas, respectively, and D_+ and D_- are called the forward and backward difference operators. Any approximation of the derivative of a function in terms of values of that function at a discrete set of points is called a *finite difference approximation.* In these equations, we have used the abbreviations $f_i = f(x_i)$ and $h_i = x_i - x_{i-1}$ in order to simplify the notation. Thus we have two distinct approximations for the derivative; it is not clear which one ought to be used. As we will later show, Eqs. (4.5) and (4.6) are equally accurate, so the choice of which one to use depends on other considerations to be discussed below.

Improvement on these formulas can be obtained by using the quadratic Lagrange polynomial:

$$\begin{aligned} f(x) \approx & \frac{(x - x_i)(x - x_{i+1})}{(x_{i-1} - x_i)(x_{i-1} - x_{i+1})} f(x_{i-1}) + \frac{(x - x_{i-1})(x - x_{i+1})}{(x_i - x_{i-1})(x_i - x_{i+1})} f(x_i) \\ & + \frac{(x - x_{i-1})(x - x_i)}{(x_{i+1} - x_{i-1})(x_{i+1} - x_i)} f(x_{i+1}) \end{aligned} \tag{4.7}$$

for the interval $x_{i-1} \leq x \leq x_{i+1}$. Differentiating this equation and evaluating the results at the central point x_i of the interval, we obtain

$$\begin{aligned} f'(x_i) \approx & -\left[\left(\frac{h_{i+1}}{h_i}\right)(h_i + h_{i+1})^{-1}\right] f_{i-1} + [h_i^{-1} - h_{i+1}^{-1}] f_i \\ & + \left[\left(\frac{h_i}{h_{i+1}}\right)(h_i + h_{i+1})^{-1}\right] f_{i+1} \end{aligned} \tag{4.8}$$

For equally spaced intervals, $h_i = h_{i-1} = h$, and this reduces to

$$f'(x_i) \approx \frac{f_{i+1} - f_{i-1}}{2h} = \tfrac{1}{2}(D_+ D_-)f \tag{4.9}$$

which is the average of the forward and backward difference approximations. Equations (4.8) and (4.9) are called *central difference approximations.*

We also give the results for the derivative at x_{i-1} and x_{i+i} for the case of equally spaced intervals

$$f'(x_{i-1}) \approx \frac{-3f_{i-1} + 4f_i - f_{i+1}}{2h} \tag{4.10}$$

$$f'(x_{i+1}) \approx \frac{f_{i-1} - 4f_i + 3f_{i-1}}{2h} \tag{4.11}$$

These are forward and backward difference approximations.

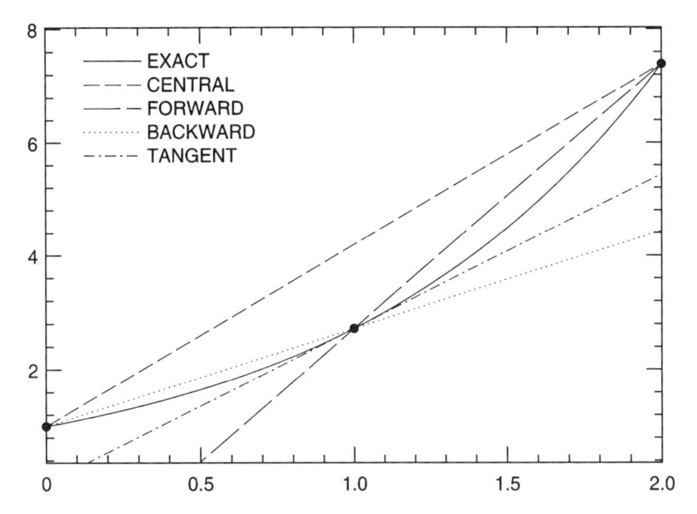

FIGURE 4.1. Graphical interpretation of the finite difference approximations applied to the function e^x. The curve represents the function and the solid dots are the data used to construct the approximations. TANGENT is the line tangent to the curve at $x = 1$; its slope is the exact derivative. The lines labeled BACKWARD, FORWARD, and CENTRAL are secant lines connecting the first and second, second and third, and first and third data points, respectively.

An interpretation of these approximations is given by Figure 4.1; the function whose derivative (at $x = 1$) is to be estimated is $f(x) = e^x$. To exaggerate the errors, the points used in the approximations, $x = 0$, 1, and 2, are rather far apart. The correct derivative is the slope of the tangent to the function curve at $x = 1$. The backward, forward, and central approximations are the slopes of the lines labeled backward, forward, and central, respectively. The figure clearly shows that the central difference approximation is more accurate than either the forward or backward approximations. Intuition tells us that this ought to be the case; a more precise estimate of the accuracy of these methods is given below.

We can also differentiate Eq. (4.7) twice to obtain a formula for the second derivative. The result is constant on the interval (x_{i-1}, x_{i+1}):

$$f'' \approx 2h_i^{-1}(h_i + h_{i+1})^{-1} f_{i-1} - \left(\frac{2}{h_i h_{i+1}}\right) f_i + 2h_{i+1}^{-1}(h_i + h_{i+1})^{-1} f_{i+1} \tag{4.12}$$

which, for equal intervals, becomes

$$f'' \approx \frac{f_{i-1} - 2f_i + f_{i+1}}{h^2} = D_+ D_- f = D_- D_+ f \tag{4.13}$$

TABLE 4.2. Coefficients in Difference Formulas

Derivative	f_{i-2}	f_{i-1}	f_i	f_{i+1}	f_{i+2}	Error
Forward formulas						
hf_i'			-1	1		$O(h)$
$2hf_i'$			-3	4	-1	$O(h^2)$
hf_i''			1	-2	1	$O(h)$
Backward formulas						
hf_i'		-1	1			$O(h)$
$2hf_i'$	1	-4	3			$O(h^2)$
hf_i''	1	-2	1			$O(h)$
Central formulas						
$2hf_i'$		-1	0	1		$O(h^2)$
$12hf_i'$	1	-8	0	8	1	$O(h^4)$
$h^2 f_i''$		1	-2	1		$O(h^2)$
$12h^2 f_i''$	-1	16	-30	16	-1	$O(h^4)$

This formula may be applied to the estimation of the second derivative at any of the three points x_{i-1}, x_i, and x_{i+1}. Depending on the point to which it is applied, it is called a forward, central, or backward difference approximation. As may be expected and as we shall show later, the formulas obtained from quadratic Lagrange interpolation are more accurate than those derived from the linear approximation.

This procedure can be extended to more accurate interpolation methods. Using cubic Lagrange polynomials, we can obtain forward and backward difference formulas; the formulas obtained at the central two of the four points do not fit into any of the categories defined above. The quartic Lagrange polynomial yields central, forward, and backward formulas as well as two formulas (at the second and fourth of the five points) that do not fit these categories. Some of these formulas are given in Table 4.2; the last column of the table gives the order of the approximation, a term defined in the next subsection.

Use of Hermite interpolation as a means of estimating first derivatives does not make sense because the approximation uses first derivatives as data, but it could be used to approximate higher-order derivatives. Methods based on these approximations have been the subject of research papers but, as they have not found widespread use, we shall not present them here.

The use of splines as sources of finite difference approximations has received some attention in recent years. The only spline treated in detail in Chapter 2 is the cubic for which a method of computing the second derivative is given by Eq. (2.30). It is interesting to note that, for equally spaced intervals, the right-hand side of Eq. (2.30) is the central difference formula derived above. The effect of the spline is to distribute the result (4.13) over the central point and its two nearest neighbors with weights $\frac{1}{6}$, $\frac{4}{6}$, and $\frac{1}{6}$. (For unequally spaced points, the right-hand side is still an approximation to the second derivative at x_i but the weights on the left-hand side are different.) Despite the fact that use of splines requires solution of sets of simultaneous algebraic equations, the increase in computational effort is small because the system is tridiagonal. The results are approximately as accurate as the formulas given above, but the error is opposite in sign, and it is possible to use this knowledge to obtain a simple formula that is better than either Eq. (4.12) or the spline result; we shall do this later.

4.1.2. Taylor Series

Another method of obtaining finite difference approximations to derivatives is through the use of Taylor series. Starting with the well-known expansion

$$f(x \pm h) = f(x) \pm hf'(x) + \frac{h^2}{2}f''(x) \pm \cdots \tag{4.14}$$

and forming linear combinations of the values of the function at various mesh points x, $x \pm h$, $x \pm 2h$, and so on, we can obtain approximations to the derivatives. For example, to construct a formula for the first derivative, we must create a linear combination of the values at the various points in which the terms involving $f(x)$ sum to zero but those involving $f'(x)$ do not. By including the values of the function at more points, we can eliminate the coefficients of some of the higher derivatives, producing a result with an error proportional to a higher power of h. The required coefficients can be found by solving a set of linear equations for them.

As an example, suppose that we wish to construct an approximation for the first derivative of the function f at the point x using the values of the function $f(x-h)$, $f(x)$, and $f(x+h)$. We could create the linear combination:

$$\begin{aligned} af(x-h) + bf(x) + cf(x+h) &= (a+b+c)f(x) + (c-a)hf'(x) \\ &\quad + (c+a)\tfrac{1}{2}h^2 f''(x) + (c-a)\tfrac{1}{6}h^3 f'''(x) + \text{HOT} \end{aligned} \tag{4.15}$$

where we have used Eq. (4.14) and HOT is an abbreviation for higher-order terms. Since we want this expression to approximate the first derivative, we want the coefficient of f' to be 1 and as many of the other coefficients as possible to be zero. Since there are three coefficients, we can require:

$$\begin{aligned} a + b + c &= 0 \\ c - a &= 1/h \\ c + a &= 0 \end{aligned} \tag{4.16}$$

Solving these equations, we find $c = -a = 1/2h$ and $b = 0$, yielding the approximation (4.9) again. We also see that the first error term is $(c-a)(h^3/6)f'''(x) = (h^2/6)f'''(x)$.

An alternative to this method is to take guidance from the results found by interpolation. Thus, using Eq. (4.14) with $x = x_i$ to compute the right-hand side of Eq. (4.5), we have

$$\frac{f(x_i) - f(x_{i-1})}{h_i} = f'(x_{i-1}) + \frac{h_i}{2} f''(x_{i-1}) + \cdots \tag{4.17}$$

Using Eq. (4.14) at $x = x_{i-1}$ to compute the right-hand side of Eq. (4.6), we have

$$\frac{f(x_i) - f(x_{i-1})}{h_i} = f'(x_i) - \frac{h_i}{2} f''(x_i) + \cdots \tag{4.18}$$

In this manner we see that Eqs. (4.5) and (4.6) are approximations to the first derivative (as we already knew) and that the errors in these approximations are $\pm hf''/2$ for small h. Thus, the error is proportional to the grid spacing if the latter is small enough that the higher-order terms can be neglected. We call an approximation whose error is proportional to h *first-order accurate*; both the forward and backward difference formulas are first-order accurate estimates of the first derivative.

The approach that we used above showed that the central difference approximation is proportional to the square of the grid size:

$$\frac{f(x_{i+1}) - f(x_{i-1})}{2h} = f'(x_i) + \frac{h^2}{6} f'''(x_i) + \cdots \tag{4.19}$$

We call this is a *second-order accurate* approximation to $f'(x_i)$. Note, however, that as an estimate for $f'(x_{i-1})$ or $f'(x_{i+1})$, it becomes identical to the forward or backward formula with h replaced by $2h$ and is therefore only first-order accurate and, in fact, less accurate than either of Eqs. (4.17) or (4.18). Thus the accuracy of a formula depends on the location at which it is asked to approximate a derivative as well as on the formula itself.

If the grid spacing h is small, a second-order formula is more accurate than a first-order formula; moreover, its error decreases more rapidly as $h \to 0$, a property that allows it to give better results for a given amount of computation. This makes second-order approximations the preferred choice when the required accuracy is higher. We shall develop approximations of still higher accuracy later.

Example 4.1: Estimating the Derivative of e^x To illustrate these ideas, we apply the finite difference approximations derived above to the estimation of the first derivative of $f(x) = e^x$ at $x = 0$ and look at the results as a function

TABLE 4.3. Errors in Finite Difference Approximations

Grid (h)	f'	Error (ϵ)	ϵ/h
	Forward Differences		
1	1.718282	0.718282	0.718282
0.5	1.297443	0.297443	0.594885
0.25	1.136102	0.136102	0.544407
0.1	1.051709	0.051709	0.517092
0.05	1.025422	0.025422	0.508439
0.025	1.012605	0.012605	0.504193
0.01	1.005017	0.005017	0.501671
0.005	1.002504	0.002504	0.500834
0.0025	1.001251	0.001251	0.500417
	Backward Differences		
1	0.632121	−0.36788	−0.36788
0.5	0.786939	−0.21306	−0.42612
0.25	0.884797	−0.1152	−0.46081
0.1	0.951626	−0.04837	−0.48374
0.05	0.975412	−0.02459	−0.49177
0.025	0.987604	−0.0124	−0.49586
0.01	0.995017	−0.00498	−0.49834
0.005	0.997504	−0.0025	−0.49917
0.0025	0.998751	−0.00125	−0.49958
	Central Differences		
1	1.175201	0.175201	0.175201
0.5	1.042191	0.042191	0.168762
0.25	1.010449	0.010449	0.167188
0.1	1.001668	0.001668	0.16675
0.05	1.000417	0.000417	0.166688
0.025	1.000104	0.000104	0.166672
0.01	1.000017	0.000017	0.166668
0.005	1.000004	4.2×10^6	0.166667
0.0025	1.000001	1.0×10^6	0.166667

of the separation distance between the points used in the formula. The exact result is, of course, $f'(0) = e^0 = 1$. The forward difference approximation,

$$f'(0;h) = \frac{f(h) - f(0)}{h}$$

when evaluated for $h = 1$, gives $f'(0;h) = 1.71828$, which is much larger than the correct value. Results for smaller h are shown in Table 4.3 along with the error (defined as the difference between the approximation and the exact value). Results for the backward and central difference approximations are also given. For the first two approximations, we also give the error divided by h; the analysis suggests that this ratio should be independent of h for small

h; this is indeed the case. For the central difference approximation, the error divided by h^2 is given because the analysis suggests that this quantity ought to be constant for small h; this is verified by the results. The second-order nature of the central difference approximation makes it much more accurate than either of the other formulas for small h. Note that the errors produced by the forward and backward approximations are nearly equal but of opposite sign.

As we shall see later, these error estimates can be applied to the calculation of the error in numerical solutions of differential equations. They also allow one to create more accurate formulas. Let the spacing between successive points be fixed at h, and let us estimate the derivative using grid size $2h$. We apply Richardson extrapolation to obtain a better estimate of the first derivative. Suppose we apply the first-order forward and backward difference formulas [Eqs. (4.5) and (4.6)] for the derivative using grid sizes h and $2h$. We can create the second-order accurate approximations 4.10 and 4.11 by applying Richardson extrapolation. For example, if we use Eq. (4.6) as the basic method, extrapolation gives

$$2f'(x,h) - f'(x,2h) = 2\frac{f_i - f_{i-1}}{h} - \frac{f_i - f_{i-2}}{2h} = \frac{3f_i - 4f_{i-1} + f_{i-2}}{2h} \tag{4.20}$$

which is Eq. (4.11). In the same manner, application of Richardson extrapolation to the central difference formula (4.8), which is second-order accurate, produces the fourth-order accurate formula given in Table 4.2. Thus the combination of Taylor series and Richardson extrapolation provides a powerful tool for constructing accurate difference schemes.

As the example shows, the error is smaller for a central difference approximation than for one that is asymmetric. Also, because there is often cancellation due to symmetry, an equally spaced mesh generally produces more accurate difference formulas than an irregular one. Finally, for a given type of difference formula, higher accuracy may be obtained by including more points in the difference formula. The inclusion of more terms introduces more coefficients, which can be chosen to cancel more terms of the Taylor series expansions. Similarly, approximation of higher-order derivatives requires the inclusion of more points in the difference formula. High-order multipoint schemes are useful, but as we shall see later, it is difficult to satisfy the boundary conditions with the same accuracy when these methods are used.

One way to avoid the need to include large numbers of points in an approximation is to use *compact schemes*. These approximations can be derived using the ideas taken from spline interpolation. If we regard Eq. (2.30) with equal spacing as a difference formula (i.e., a means of computing f''), then we can use Taylor series to show that it is second-order accurate and that the error is precisely equal and opposite to that of Eq. (4.13). Consequently, the average of these two equations, which is

$$\tfrac{1}{12}[f''(x_{i-1}) + 10f''(x_i) + f''(x_{i+1})] = \frac{f(x_{i-1}) - 2f(x_i) + f(x_{i+1})}{h^2} \tag{4.21}$$

is both fourth-order accurate and compact in the sense that it uses data at only three mesh points. The fact that it requires the solution of a tridiagonal set of equations to compute the derivatives is, generally, only a slight inconvenience. One can derive a similar scheme for the first derivative:

$$\tfrac{1}{6}[f'(x_{i-1}) + 4f'(x_i) + f'(x_{i+1})] = \frac{f(x_{i+1}) - f(x_{i-1})}{h} \tag{4.22}$$

We shall apply compact schemes in the next chapter.

4.1.3. Numerical Integration

It should be no surprise that numerical quadrature methods can be used as the basis of techniques for solving ordinary differential equations. Since the quadrature methods developed in Chapter 3 were based on interpolation methods, however, we expect that any method derived from a quadrature scheme can also be found by the approaches considered above. It is, nonetheless, interesting to see where some of the quadrature methods lead in the way of differential equation solvers.

Suppose that we wish to solve the differential equation

$$y'(x) = f(x, y) \tag{4.23}$$

and that we have the value of $y(x)$ at the point x_n. The objective is to find a procedure for computing $y(x_{n+1}) = y(x_n + h)$. Having that, we can use it as a new initial point to compute $y(x_{n+2}), y(x_{n+3}), \ldots$ and thus generate an approximate solution to the differential equation.

To accomplish this, we can simply integrate Eq. (4.23) from x_n to x_{n+1} to get:

$$y_{n+1} - y_n = \int_{x_n}^{x_{n+1}} f(x, y(x))\, dx \tag{4.24}$$

where we have explicitly indicated that, in performing the integration, y must be considered a function x.

We can apply any quadrature formula to the estimation of this integral. The first two methods we shall consider are so simple that we did not even use them in Chapter 3. These are the forward and backward formulas:

$$\int_{x_n}^{x_{n+1}} f(x, y(x))\, dx \approx hf(x_n, y(x_n)) \tag{4.25}$$

and

$$\int_{x_n}^{x_{n+1}} f(x,y(x))\,dx \approx hf(x_{n+1},y(x_{n+1})) \tag{4.26}$$

where again $h = x_{n+1} - x_n$. These formulas are equivalent to applying the forward and backward difference approximations to the differential equation. As methods for solving ordinary differential equations, they are known as the forward (or explicit) and backward (or implicit) Euler methods. We shall investigate them in some detail later.

The other two formulas that we shall consider here are based on the midpoint rule, Eq. (3.12). Applying the midpoint rule to the the approximation of the integral in Eq. (4.24), we have

$$y_{n+1} - y_n = hf(x_{n+1/2},y_{n+1/2}) \tag{4.27}$$

This method would be satisfactory if we had a means of evaluating $f(x_{n+1/2}, y_{n+1/2})$. Since doing so requires a further approximation, we also look at the result obtained by applying the midpoint rule to the interval (x_{n-1},x_{n+1}):

$$y_{n+1} - y_{n-1} = 2hf(x_n,y_n) \tag{4.28}$$

This is known as the *leapfrog method* and will be analyzed in more detail later.

Application of the trapezoid rule to the integration of the differential equation (5.11) from x_n to x_{n+1} produces

$$y_{n+1} - y_n = \tfrac{1}{2}h[f(x_n,y_n) + f(x_{n+1},y_{n+1})] \tag{4.29}$$

This formula may also be obtained by approximating $f(x_{n+1/2},y_{n+1/2})$ in the midpoint rule formula (4.27) by the average of its values at n and $n+1$. This method, which is used very commonly, goes by various names. As a method for solving ordinary differential equations, it is known simply as the *trapezoid rule*.

As a final example we apply Simpson's rule to Eq. (5.11). The nature of Simpson's rule requires using the interval (x_{n-1},x_{n+1}) and we have

$$y_{n+1} - y_{n-1} = \tfrac{1}{3}h[f(x_{n-1},y_{n-1}) + 4f(x_n,y_n) + f(x_{n+1},y_{n+1})] \tag{4.30}$$

which is very accurate but is not used often by itself because it is unstable. Instability is an important issue that we shall discuss later.

4.2. NONUNIFORM GRIDS

When the grid size is not constant, it is more difficult to construct finite difference approximations to derivatives and the errors tend to be larger. It is natural to ask why one would ever use a nonuniform grid. The answer is that we want the error to be uniformly distributed over the domain and, usually, the solution changes rapidly in some parts of the solution domain more gradually in others. Since the error is proportional to the product of the grid size and a derivative of the solution, unless we use a high-order approximation (there are difficulties with that), the only other way to reduce the error at reasonable cost is to use a grid that is more concentrated where the derivatives of the function are larger.

The approximation (4.8) is second-order accurate. Applying the kind of Taylor series analysis that we used above, we find that the error in that formula is $(h_n h_{n+1}/6)f'''$. However, as the formula (4.8) is a bit complicated, many people prefer to use the simpler approximation:

$$f'(x_i) \approx \frac{f(x_{i+1} - f(x_{i-1})}{x_{i+1} - x_{i-1}} \tag{4.31}$$

The error in this approximation is

$$\frac{h_{i+1} - h_i}{2} f''(x_i) - \frac{h_{i+1}^3 + h_i^3}{6(h_{i+1} + h_i)} f''' \tag{4.32}$$

Formally, this method has to be called first-order accurate as there is an error term that decreases linearly with the grid size. However, this error depends on the ratio of the grid sizes $\alpha = h_{i+1}/h_i$. For a uniform grid, for which the ratio is one, the first-order error is zero and the method is second-order accurate, as we have already seen. When a grid is refined, the ratio α is usually reduced toward unity and, despite the formal first-order nature of the error, it actually behaves more like a second-order error. For this reason, this method is very commonly used.

4.3. EULER EXPLICIT METHOD

We now come to the principal topic of this chapter—methods for solving initial value problems for ordinary differential equations. As we noted earlier, what is needed is a method for computing the solution at $x + h$ given the solution at x. Once the solution at $x + h$ has been found, we can use it as a new initial condition to compute the solution at $x + 2h$. Continuing in this way, it is possible to generate the solution for as long a range of x as desired. This is the marching method defined earlier.

Our overall strategy will be to learn about the properties by a thorough study of some of the simplest methods. This will tell us what properties are important. We then study more complex (but more accurate) methods in less detail. We start with the simplest of all methods, the explicit Euler method.

This method was given above and is based on the forward difference formula (4.5). As we will see, it is not very accurate so it has been largely superseded by better methods. It is studied here because it is simple to use. The basic equation of the method was given above and is repeated here:

$$y_{n+1} = y_n + hf(x_n, y_n) \tag{4.33}$$

where $y' = f(x,y)$. We shall investigate a number of properties of this method.

First, we note that, if y_n is known, y_{n+1} can be computed by evaluating f, multiplying it by h, and adding the result to y_n; a method having this property of allowing direct calculation of y_{n+1} is called *explicit*.

We shall use y_n to represent the computed value of the function at x_n and $y(x_n)$ to represent the exact value. For convenience, we assume that $h = x_{n+1} - x_n$ is independent of n. Comparing Eq. (4.33) with the exact Taylor series

$$y(x_{n+1}) = y(x_n) + hy'(x_n) + \tfrac{1}{2}h^2y''(x_n) + \cdots \tag{4.34}$$

we see that if $y_n = y(x_n)$, that is, if the solution at x_n is exact, then

$$y(x_{n+1}) - y_{n+1} \approx \tfrac{1}{2}h^2y_n'' \tag{4.35}$$

Thus the error introduced in a single step is approximately $h^2y''/2$. The nomenclature of numerical analysis calls this a *first-order method* because the approximation to the derivative is first-order accurate, as was shown in the preceding section. Also, if we wish to compute the solution from $x = x_0$ to $x = x_f$, the number of steps of size h required is proportional to $1/h$. Ignoring possible error amplification (see below), the error in $y(x_f)$ will be the product of the number of steps and the error per step, and thus be proportional to h.

To understand the global behavior of this method, we note that the solutions of the differential equation (5.11) are a one-parameter family of curves in the $x-y$ plane. Suppose we compute in the direction of increasing x. Starting from a point $x_0, y(x_0)$ on the desired curve, we use Eq. (4.33) to compute y_1. If the solution curve is concave upward, the slope at the initial point is smaller than the average slope of the solution curve between x_0 and x_1 so y_1 is smaller than the correct value $y(x_1)$. If the slope of the solution curve at (x_1, y_1) is smaller than the slope at $(x_1, y(x_1))$, an even larger error is produced at the next step. The error can increase exponentially.

On the other hand, if the slope of the solution curve passing through (x_1, y_1) is greater than the slope of the exact solution curve, the error due to being off the solution curve opposes the error due to the finite difference approximation and the error growth is less disastrous.

What distinguishes the two cases from each other is that in the first case $dy/dx = f$ increases with increasing y for fixed x, while, in the second case, it decreases. Thus, the behavior is determined by the sign of $\partial f/\partial y$. If $\partial f/\partial y > 0$, the solution tends to behave in an unstable manner; but if $\partial f/\partial y < 0$, better results can be expected.

One can obtain an estimate for the global error, in other words, the error after many steps. Not surprisingly, it depends strongly on $\partial f/\partial y$. The detailed analysis can be found in advanced numerical analysis texts (see, e.g., Isaacson and Keller, 1966) and the result is

$$|y(x_n) - y_n| \leq (e^{Lnh} - 1)\frac{hN}{2L} + |\epsilon_0||1 + hL| \tag{4.36}$$

where ϵ_0 is the initial error

$$\epsilon_0 = y(0) - y_0 \tag{4.37}$$

and L and N are upper bounds on $\partial f/\partial y$ and y'', respectively:

$$\partial f/\partial y < L \qquad |y''| < N \tag{4.38}$$

Equation (4.36) tends to be a very generous estimate and may not be very useful. Its importance lies in showing that it is possible for the error to grow exponentially.

We see from Eq. (4.36) that one means of reducing the error is to make the step size h smaller. For the Euler explicit method this results in a decrease in the error that is approximately linear with h. As h is reduced, however, the number of steps to cover a given distance increases. This results in an increase in both cost and round-off error. Round-off error is the error produced because the representation of any number in a computer is finite and therefore must be rounded off after every numerical operation that is done. When the number of operations is large, the round-off error can accumulate and may become sufficiently large to make the results worthless.

Example 4.2: Round-off Error in Solution of Differential Equations To demonstrate the importance of round-off error, we shall solve the ordinary differential equation:

$$y' = y$$

with the initial condition $y(0) = 1$. The exact solution is $y(x) = e^x$. For this equation, Eq. (4.33) reduces to

$$y_{n+1} = (1 + h)y_n$$

This method was programmed and the error in the resulting solution at $x = 1$ is shown in Figure 4.2. We see that the error decreases with decreasing h up to a point but then increases so that there is a step size that produces the minimum error. When h is decreased beyond this value, round-off error becomes larger than the approximation error and causes the total error to increase. The second curve is the error due to the numerical approximation (truncation error) and is linear in h; the difference between the two curves

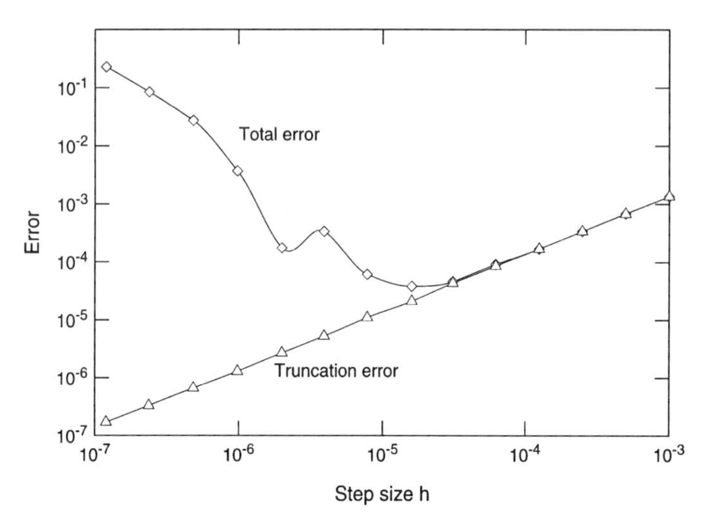

FIGURE 4.2. Error versus step size in the Euler solution of $y' = y$.

is the round-off contribution. The round-off error should be proportional to $h^{-1/2}$ but it is statistical in nature; this explains the irregularity in the error at one point. It is likely that a different program or the same program run on another machine or with another compiler would give different results in detail but not a different trend. In practice, round-off error is not often significant in ordinary differential equation solutions.

4.4. STABILITY

Accuracy, as measured by the error produced by a numerical scheme, is clearly one of its most important properties. There is another and equally important property of a scheme, its stability, that must be understood and appreciated. That is the subject of this section.

To study stability, we first note that, just as some ordinary differential equations can be solved exactly, it is sometimes possible to find exact solutions to difference equations. Generally, if the differential equation has a known exact solution, there is a good chance that a finite difference approximation to it can also be solved analytically. Since we are considering single first-order differential equation of the form of Eq. (4.23), the ideal choice of model equation on which to test this idea is the simplest nontrivial equation of that kind:

$$y' = \alpha y \tag{4.39}$$

Before presenting the solution of this equation, it is important to note that it can serve as a model for the generic equation (4.23). To see this, we expand the right-hand side of Eq. (4.23) in a Taylor series about some point (x_0, y_0):

$$f(x,y) = f(x_0, y_0) + (x - x_0)\frac{\partial f}{\partial x}(x_0, y_0) + (y - y_0)\frac{\partial f}{\partial y}(x_0, y_0) + \cdots \tag{4.40}$$

The higher-order terms may be neglected if $(x - x_0)$ and, thus, $(y - y_0)$ are small. Substituting Eq. (4.40) into the ordinary differential equation (4.23), lumping all of the terms that can be evaluated into a single term, and introducing new names for the coefficients, we have

$$y' = \alpha y + (\beta_0 + \beta_1 x) \tag{4.41}$$

The solution to this equation consists of two parts, a particular solution of the inhomogeneous equation and a solution of the homogeneous equation. Both parts of the solution are equally well (or badly) approximated by the difference method, but the stability characteristics are governed almost entirely by the homogeneous part of the equation. This is why the study of the behavior of numerical methods applied to Eq. (4.39) is important; most of what we need to know about the full equation (4.23) can be discovered through study of the simpler equation. This will be demonstrated in the examples. For later reference, we note that $\alpha = \partial f/\partial y$.

Let us now return to Eq. (4.39); its exact solution is well known:

$$y(x) = y_0 e^{\alpha x} \tag{4.42}$$

If we solve the differential equation (4.39) by Euler's method with step size h, the finite difference approximation is

$$y_{n+1} = y_n + \alpha h y_n \tag{4.43}$$

which is easily rearranged to

$$y_{n+1} = y_n(1 + \alpha h) \tag{4.44}$$

Using this result inductively, we find that the finite difference solution to the equation (4.39) at point x_n is

$$y_n = y_0(1 + \alpha h)^n \tag{4.45}$$

Thus the solution to the difference equation is exponential. In fact Eq. (4.42) is identical to Eq. (4.45) if $e^{\alpha h}$ is replaced by $(1 + \alpha h)$. Since

$$e^{\alpha h} = 1 + \alpha h + \tfrac{1}{2}(\alpha h)^2 + \cdots \tag{4.46}$$

we see that Euler's method reproduces the first two terms of the Taylor series of the exponential. This is another way of saying that the Euler explicit method is first-order accurate.

Important things can be learned from this solution. Suppose the coefficient α in Eq. (4.39) is complex. It is essential to allow this possibility because higher-order differential equations (or systems of equations) may have oscillatory solutions; these functions also satisfy first-order equations with complex coefficients. Recall that, if $\alpha = \beta + i\gamma$, then

$$e^{\alpha x} = e^{\beta x}(\cos\gamma x + i\sin\gamma x) \tag{4.47}$$

If β, the real part of α, is positive, both the exact solution (4.47) and the approximate solution (4.45) grow with increasing x. For large values of βh, the approximate solution is badly in error, but it grows with increasing x, as does the exact solution. This is an issue of accuracy rather than stability.

On the other hand, if the real part of α is negative, there can be a serious difficulty. First, consider the case of real α. The solution of the differential equation decays with increasing x, but the solution found from the Euler explicit method may not decay at all. In fact if $|1+\alpha h| > 1$, the magnitude of the numerical solution grows with increasing x. In particular, if $\alpha h < -2$, the numerical solution grows in magnitude while the exact solution decays; it does this in an oscillatory manner because $1+\alpha h$ is negative. Clearly, this is unacceptable and is a situation to be avoided at all costs. It is an example of one of the greatest dangers in numerical differential equation solvers—*instability*.

The problem also occurs for complex α. If the real part of α is negative and $|1+\alpha h| > 1$, we obtain a growing solution when we should find a decaying one. (The only difference is that now both solutions also oscillate in sign.) The region in the complex αh plane (see Figure 4.3) in which $|1+\alpha h| < 1$ is a disk of unit radius centered at the point -1. For any value of αh in the left-half plane outside this circle Euler's method yields an increasing solution when it should produce a decaying one, that is, the method is unstable.

The issue of stability of numerical methods for solving both ordinary and partial differential equations is extremely important. A number of definitions of stability are in use. The type of stability we have defined is called *A-stability* in the numerical analysis literature and is the one most often used. Simply put, a method is stable if it produces a bounded solution when the solution of the differential equation is bounded; otherwise, it is unstable. A-stability is equivalent to this property for Eq. (4.39). Euler's method is said to be *conditionally stable* because it is stable for some, but not all, values of αh. A method that is stable for all values of this parameter is said to be *unconditionally stable*; one that is not stable for any value of αh is called *unconditionally unstable*.

One might think that basing our analysis of stability on an equation as simple as Eq. (4.39) is not meaningful. As shown above, any equation can be linearized locally; α represents the local value of $\partial f/\partial y$. Since the stability

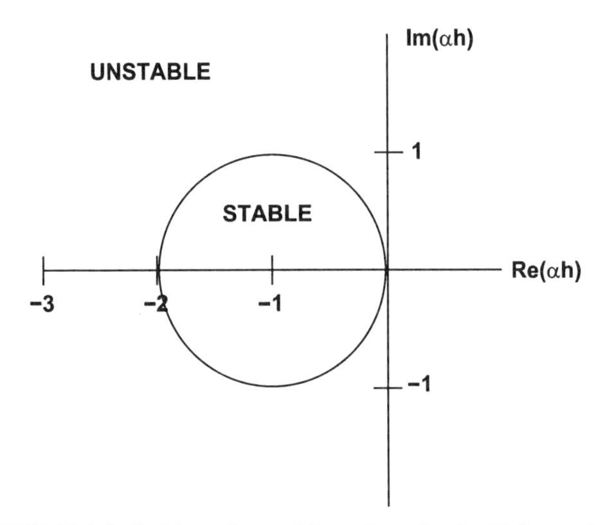

FIGURE 4.3. Stable and unstable regions for the Euler method.

condition for Eq. (4.39) always takes the form of restriction on αh, we surmise that a method will be stable if $h\partial f/\partial y$ lies within the stability bound on αh for Eq. (4.39). This turns out to be very nearly correct; if this condition is obeyed, the solution will not grow when it should not. Note that the question of stability has nothing whatever to do with the accuracy of the solution. It is quite possible for a method to be stable and highly inaccurate; it is also possible for a method to be accurate and unstable. For methods of the latter type, see the section on multistep methods; a method with these characteristics is of little value by itself but may be useful as part of a combination of methods; we demonstrate this later in this chapter.

Euler's method is unstable if $(1 + \alpha h)$ is greater than unity in magnitude. When Euler's method is unstable, it usually produces a result that alternates in sign at every step. A growing solution that changes sign at every step is a sure indicator that something is amiss. The only cures for the problem are to reduce the step size so that αh is within the stability bounds or to change to another, more stable, method.

We mention that spreadsheet programs for personal computers provide a simple and effective way to implement and test methods for solving ordinary differential equations. To use them, the parameters are entered into the top of the spreadsheet and can be given names for convenience in referring to them. The initial values of the independent and dependent variables are then entered into a row, and the formulas for advancing the solution one step (and for updating the independent variable) are entered into the next row. Copying these formulas to a number of lower rows then produces the solution for a large range of the independent variable. The graphics capabilities of these programs allows a graph of the solution to be obtained very quickly. Note that the accuracy of these programs is usually greater that the inherent accuracy of

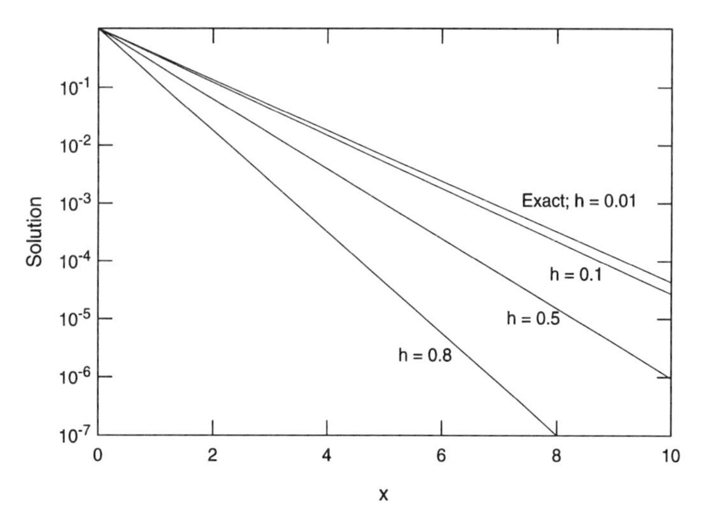

FIGURE 4.4. Euler solution of $y' = -y$ for various step sizes.

the machine discussed earlier; this is because they are designed for financial calculations. Programs designed for scientific computation, such as Matlab, Mathematica, and Maple are also useful tools for these problems.

Example 4.3: Euler Solution of $y' = -y$ Consider Eq. (4.39) with $\alpha = -1$, that is, $y' = -y$. The behavior of the solution is shown in Figure 4.4 for small h. Both the exact and computed solutions are exponential. In fact, the error is easily computed. The computed solution is

$$\begin{aligned} y_n &= (1-h)^n \approx (e^{-h} - \tfrac{1}{2}h^2)^n \approx e^{-nh} - (\tfrac{1}{2}nh^2)e^{-(n-1)h} \\ &= e^{-nh}[1 - (\tfrac{1}{2}nh^2)e^h] \approx e^{-x}[1 - \tfrac{1}{2}hx] \end{aligned} \tag{4.48}$$

since $e^h \approx 1$ for small h. Thus the error is approximately $-hxe^{-x}/2$ and is proportional to h as expected. The relative error, that is, the error divided by the actual solution, increases linearly with x. These results are quite accurate up to about $h = 0.1$ ($\alpha h = 0.1$ in the more general case). The error divided by h, which should be independent of h, is given in Table 4.4 and is indeed independent of h up to $h \approx 0.2$.

For larger values of h, the results are shown in Figure 4.5; the computed result behaves reasonably for h less than about 0.5. At $h = 1$ the solution becomes zero at the first step and stays there. For $1 < h < 2$, the solution oscillates with decaying amplitude. Finally, for $h > 2$, the instability is obvious and is manifested as a growing oscillation.

Example 4.4: Euler Computation of an Oscillating Solution As the next example, we take a differential equation with an oscillating solution. This situa-

TABLE 4.4. Error in Euler Solution of $y' = -y$

Grid (h)	$y(1)$	Error (ϵ)	ϵ/h
0.001	0.367695	1.84×10^{-4}	0.184
0.01	0.366032	1.84×10^{-3}	0.184
0.02	0.364170	3.71×10^{-3}	0.185
0.05	0.358496	9.39×10^{-3}	0.188
0.1	0.348678	1.92×10^{-2}	0.192
0.2	0.327680	4.02×10^{-2}	0.201
0.5	0.250000	1.18×10^{-1}	0.236

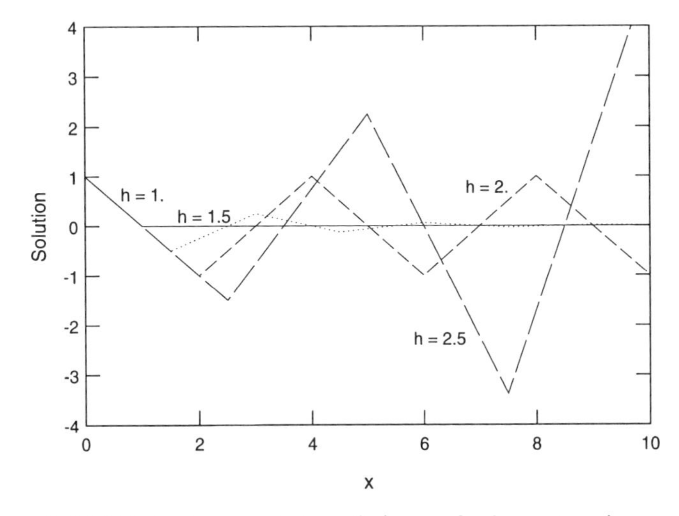

FIGURE 4.5. Euler solution of $y' = -y$ for large step sizes.

tion arises when α is complex (or pure imaginary). The same problem will be done two ways to illustrate the use of the Euler explicit method for systems of equations. The complex form is simpler for analytical treatment, so we take it first:

$$y' = i\pi y \tag{4.49}$$

The factor π has been inserted to make the solution easier to interpret. With the initial condition $y(0) = 1$, the exact solution is

$$y(x) = e^{i\pi x} = \cos \pi x + i \sin \pi x \tag{4.50}$$

We are usually interested in the real part of the solution.

Applying Euler's method to this problem, we find

$$y_{n+1} = (1 + i\pi h)y_n \tag{4.51}$$

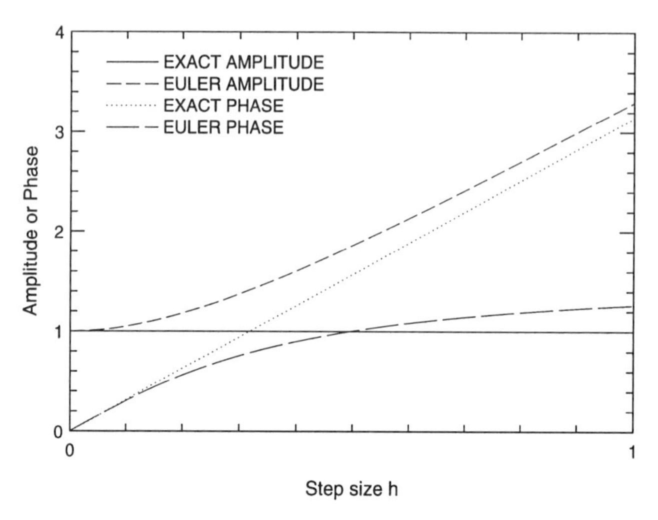

FIGURE 4.6. Amplitude and phase of the solution of $y' = -i\pi y$ obtained with the explicit Euler method.

and, by induction, the solution is

$$y_n = (1 + i\pi h)^n \tag{4.52}$$

This is more easily compared with the exact result if we use the polar form $(1 + i\pi h) = Ae^{i\theta}$ in which the amplitude is

$$A = (1 + \pi^2 h^2)^{1/2} \tag{4.53}$$

while the phase angle is

$$\theta = \tan^{-1} \pi h \tag{4.54}$$

Thus Euler's method results in the replacement of the exact solution $e^{i\pi x} = e^{i\pi nh}$ by $A^n e^{in\theta}$. The amplitude, which is unity in the exact solution, has been replaced by $A^n = (1 + \pi^2 h^2)^{n/2}$, which is greater than unity. Thus a solution that remains bounded has been replaced by a growing one, that is, Euler's method applied to Eq. (4.49) is unstable for any h. This is not surprising because Figure 4.3 clearly shows that the entire imaginary axis lies outside the domain of stability of the Euler method. Despite this, the growth is fairly weak if h is small. In fact

$$(1 + \pi^2 h^2)^{1/2} \approx 1 + \tfrac{1}{2}\pi^2 h^2 \tag{4.55}$$

is only slightly greater than unity if h is not too large. Thus the instability is rather mild and Euler's method may be useful even though it is technically unstable. The amplitude as a function of step size is shown in Figure 4.6.

We next look at the phase of the solution. Euler's method replaces the phase of the exact solution $n\pi h$ by $n\tan^{-1}\pi h$. For small h, the Taylor series expansion

$$\tan^{-1}\pi h \approx \pi h - (\tfrac{1}{3}\pi h)^3 \tag{4.56}$$

shows that the phase error is quite small for small values of πh. For larger values of πh, the error is much larger. The phase as a function of step size is also shown in Figure 4.6.

This example illustrates a number of points that arise with oscillatory solutions. The order of a method may not be a good indicator of its quality nor is the error that was defined earlier. The latter may behave very erratically. Consequently, when one is solving problems with oscillatory solutions, it is more appropriate to discuss the accuracy of numerical methods in terms of amplitude and phase errors rather than the apparently simpler truncation error.

Example 4.5: Euler's Method for Two ODEs In this example, we treat the problem of the preceding example in terms of real equations and compute the solution numerically. If we write

$$y = y_1 + iy_2 \tag{4.57}$$

then the differential equation of Example 4.4 can be replaced by the pair of equations:

$$\begin{aligned} y_1' &= -\pi y_2 \\ y_2' &= \pi y_1 \end{aligned} \tag{4.58}$$

The initial conditions are $y_1(0) = 1$, $y_2(0) = 0$.

Euler's method can be applied to a system of differential equations just as easily as it is to a single equation. For the system (4.58), we may write

$$\begin{aligned} y_{1,n+1} &= y_{1,n} - \pi h y_{2,n} \\ y_{2,n+1} &= y_{2,n} + \pi h y_{1,n} \end{aligned} \tag{4.59}$$

This problem can be solved with the program EULER, which can be found on the Internet site described in Appendix A. In fact, the subroutine solves the system of equations

$$y_i' = f_i(x, y_1, y_2, \ldots, y_m) \qquad m = 1, 2, \ldots, n \tag{4.60}$$

for any number of equations up to 10 (this limit comes from the declared dimension and is easily increased).

TABLE 4.5. Solution of $y' = -2xy$ by Explicit Euler Method with $h = 0.4$

x	y	Exact Solution
0.0	1.0000	1.0000
0.4	1.0000	0.8521
0.8	0.6800	0.5273
1.2	0.2448	0.2369
1.6	0.0098	0.0773
2.0	−0.0027	0.0183
2.4	0.0016	0.0032
2.8	−0.0015	3.9×10^{-4}
3.2	0.0019	3.6×10^{-5}
3.6	−0.0029	2.4×10^{-6}
4.0	0.0055	1.1×10^{-7}
4.4	−0.0121	3.9×10^{-9}
4.8	0.0305	9.9×10^{-11}

The results obtained are, of course, identical to those obtained earlier, so they are not given again.

Example 4.6: An Equation with Variable Coefficients Next we consider an equation with nonconstant coefficients:

$$y' = -2xy \tag{4.61}$$

with $y(0) = 1$. The exact solution is $y(x) = e^{-x^2}$.

Computing the solution with $h = 0.4$ gives the results shown in Table 4.5. The results are presented in tabular form because it is easier to see the behavior of the solution in this way. Due to the large step size, the solution is not very accurate but, at least up to $x = 1.6$, it bears some resemblance to the correct solution. Beyond $x = 1.6$, the point at which $|h\partial f/\partial y| = 2xh$ first exceeds unity, the numerical solution oscillates, while the exact solution remains monotonic. Beyond $x = 3.2$ (where $h\partial f/\partial y$ first exceeds 2), the solution not only oscillates but grows in amplitude as well. This accords with the conjecture made about the application of the stability criterion to differential equations other than the model equation earlier; these results show that it is valid for equations with nonconstant coefficients. It holds for nonlinear equations as well.

For this equation, the instability occurs in a region in which the solution is nearly zero and very smooth, a region where one would least expect a problem. This behavior is a typical consequence of instability; the problem often occurs where one least expects it. It is due entirely to the numerical method and not the differential equation or its solution. We shall later see that instability can also wreak havoc with solutions of systems of equations. The instability arises not where the solution is difficult to compute but where it is

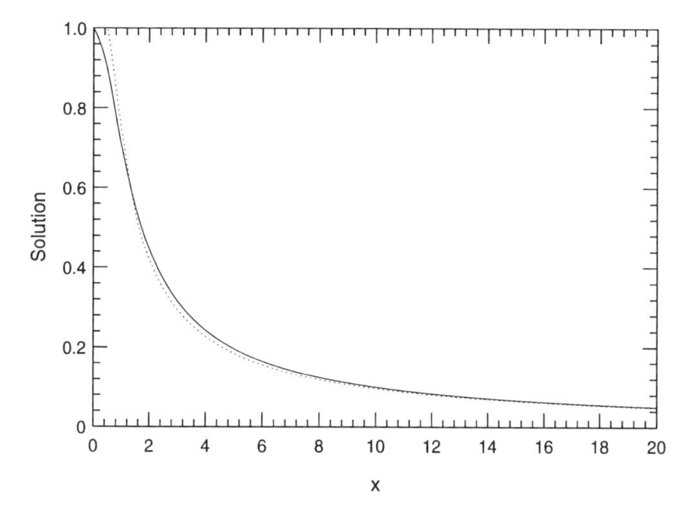

FIGURE 4.7. Solution of the nonlinear equation $y' = -xy^3$ obtained with the Euler method with $h = 0.5$. The solid line is the numerical solution, the dashed line, the exact solution.

well behaved. When this problem occurs for systems of equations, it is usually due to a property called stiffness.

Example 4.7: A Nonlinear Equation Nonlinear equations pose no special difficulty for explicit methods other than the ones we encountered for linear equations. To demonstrate this, let us try solving the equation:

$$y' = -xy^3 \tag{4.62}$$

with the initial condition $y(0) = 1$. The exact solution of this problem is

$$y(x) = \frac{1}{(1 + x^2)^{1/2}} \tag{4.63}$$

The explicit Euler method with step size h applied to Eq. (4.62) gives

$$y_{n+1} = y_n - hx_n y_n^3 \tag{4.64}$$

The solution obtained is shown in Figure 4.7. A relatively large step size, $h = 0.5$, was used to demonstrate the fact that the nonlinearity poses no special difficulty for the Euler method. In fact, for this equation, $\partial f/\partial y = -3xy^2$ and, even though x increases (which caused difficulty in the preceding example), y decreases and the stability condition is never violated. Despite the rather large step size, a reasonably good solution is obtained.

4.5. BACKWARD OR IMPLICIT EULER METHOD

In the last two sections we investigated the properties of one method, the Euler explicit method. The properties we found are similar to those of other explicit methods. In this section we will look into properties of methods belonging to the other major class of numerical methods for ordinary differential equations—implicit methods.

The method we shall study is similar in many respects to the explicit Euler method but the differences are important and interesting. This method is known as either the *backward* or *implicit Euler* method and is based on the backward difference formula (4.6). The difference equation for this method is

$$y_{n+1} - y_n = hf(x_{n+1}, y_{n+1}) \tag{4.65}$$

Like the explicit Euler method presented above, this method is first-order accurate; we will present an estimate of the error shortly. Before doing so, we note an important difference between this method and the preceding one. Equation (4.65) cannot be solved for y_{n+1} without considerable effort unless the function f is extremely simple; in general, it is a nonlinear algebraic equation that must be solved with iterative methods such as the Newton or secant methods described in Appendix C. Any method that requires the solution of such an equation for the value of the function at the new step is called *implicit*; the backward Euler method is the simplest member of this class of solution methods.

One might be tempted to reject implicit methods outright because they require more computation per step than explicit methods, but they have the important redeeming virtue of stability. We will therefore spend some time studying the backward Euler method as a representative member of the class.

The error analysis for this method is more complicated than for the explicit Euler method, so we do it only for the case in which the method is applied to the model equation (4.39). For this equation, the difference equation (4.65) becomes

$$y_{n+1} - y_n = \alpha h y_{n+1} \tag{4.66}$$

which, because it is linear, is easily solved; we have

$$y_{n+1} = (1 - \alpha h)^{-1} y_n \tag{4.67}$$

so that

$$y_n = (1 - \alpha h)^{-n} y_0 \tag{4.68}$$

We emphasize that the solution was found so easily only because the differential equation is linear and that this is a special case. The solution is not so easily found for nonlinear equations.

Use of the backward Euler method is equivalent to replacing $e^{\alpha h}$ by $(1-\alpha h)^{-1}$. Comparing the Taylor series expansion of $(1-\alpha h)^{-1}$

$$(1-\alpha h)^{-1} = 1 + \alpha h + (\alpha h)^2 + \cdots \tag{4.69}$$

with the Taylor series expansion (4.46) for $e^{\alpha h}$, we see that this method reproduces only the first two items of the Taylor series correctly and is, therefore, first-order accurate. It is interesting to note, however, that the error to lowest order is

$$y(x_n) - y_n = -\tfrac{1}{2}(\alpha h)^2 y(x_n) \tag{4.70}$$

which is equal in magnitude to the error of the explicit Euler method but of opposite sign.

Next, let us consider the stability of this method. It is not difficult to see that, if the real part of αh is negative, then the real part of $(1-\alpha h)$ is always greater than unity and, no matter what the imaginary part of αh may be, the magnitude of $(1-\alpha h)$ is greater than 1. Consequently, the magnitude of $(1-\alpha h)^{-1}$ is less than 1 for all αh in the left-half complex αh plane and, according to the definition introduced above, the backward Euler method is *unconditionally stable*. We emphasize again that stability does not imply anything about the accuracy of the method or its overall usefulness.

This result is somewhat general. The only methods that are unconditionally stable are implicit methods; implicit methods, therefore, become the methods of choice when stability is the overriding consideration. Not all implicit methods are unconditionally stable, but they are almost always (if not always) more stable than explicit methods of similar accuracy and type. This brings us to the major tradeoff that plays an important role in much of the remainder of this chapter. Explicit methods offer more speed per step, but, in some problems, the step size required for stability may be so small that the advantage is lost. Implicit methods place much weaker restrictions on the size of the step allowed by stability, but the computation cost per step is higher; they can be less expensive overall if they can be run with a step size large enough to compensate for the additional cost of each step.

Example 4.8: Implicit Euler Solution of $y' = -y$ We repeat the problem of Example 4.1 using the implicit Euler method. The resulting solution is

$$y_n = (1+h)^{-n} \tag{4.71}$$

and is plotted in Figure 4.8. It is clear that the error is opposite to that produced by the explicit Euler method, as expected from the truncation error formula given above. More importantly, we see that for large step sizes the error continues to increase but the solution remains quite smooth. There is no tendency for the solution to oscillate or amplify.

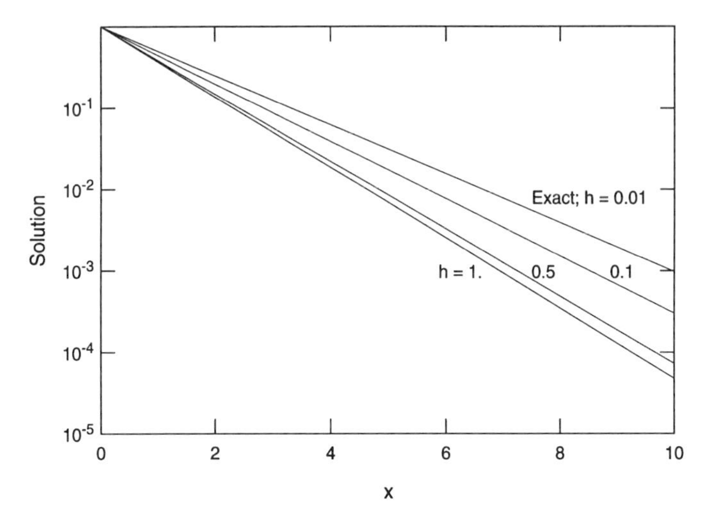

FIGURE 4.8. Solution of $y' = -y$ obtained with the implicit Euler method with various step sizes.

A warning should be issued here. The inaccurate solution produced by the backward Euler method with a large step size can be a curse rather than a blessing. Because the solution seems "reasonable," there is a temptation to accept it as correct even though it contains considerable error; the consequences can be very serious indeed. One can never question the validity of a numerically obtained solution carefully enough. The greatest disaster that can occur when problems are solved on computers is *not* instability. Instability produces solutions that are so far from correct that there is little possibility of accepting them; any sensible engineer would know that they are incorrect and search for a cause or try another approach. A numerical solution that is incorrect by 10, 20, or 50% or even a factor of 2 poses a greater danger because one might believe it to be correct; the consequences of doing so can be very severe.

The easiest way to assure that the solution obtained is accurate enough is to repeat the procedure using half the step size. This (for a first-order method) should halve the error, and it is then possible to estimate the error to a sufficiently accurate approximation using the Richardson method; we shall demonstrate how this can be done in more detail in Section 4.6.

Example 4.9: Implicit Euler Solution of $y' = i\pi y$ Solving the problem of Example 4.4 with the backward Euler method gives the solution

$$y_n = (1 - i\pi h)^{-n} \tag{4.72}$$

We can write

$$(1 - i\pi h)^{-1} = Ae^{i\theta} \tag{4.73}$$

with $A = (1 + \pi^2 h^2)^{-1/2}$ and $\theta = \tan^{-1} \pi h$. The phase is exactly the same as the phase produced by the explicit Euler method, while the amplitude is always less than unity. Thus the implicit Euler method is also stable for oscillating solutions, as expected from the argument made above. For small step sizes, the implicit method produces an amplitude error of approximately the same size as the explicit method but in the opposite direction. The "stable" solution produced is thus no better than the "unstable" one that the explicit method gives. In fact, the choice between the two methods is a matter of the user's taste and judgment rather than any objective criterion.

Solving the problem of Example 4.5 with the implicit method requires the solution of a system of two equations for the two dependent variables at the new step. For this linear problem we can solve the system of equations by hand or through the use of a standard linear equation solver at fairly modest cost.

The problem of Example 4.6 also offers no challenge to the implicit Euler method. The problem is linear and a well-behaved solution is produced. At large x, however, $\partial f/\partial y$ becomes large and accuracy will be sacrificed unless the step size is reduced.

Finally, the nonlinear equation of Example 4.7 poses a serious problem for the implicit method. The equation from which y_{n+1} must be obtained is

$$y_{n+1} - y_n = -hx_{n+1} y_{n+1}^3 \tag{4.74}$$

and is a cubic equation for the desired quantity. It may be solved either with the explicit formula for the solution of cubic algebraic equations or by some method such as Newton–Raphson. In either case, the cost of obtaining the solution is considerably larger than it was for the explicit method.

For other nonlinear equations there is no choice but to use an iterative solver. If the step size is small enough that the error created by the backward Euler approximation is not too large, the change in the solution from one step to the next is fairly small and the following strategy is usually successful. For the initial guess at the new step, we take the converged value at the preceding step. It is then sufficient to perform just two Newton–Raphson iterations to find the solution at the new step with an error less than the error inherent to the solution method. With this strategy, the cost of the implicit Euler method is only about double that of the explicit method. However, the relative cost increase is larger for systems of equations.

4.6. ERROR ESTIMATION AND ACCURACY IMPROVEMENT

In most cases, the desired accuracy of the solution fixes the step size. Typically, the user wants the solution for some range of the independent variable with a specified accuracy. It is difficult to predict (and, therefore, to control) the global accuracy of a method, but we can make an educated guess based on the single-step error. More sophisticated approaches to error control will be discussed later. For the first-order methods considered so far, the magnitude of the single-step error is approximately $h^2 y''/2$. As the step size is decreased, the number of steps required to reach a particular final value of x increases in proportion to $1/h$. If the sum of the single-step errors is taken as an estimate of the global error [rather than the more accurate, but harder to apply, Eq. (4.36)], the global error is proportional to h. Thus the step size must be reduced in proportion to the error tolerance, which can be very expensive if high accuracy is required. On the other hand, with a more accurate method, for example, one in which the error scales like h^p, the problem would be much less severe because the required step size would be proportional to $\epsilon^{1/p}$ where ϵ is the desired error. If ϵ is small, it is desirable that p be fairly large. Thus there is considerable incentive to seek more accurate methods and we shall spend much of the rest of this chapter looking at some possibilities. Naturally, increased accuracy comes at a price.

There are a number of ways in which one can obtain increased accuracy. In this section we discuss some of the simpler ones; other methods will be described later. We begin by discussing methods for estimating errors.

4.6.1. Error Estimation

The fact that the error is proportional to the step size h (or some higher power of h in the methods to be introduced below) allows us to estimate the error with little effort. The concept is similar to the one used in Richardson extrapolation and is therefore sometimes called the *Richardson estimator.*

We know that, for a first-order method, the error is proportional to the step size h times some unknown function of the independent variable x. This knowledge can be represented mathematically as:

$$f(x;h) \approx f(x) + hf_1(x) \tag{4.75}$$

where $f(x;h)$ is the solution computed numerically with step size h, $f(x)$ is the exact solution (which is not available to us), and $f_1(x)$ is an unknown function that can be shown to be independent of h.

If a second solution is computed with the same method using half the step size, the solution obtained is

$$f(x;\tfrac{1}{2}h) \approx f(x) + \tfrac{1}{2}hf_1(x) \tag{4.76}$$

If we subtract the two results, we obtain

$$f(x;\tfrac{1}{2}h) - f(x;h) \approx -\tfrac{1}{2}hf_1(x) \approx f(x) - f(x;\tfrac{1}{2}h) \tag{4.77}$$

which is approximately the error in the solution obtained with the smaller step size or half the error in the solution obtained with the larger step size. This provides a simple means of estimating the error.

If one cannot afford to repeat the calculation with half the step size, the estimation can be based on calculations with step sizes h and $2h$.

Of course, expressions (4.75) and (4.76) contain higher-order terms that we have ignored. When h is small, the first term (the one we kept) is much larger than the others, and this estimate is fairly accurate. When the step size h is large, this estimate may not be very accurate, but it will be large and that tells the user that the solution obtained is not very accurate, which is often all we need to know.

4.6.2. Richardson Extrapolation

The idea of using Richardson extrapolation to improve the accuracy of a solution is a natural one. Suppose we use Euler's method to compute y_{n+1}:

$$y_{n+1}^{(1)} = y_n + hy'_n \tag{4.78}$$

The superscript (1) indicates that this value will not be accepted as the final result. We repeat the calculation with step size $h/2$, using two steps to go the same distance:

$$\begin{aligned} y_{n+1/2} &= y_n + \tfrac{1}{2}hy'_n \\ y_{n+1}^{(2)} &= y_{n+1/2} + \tfrac{1}{2}hy'_{n+1/2} \end{aligned} \tag{4.79}$$

Then we combine these results using the Richardson extrapolation method of Section 3.2; the result is

$$y_{n+1} = 2y_{n+1}^{(2)} - y_{n+1}^{(1)} \tag{4.80}$$

Applied to the test differential equation $y' = \alpha y$, this gives

$$y_{n+1} = 2(1 + \tfrac{1}{2}\alpha h)^2 y_n - (1 + \alpha h)y_n = (1 + \alpha h + \tfrac{1}{2}(\alpha h)^2)y_n \tag{4.81}$$

which is a second-order accurate approximation to the exact result. Assuming that evaluation of y' is the most expensive part of the calculation, the total computational cost is only a little more than the cost of doing the calculation on the finer grid since the value of y'_n needed in Eq. (4.79) is the same one needed by Eq. (4.78). This procedure can be applied to any of the other methods presented in this chapter.

This method has not found widespread popularity. Apparently one reason is that application of Richardson extrapolation to solving ordinary differential equations was not given much attention until after some of the more popular methods given below were already in common use. The particular method we have just given turns out to be a rearrangement of the second-order Runge–Kutta method presented in Section 4.8.

4.6.3. Trapezoid Rule

Another way to obtain increased accuracy is to use a higher-order method, but, as we have seen, accurate estimation of derivatives requires using more data points in the finite difference approximation. Methods based on the use of data from multiple points are among the best methods available, but they require a somewhat more involved analysis, so we defer discussion of them until later.

If we are restricted to using data only at points x_n and x_{n+1}, the simplest and best means of obtaining a second-order formula is to use the trapezoid rule. This method was derived using the trapezoid rule for integration in Section 4.1.3. It can also be derived by using the second-order accurate central difference formula. The basic equation for the method is

$$y_{n+1} - y_n = \tfrac{1}{2}h[f(x_n, y_n) + f(x_{n+1}, y_{n+1})] \tag{4.82}$$

This method is implicit and therefore suffers the same difficulties as the backward Euler method. That is, if f is a nonlinear function of y, we need to iterate to find y_{n+1}. To obtain a better understanding of this method, let us apply it to the test differential equation (4.39). With little difficulty we find

$$y_n = \left(\frac{1 + \alpha h/2}{1 - \alpha h/2}\right)^n y_0 \tag{4.83}$$

Thus the trapezoid rule method is equivalent to approximating $e^{\alpha h}$ by $(1 + \alpha h/2)/(1 - \alpha h/2)$. The Taylor series expansion of this function is

$$\frac{1 + \alpha h/2}{1 - \alpha h/2} = 1 + \alpha h + \tfrac{1}{2}(\alpha h)^2 + \tfrac{1}{4}(\alpha h)^3 + \cdots \tag{4.84}$$

and is more accurate than the corresponding results for either the forward or backward Euler method. The first three terms of the Taylor series for $e^{\alpha h}$ is reproduced so the method is second-order accurate.

Because the trapezoid rule provides more accuracy than backward Euler method at almost no increase in cost, it is almost always the preferred choice. There is, however, one property that makes the backward Euler method attractive. When αh is large and negative, the trapezoid rule approximation to $e^{\alpha h}$ becomes -1, while that for the backward Euler method goes to zero as $(\alpha h)^{-1}$. Although $(\alpha h)^{-1}$ is very different from the exponential function it is supposed to approximate, its asymptotic limit is 0, which is better than the limiting value of -1 produced by the trapezoid rule. This property makes use of the backward Euler method worthwhile when large step sizes are required. Examples of where this situation arises are in stiff problems (see Section 4.11)

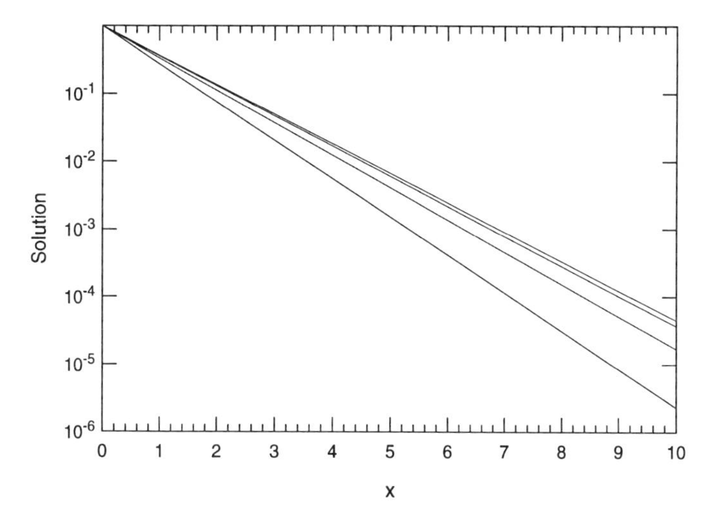

FIGURE 4.9. Solution of $y' = -y$ obtained with the trapezoid rule method with step sizes $h = 0.1$, 0.5, 1.0, and 1.5.

and when uses implicit methods with large time steps are used to find steady-state solutions (this is a common method for partial differential equations).

Example 4.10: Trapezoid Rule Solution of $y' = -y$ The trapezoid rule is much more accurate than the first-order methods discussed earlier. Results for the test problem are shown in Figure 4.9. The results for $h < 1$ are very close to the exact solution. For $h = 2$ (not shown) the trapezoid rule produces a solution that decays to zero on the first step and stays there. For $h > 2$ the result is a damped oscillation with the decay rate decreasing as h increases. Hence, even though the method is stable, it should not be used with too large a step size if an accurate solution is to be obtained.

Comparing the trapezoid method with the first-order methods, we note that, for small step sizes, the magnitude of the error produced by either of the Euler methods with step size h_e, is $h_e y''/2 = h_e y/2$ for the differential equation $y' = -y$. The error produced by the trapezoid rule with step size h_t is $h_t^2 y'''/12 = -h_t^2 y/12$ for the same differential equation. So, to equal the accuracy of the trapezoid rule for this equation, the Euler methods would have to use a step size $h_e = h_t^2/6$, which is much smaller than h_t if the latter is small. The higher the required accuracy, the greater the relative cost of the Euler methods.

Example 4.11: Trapezoid Rule Solution of $y' = i\pi y$ Applying the trapezoid method to $y' = i\pi y$ with $y(0) = 1$ produces the solution

$$y_n = \left(\frac{1 + i\pi h/2}{1 - i\pi h/2} \right)^n \tag{4.85}$$

One can show that

$$\left(\frac{1 + i\pi h/2}{1 - i\pi h/2}\right) = e^{i\theta} \tag{4.86}$$

with $\theta = 2\tan^{-1}\pi h/2$. The amplitude is exactly unity, so there is no amplitude error. The phase error is considerably smaller than for the Euler methods. The lack of amplitude error makes the trapezoid rule an excellent choice for solving problems with oscillatory solutions. This is the reason why the trapezoid rule is the basis for some of the best methods of solving partial differential equations. For the application of this method to partial differential equations (for which it is called the Crank–Nicolson method), see Section 2 of Chapter 5.

Applying the trapezoid rule to systems of equations or nonlinear equations poses difficulties similar to those described for the implicit Euler method. Since these methods are only occasionally used in solving ordinary differential equations, details of these applications are not given here.

4.6.4. Other Approaches

Another approach to finding higher-order methods makes direct use of the Taylor series:

$$y_{n+1} = y_n + hy_n' + h^2 y_n'' + h^3 y_n''' + \cdots \tag{4.87}$$

In fact all high-order methods can be derived from this equation by using appropriate approximations for derivatives higher than the first. Alternatively, the derivatives can be obtained directly by differentiating the differential equation. One need only be careful to take the total (as opposed to the partial) derivative of f with respect to x, that is,

$$y'' = \frac{df}{dx} = \frac{\partial f}{\partial x} + \frac{\partial f}{\partial y}y' = \frac{\partial f}{\partial x} + f\frac{\partial f}{\partial y} \tag{4.88}$$

In some cases the differentiation is easy to perform and the evaluation of y'' is no more difficult than the evaluation of y'. For these problems, this may be a useful approach. More commonly, evaluation of the higher derivatives becomes successively more difficult so this method is not practical. One must then use a method that estimates the higher derivatives by a finite difference approximation. Some of these, not necessarily derived in this way, are studied in the sections below.

As we saw in Section 4.1, estimation of higher-order derivatives (or the higher-order accurate estimation of first derivatives) requires data at more than two points. For this reason, all explicit methods of order higher than 1 (and implicit methods of order higher than 2) must use data at points other than x_n and x_{n+1}. There are two approaches to including more points; they lead to two major types of high-order methods. In *multistep methods* the value of y_{n+1} is computed from estimates of the solution that have already been computed, that is, $y_n, y_{n-1}, y_{n-2}, \ldots$. These methods have the disadvantage that other methods must be used at the first few points, but they are simple to program and fast. In the other category of methods, *Runge–Kutta* methods, temporary use is made of the values of the function at points between x_n and x_{n+1}. Thus we might first compute the value of y at $x_{n+1/2} = x_n + h/2$, then use this value to compute y_{n+1}. Runge–Kutta methods have no problem in getting started and can be very accurate, but they tend to be more expensive in terms of computation time. They are also closely related to *predictor–corrector* methods, which are discussed in Section 4.7.

4.7. PREDICTOR–CORRECTOR METHODS

When the differential equation $y' = f(x,y)$ is nonlinear, implicit methods require the solution of a nonlinear algebraic equation. This must be done with an iterative method such as Newton–Raphson. If the iteration process is not carried far enough, an additional error is introduced; it is the difference between the current iterate and the solution obtained if iteration is continued until the solution converges to the precision of the computer. In fact, it does not make sense to continue the iterative method beyond the point at which the error due to incomplete convergence of the iterative method is much smaller than the discretization error, that is, the error made in approximating the differential equation. Complete convergence of the iterative method at each step is required only when stability is crucial.

We are thus led to consider a compromise between the simplicity of explicit methods (with attendant lack of stability) and the stability of the implicit approach (with more difficult computation). This compromise, which can be thought of as an approximate iterative solution to the implicit equation, is the *predictor–corrector* approach. As the name suggests, the method consists of the application of an explicit method to estimate (predict) the new value of the dependent variable, and the subsequent application (or applications) of an implicit method to improve (correct) it. There are many predictor–corrector methods. It is not our purpose to survey this class of methods exhaustively; we examine a few to discover their general properties and give a few of the better methods of the class.

The simplest predictor–corrector scheme uses the explicit Euler method as a predictor

$$y^*_{n+1} = y_n + hf(x_n, y_n) \tag{4.89}$$

where the asterisk (*) indicates that this is a predicted value that is not accepted as the final result. To correct this value, we use the trapezoid rule:

$$y_{n+1} = y_n + \tfrac{1}{2}h[f(x_n, y_n) + f(x_{n+1}, y^*_{n+1})] \tag{4.90}$$

This method has been so commonly used that it is sometimes called *the* predictor–corrector method. It is also known as *Heun's* method or, sometimes, as the improved Euler method.

Applying this method to the differential equation (4.39) we find

$$\begin{aligned} y_{n+1}^* &= (1 + \alpha h)y_n \\ y_{n+1} &= (1 + \tfrac{1}{2}\alpha h)y_n + (\tfrac{1}{2}\alpha h)y_{n+1}^* = [1 + \alpha h + \tfrac{1}{2}(\alpha h)^2]y_n \end{aligned} \tag{4.91}$$

Thus this method inherits the second-order accuracy of the trapezoid rule corrector. This result is fairly general; predictor–corrector methods, when properly designed, have the accuracy of the final corrector step. Also note that when applied to the test equation $y' = \alpha y$, it is identical to the Richardson extrapolation method of the preceding section. For nonlinear equations or equations with nonconstant coefficients, the two methods differ.

The stability limit of this method is not easily found since the boundary curve is a quartic (a polynomial of degree 4). For almost all higher-order methods the stability limit is best found by direct calculation. The problem is to find the complex values of αh for which the characteristic function [in the present case, the polynomial multiplying y in the last part of Eq. (4.91)] has unit magnitude. We are interested only in the left-half complex αh plane. Also, since the coefficients of the characteristic function are real (as they are in most cases), the boundary curve is symmetric about the real axis, and it is sufficient to consider the second quadrant of the complex αh plane. The calculation may be done by choosing various phases of αh between 90° and 180°; for each phase, the magnitude of αh is increased until the magnitude of the characteristic function is unity; other methods can be used but there is no need to give them here. Using this method, we find the stability of the predictor–corrector method; the result is given in Figure 4.10; the results are presented in a different form from that used in Figure 4.3 to familiarize the reader with different methods of representing the results. For real values of αh, the predictor–corrector method has the same stability limit as Euler's method, but the stability is improved for complex values of αh.

We can try to further improve the method by treating the result (4.90) as a new predicted value and using the trapezoid rule again as a second corrector. Applying this method to Eq. (4.39) yields

$$y_{n+1} = [1 + \alpha h + \tfrac{1}{2}(\alpha h)^2 + \tfrac{1}{4}(\alpha h)^3]y_n \tag{4.92}$$

in place of Eq. (4.91). This result is still only second-order accurate, but it has a smaller error than Eq. (4.91). The coefficient of y_n in Eq. (4.92) is the sum of the first four terms of the Taylor series expansion of $(1 + \alpha h/2)/(1 - \alpha h/2)$, the characteristic function of the trapezoid rule. One might think that this method of iteration would produce the trapezoid rule if continued, but the Taylor series for $(1 + \alpha h/2)/(1 - \alpha h/2)$ converges only for $\alpha h < 2$. For values of αh larger

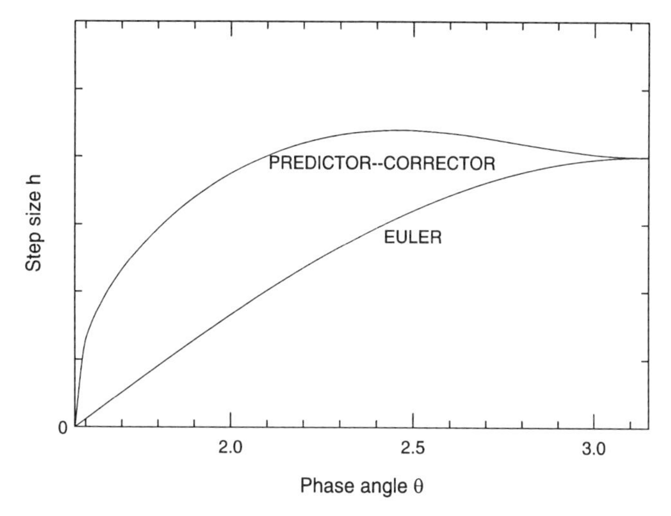

FIGURE 4.10. Maximum step size for which the explicit Euler and predictor–corrector methods are stable as functions of the phase angle of the coefficient.

than this, the method does not converge, so there is no point in continuing to iterate, as we will achieve neither increased accuracy nor improved stability. Furthermore, each iteration requires another evaluation of the function $f(x, y)$ and increases the cost.

4.8. RUNGE–KUTTA METHODS

An alternative approach to obtaining high accuracy uses points that lie between x_n and x_{n+1}; the values of the solution at these points have no importance as they are only temporary. This type of scheme is similar to a predictor–corrector method, but these methods are usually regarded as belonging to a separate class called Runge–Kutta methods. There are many Runge–Kutta methods (indeed, there is ongoing development of new ones); only a few will be presented here. The simplest method of this class uses the explicit Euler method to predict the value at $x_{n+1/2} = x_n + h/2$:

$$y^*_{n+1/2} = y_n + \tfrac{1}{2}hf(x_n, y_n) \tag{4.93}$$

This is then corrected using the midpoint rule:

$$y_{n+1} = y_n + hf(x_{n+1/2}, y_{n+1/2}) \tag{4.94}$$

Application of this method to Eq. (4.39) shows that it has the same characteristic function as the predictor–corrector method presented above; the two methods therefore share the same stability properties. In fact, this method is completely identical to the Richardson extrapolation method presented earlier. It is not at all uncommon that two apparently different methods are actually different ways of writing the same method.

The most commonly used Runge–Kutta methods are fourth-order accurate; there are a number of these. The best known such method (sometimes called *the* fourth-order Runge–Kutta method) is

$$\begin{aligned} y^*_{n+1/2} &= y_n + \tfrac{1}{2}hf(x_n, y_n) \quad \text{(Euler half-step predictor)} \\ y^{**}_{n+1/2} &= y_n + \tfrac{1}{2}hf(x_{n+1/2}, y^*_{n+1/2}) \quad \text{(backward Euler half-step corrector)} \\ y^{***}_{n+1} &= y_n + hf(x_{n+1/2}, y^{**}_{n+1/2}) \quad \text{(midpoint rule full-step predictor)} \\ y_{n+1} &= y_n + \tfrac{1}{6}h[f(x_n, y_n) + 2f(x_{n+1/2}, y^*_{n+1/2}) \\ &\qquad + 2f(x_{n+1/2}, y^{**}_{n+1/2}) + f(x_{n+1}, y^{***}_{n+1})] \\ &\qquad\qquad \text{(Simpson's rule full-step corrector)} \end{aligned} \tag{4.95}$$

where an interpretation of each step has been given.

By looking at this method, one can see that derivation of Runge–Kutta methods is not an easy task. Analysis of this method for general nonlinear differential equations is also difficult. It is not difficult to analyze, however, when applied to Eq. (4.39). We find

$$y_{n+1} = (1 + \alpha h + \tfrac{1}{2}(\alpha h)^2 + \tfrac{1}{6}(\alpha h)^3 + \tfrac{1}{24}(\alpha h)^4)y_n \tag{4.96}$$

which demonstrates that the method is indeed fourth-order accurate and the leading error term is $(\alpha h)^5/120$. It is interesting to note that the steps that comprise this method are of orders 1, 1, 2, and 4, respectively, and the method has inherited the accuracy of the final corrector.

The stability limit for this method is shown in Figure 4.11. We see that it is both considerably more accurate and more stable (for a wide range of the phase of α) than any of the methods described earlier. This property and the ease of programming the method are the reasons why it has been and still is a very popular method. Its only disadvantage is that it requires the evaluation of the derivative function four times on each step. This makes it relatively expensive and is the reason why it has been partially supplanted by the multipoint methods given below. One problem with Runge–Kutta methods of the past was the difficulty in estimating the error in the results during the calculation, making automatic error control difficult. A number of methods have been developed that overcome this deficiency; they use an additional calculation at each step to estimate of the error. Among these popular methods of this type are the Merson, Scraton, and Fehlberg variations.

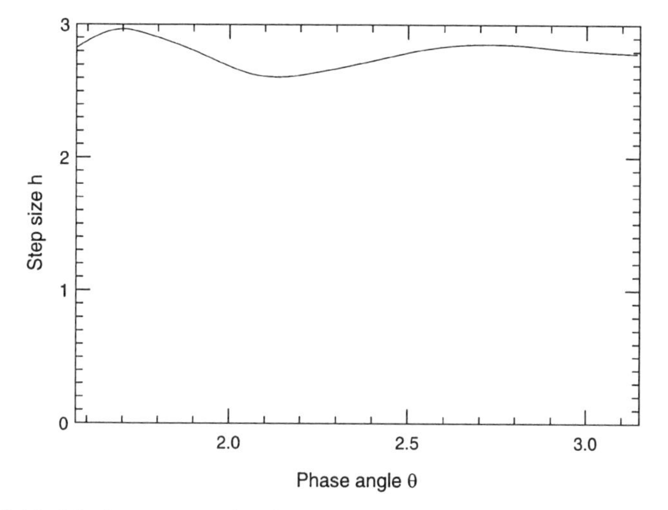

FIGURE 4.11. Maximum step size for which the fourth-order Runge–Kutta method is stable as a function of the phase angle of the coefficient.

The Runge–Kutta–Merson method actually has five steps and is

$$y^*_{n+1/3} = y_n + \tfrac{1}{3}hf(x_n, y_n) \quad \text{(Euler one-third-step predictor)}$$

$$y^{**}_{n+1/3} = y_n + \tfrac{1}{6}h[f(x_n, y_n) + f(x_{n+1/3}, y^*_{n+1/3})]$$

(trapezoid rule one-third-step corrector)

$$y^{***}_{n+1/2} = y_n + \tfrac{1}{8}h[f(x_n, y_n) + 3f(x_{n+1/3}, y^{**}_{n+1/3})]$$

(Adams–Bashforth half-step predictor) (4.97)

$$y^*_{n+1} = y_n + \tfrac{1}{2}h[f(x_n, y_n) - 3f(x_{n+1/3}, y^{**}_{n+1/3}) + 4f(x_{n+1/2}, y^{***}_{n+1/2})]$$

(Adams–Bashforth full-step predictor)

$$y^{**}_{n+1} = y_n + \tfrac{1}{6}h[f(x_n, y_n) + 4f(x_{n+1/2}, y^{***}_{n+1/2}) + f(x_{n+1}, y^*_{n+1})]$$

(Simpson's rule full-step corrector)

One can show [the easiest way is by application to Eq. (4.39)] that both y^*_{n+1} and y^{**}_{n+1} are fourth-order accurate. In fact, if $y_n = y(x_n)$ is the exact value of y at x_n, then one can show that

$$y^*_{n+1} = y(x_{n+1}) - \frac{h^5}{120}y^{(v)} + O(h^6) \tag{4.98}$$

$$y^{**}_{n+1} = y(x_{n+1}) - \frac{h^5}{720}y^{(v)} + O(h^6) \tag{4.99}$$

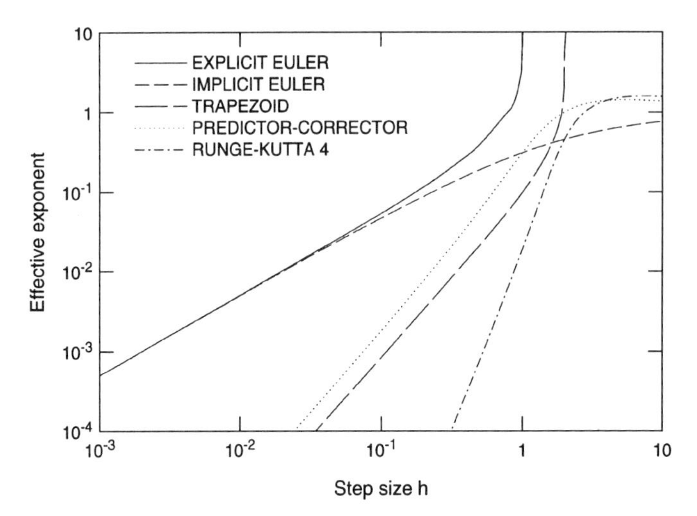

FIGURE 4.12. Reduced effective exponent ($\gamma - 1$) of five methods of solving $y' = -y$.

Thus y^{**}_{n+1} contains approximately one-sixth the error of y^{*}_{n+1}. The error in the former may therefore be estimated as one-fifth the difference between y^{*}_{n+1} and y^{**}_{n+1}. This knowledge allows the routine to adjust the step size automatically to maintain the desired accuracy. The cost of this improvement is an extra evaluation of the derivative.

Example 4.12: Runge–Kutta Solution of $y' = -y$ The predictor–corrector method, when applied to the equation $y' = -y$ yields results similar to ones obtained earlier. The error is opposite in sign to that of the trapezoid rule and approximately twice as large. For step sizes larger than approximately 1.23 the solution begins to oscillate, and for steps larger than 2 oscillatory growth similar to that produced by the Euler method is found.

Since the results for predictor–corrector methods are similar to those for methods considered earlier, we shall introduce another way to compare them. All of the methods considered so far, when applied to $y' = -y$, produce solutions of the form $y_n = \beta^n$ where β is an approximation to e^{-h}, which can be written as $\beta = e^{-\gamma h}$. Therefore $\gamma - 1$ is a measure of the accuracy of a method. We call $\gamma - 1 = -h^{-1} \ln \beta(h) - 1$ the *reduced effective exponent*; it is plotted for all of the methods discussed so far in Figure 4.12.

For small h, $\gamma - 1$ is proportional to h^n for an nth-order method so all of the curves are straight lines whose slopes are the order of the method to which they pertain. At larger h the curves deviate from straight-line behavior. For the explicit Euler and trapezoid methods, which give $\beta = 0$ at some value of h, the inaccuracy is very obvious on this plot.

The fourth-order Runge–Kutta method goes unstable by producing exponential growth rather than decay. This contrasts with the oscillatory behavior

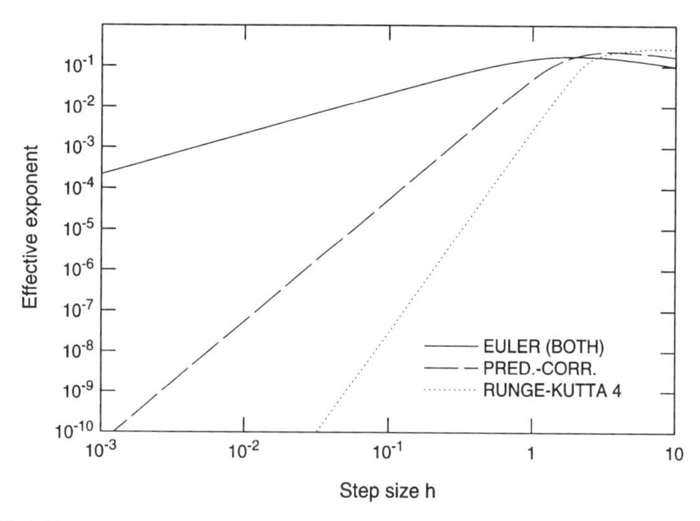

FIGURE 4.13. Effective exponent for the amplitude for several methods of solving $y'' + \pi^2 y = 0$.

of the other methods and is the reason for the behavior of the effective exponential of the Runge–Kutta method seen in Figure 4.12.

More importantly, Figure 4.12 can be used to determine the most effective method for a given accuracy. We note that $\gamma - 1$ is a measure of the error produced by a method, and we can use it to provide a way to compare methods. Thus, for $\gamma - 1 \approx 0.01$, the allowable step size for the Runge–Kutta method is approximately 2.5 times that for the trapezoid rule, 4 times the predictor–corrector step size, and 30 times the step size the Euler method would require. Although the Runge–Kutta method requires twice as much computation per step as the predictor–corrector method and four times as much effort as the Euler method, it actually produces a solution of the required accuracy at the lowest cost. The advantage is still greater if higher accuracy is required. This combination of accuracy, relatively low cost, and ease of programming (see the next example) have made Runge–Kutta a very popular method.

Example 4.13: Effective Exponents for $y' = i\pi y$ It is also interesting to compute the solution to $y' = i\pi y$ by the various methods. Since the amplitude error, when there is any, is exponential, we present the results for the amplitude in a manner similar to the one used in the preceding example. For the methods introduced so far, the numerical solution has the form $y_{n+1} = \beta y_n$ where β is complex. In Figure 4.13 we plot $h^{-1}|\ln|\beta(h)||$ as a function of h.

Several points are worthy of note in Figure 4.13. The trapezoid rule gives no amplitude error and does not appear in the plot. The two Euler methods have errors in opposite directions but of equal magnitude. The Runge–Kutta method has a negative (dissipative) amplitude error and is stable until $h = 2\sqrt{2}/\pi = 0.90$; beyond that it is unstable. Due to the change in sign of $\ln \beta(h)$, the result could be plotted only up to $h = 0.9$. It is also interesting to note that the slopes of the curves in Figure 4.13 are larger than those of the corresponding curves in Figure 4.12—in other words, the amplitude error of these methods is of higher order than the formal order of the method.

Phase errors are not plotted since both the predictor–corrector and Runge–Kutta phase errors change sign as h increases. The Euler methods give negative phase errors, indicating that the solution lags the exact solution and the phase error is second order for small step size. The trapezoid rule produces a lagging phase error that, for small step sizes, is about one-quarter of the phase error of the Euler methods. The predictor–corrector method yields a lagging error whose magnitude is half that of the Euler methods at small step sizes and changes to a leading error at large step sizes. Finally, the fourth-order Runge–Kutta method gives very small lagging phase errors at small step sizes but leading errors for $h > 0.5$.

The Runge–Kutta method shows itself to be superior. It is capable of handling solutions of both the exponential and oscillatory types with great ease. For this reason we give a Runge–Kutta subroutine (RUNKUT), which is available at the Internet site described in Appendix A. It can be used in the same way as the Euler subroutine; one need only change the name in the calling statement.

4.9. MULTISTEP METHODS

Runge–Kutta methods were the methods of choice for solving ordinary differential equations for a long time; they are still used often. Some reasons for increased use of other methods are the availability of fast computers and the extreme accuracy required in a number of applications. It is possible to develop Runge–Kutta methods of very high accuracy; in fact, the best way to do so is to use with the aid of a computer. However, Runge–Kutta methods have another disadvantage—they require a large number of function evaluations per step. In this section, we consider methods that provide high accuracy without a large number of function evaluations; naturally, the advantages come at a price.

It is not difficult to construct accurate methods that do not require many evaluations of the function; many exist and more are being developed. Suppose that the solution has been computed from $x = x_0$ to $x = x_n$. We can then construct accurate methods for finding the solution at x_{n+1} by using the Taylor series (4.14) about the point x_n and approximating the second and higher derivatives by the kind of difference formulas presented in Section 4.1. We may use data at mesh points $1, 2, \ldots, n-1, n$ in these difference formulas; data at the new point $n+1$ may be used in implicit methods. Both the function and the derivative at these points can be used, so an enormous range of possibilities exists. Methods which use data from more than two points are, sensibly enough, called *multipoint* or *multistep* methods.

We shall treat one of the simplest multistep methods in some detail in order to discover some of the properties of this class of methods. Then some of the more important members of this class are given with minimal derivation; finally, means of using them in practical computations are given.

As the method to examine in detail we select the *leapfrog method* for which the basic formula is

$$y_{n+1} = y_{n-1} + 2hf(x_n, y_n) \tag{4.100}$$

As noted earlier, this method can be regarded as the midpoint rule applied to the interval (x_{n-1}, x_{n+1}). It is second-order accurate and is a two-step formula because data from the last two data points are required to compute the value of y_{n+1}.

The major difficulty shared by all multipoint schemes is immediately obvious; one cannot start a calculation with this method if data are given at only a single point. For solution of the differential equation, only the initial value y_0 need be given. However, Eq. (4.100) does not allow computation of y_1 if that is the only information available. On the other hand, if both y_0 and y_1 are known, there is no problem computing y_2. Thus, another method (which could be any of the methods described earlier) must be used to provide y_1 and thus allow use of the leapfrog technique. The choice of starting method will turn out to be important and can have a significant bearing on the results as we will see later.

For now, let us assume that values of y_0 and y_1 are available and see what the leapfrog method does, using Eq. (4.39) as a test case. The difference equation for this method is not as easy to solve as the difference equations for the methods studied earlier. Recall that, for the earlier methods, the solution to the difference equations derived from Eq. (4.39) were of the form $y_n = y_0\rho^n$; that is not the case for the leapfrog method. Part of the reason is that, since the method requires the values of both y_0 and y_1 to get started, there must be at two parameters in the solution. On the other hand, since the method is expected to be second-order accurate (because the midpoint rule is second-order accurate as a quadrature method), it should be capable of producing an exponential solution.

This leads us to guess that the solution to the difference equation for the leapfrog method applied to Eq. (4.39):

$$y_{n+1} = y_{n-1} + 2\alpha h y_n \tag{4.101}$$

might be of exponential form:

$$y_n = \rho^n \tag{4.102}$$

[There are other methods of solving Eq. (4.101). This one is chosen for its close resemblance to methods used for differential equations.] Plugging

(4.102) into the difference equation and dividing by ρ^{n-1}, we find that ρ satisfies a quadratic equation:

$$\rho^2 - 2\alpha h \rho - 1 = 0 \tag{4.103}$$

that has two solutions:

$$\rho_{1,2} = \alpha h \pm \sqrt{\alpha^2 h^2 + 1} \tag{4.104}$$

in contrast with the methods presented earlier, which had only a single exponential solution.

To see what these roots represent, let us expand the square root in a Taylor series in αh. We find

$$\begin{aligned} \rho_1 &\approx 1 + \alpha h + \tfrac{1}{2}(\alpha h)^2 - \tfrac{1}{8}(\alpha h)^3 + \cdots \\ \rho_2 &\approx -1 + \alpha h - \cdots \end{aligned} \tag{4.105}$$

The first root is a second-order accurate approximation to $e^{\alpha h}$ and is the one we expected to find. The second root bears no relation to the solution of the differential equation; it is called the *computational* or *parasitic* root and does nothing but cause trouble. In fact if αh is real and negative, ρ_2 has magnitude greater than 1 and causes the method to be unstable; in fact, the leapfrog method is unconditionally unstable. Despite this, it has some virtues so let us look a little deeper into the solution of the difference equation.

Since the difference equation (4.101) is linear, the general solution is a linear combination of the two solutions, that is, it has the form:

$$y_n = a_1 \rho_1^n + a_2 \rho_2^n \tag{4.106}$$

as can be verified by substituting Eq. (4.106) into the difference equation. The presence of two coefficients (a_1 and a_2) provides the freedom needed to match the starting values y_0 and y_1. Setting Eq. (4.106) with $n = 0$ and $n = 1$ equal to these values, and solving the resulting two linear algebraic equations, we find

$$a_1 = \frac{y_1 - \rho_2 y_0}{\rho_1 - \rho_2} \qquad a_2 = \frac{y_1 - \rho_1 y_0}{\rho_1 - \rho_2} \tag{4.107}$$

Note that if $y_1 = \rho_1 y_0$, then $a_2 = 0$, and the computational root plays no role in the solution. In other words, for the test equation, it is possible to choose starting values that totally suppress the parasitic root. This can be achieved only for this equation and, possibly, a few others.

If y_1 is a second-order accurate approximation to $y(x_1)$, the exact solution of the differential equation, it is also a second-order accurate approximation to $\rho_1 y_0$. Then a_2 must be of order h^3 and, for small h, it is very small and may not cause much of a problem. However, this shows that the method used to start the computation can have an important effect on the results.

Even if the parasitic component of the solution is initially zero, there may be a problem. As we saw, if the real part of αh is negative, the magnitude of ρ_2 is greater than 1. So, even if the starting values totally suppress the parasitic root, the inevitable round-off error may be interpreted as adding a small parasitic component to the solution. This component can then be amplified and may eventually overwhelm the accurate part of the solution. In a calculation with the leapfrog method, the sign that this is occurring is a solution that oscillates about the correct solution with a period of two steps; this is a consequence of the parasitic root being negative.

The problem can be corrected in a number of ways. The objective is to remove the parasitic component of the solution before it grows large enough to be a serious problem. One way to do this is to perform a step with another method every so often; to be consistent, it is best to use a second-order method such predictor–corrector or second-order Runge–Kutta. Another method is to average the results at successive steps every once in a while. Both of these methods prevent the parasitic component of the solution from growing too large. So, although the leapfrog method is unconditionally unstable, the difficulty due to the parasitic root may be suppressed and the method has seen considerable service. Another reason the leapfrog method has been popular will be pointed out in the examples that follow.

Example 4.14: Leapfrog Solution of $y' = -y$ We shall solve $y' = -y$ using the leapfrog method. With $h = 1$, a large step size, the two roots of Eq. (4.103) are

$$\rho_1 = 0.414 \qquad \rho_2 = -2.414 \tag{4.108}$$

The first root, ρ_1, is an approximation to the correct solution, e^{-h}; the exact solution is 0.368; the approximation is reasonably accurate given the step size. The second root is the troublemaker. With starting values taken from the exact solution of the differential equation, that is, $y_0 = 1$, $y_1 = e^{-1}$, we obtain the results shown in Table 4.6. It is obvious that, despite the accurate starting values, the results are worthless. Once the instability takes over, $y_{n+1} \approx -2.414 y_n$ and the computed "solution" grows rapidly in an oscillatory manner.

This case is extreme, of course, as the step size is larger than one would normally use for this problem. If the step size were smaller, the instability, although present, would not grow as fast and its presence would not be obvious for some time. It is possible to use the leapfrog method if the range of integration is not large, but caution is needed.

We next solve the same problem with the starting value $y_1 = 0.414$, the exact "physical" root of the leapfrog method for the differential equation. This calculation was done in double precision (to reduce round-off error) with a step size of 0.1. With these starting values, the leapfrog method behaves like a stable second-order method and the error at $x = 10$ is only approximately

TABLE 4.6. Leapfrog Solution of $y' = -y$ with Exact Starting Values

x	y
1	0.368
2	0.264
3	−0.161
4	0.585
5	−1.33
6	3.24
7	−7.28
8	18.91
9	−45.64
10	110.2

8×10^{-7}, or 2%. In other problems, for example, ones with coefficients that are not constant, it is not possible to make leapfrog behave this well. Round-off error will eventually creep in and destroy the results. Using a good starter method may delay the onset of disaster but cannot entirely prevent it.

This example displays the most important features of multistep methods: the importance of the method of starting the calculation and the trouble caused by the parasitic root. The next example shows that a method that has difficulty in problems with exponential-like solutions may do very well when the solutions are oscillatory.

Example 4.15: Leapfrog Solution of $y' = i\pi y$ When the leapfrog method is applied to the solution of $y' = i\pi y$, the results are excellent. However, the existence of the parasitic solution makes it impossible to express the results simply in terms of amplitude and phase error. We therefore discuss them in a qualitative manner. The reason excellent results are obtained is that when α is pure imaginary, both roots of Eq. (4.103) have unit magnitude so the leapfrog method is not unstable; in fact, it is neutrally stable, which is the ideal situation. Furthermore, as we saw in Example 4.14, a good choice of starting method produces a coefficient of the parasitic component of the solution that is quite small. Consequently, the method behaves very much like the second-order methods discussed earlier.

As noted, for problems with oscillatory solutions, the leapfrog method is very nearly neutrally stable. Neutral stability means that the method produces neither unwanted growth of the solution (instability) nor artificial damping (dissipation). This is a desirable property that is possessed by few methods; the combination of near-neutral stability, the ease of use, and the speed resulting from needing only one evaluation of the derivative per step have made the leapfrog method the basis for some important methods for solving partial differential equations.

The leapfrog method is not exactly neutrally stable because the parasitic term, although small, has a different phase from the physical term and the interference between the two produces small oscillating amplitude errors.

All multistep methods, when applied to $y' = \alpha y$ have more than one solution. In fact, the number of solutions is equal to the number of steps that the method uses. One of these roots is an approximation to the solution of the differential equation; all of the others are parasitic and an inconvenience. All explicit multistep methods are at most conditionally stable, but, unlike the leapfrog method, many of them have regions in which they are truly stable. Implicit multistep methods may, of course, be unconditionally stable.

The most general k step method is

$$\sum_{m=0}^{k} \alpha_m y_{n+1-m} + \sum_{m=0}^{k} \beta_m f(x_{n+1-m}, y_{n+1-m}) = 0 \tag{4.109}$$

A particular method is defined by the number of steps k and the parameters α_m and β_m. We can let $\alpha_0 = 1$ with no loss of generality. If $\beta_0 = 0$, the method is explicit; otherwise it is implicit. It is clear that there are many methods of this type; only a few classes of multistep methods will be presented here.

Methods in which $\alpha_0 = 1$ and $\alpha_1 = -1$ and $\alpha_m = 0$ for $m > 1$ are called *Adams* methods and are among the most widely used members of this class. The explicit methods of this type are called *Adams–Bashforth* methods and the implicit ones are called *Adams–Moulton* methods. They can be derived in a number of ways, one of which is from difference formulas. For the Adams–Bashforth methods backward formulas are needed, while for the Adams–Moulton case formulas with one forward point are required. They can also be derived through the use of interpolation formulas. Fitting a Lagrange polynomial to the derivative y' at the points $x_{n-k+1}, x_{n-k}, \ldots, x_n$ and integrating the result from x_n to x_{n-1} yields the Adams–Bashforth formulas. Doing the same calculation with y'_{n+1} included in the interpolation yields the Adams–Moulton methods. The various derivations lead, of course, to the same results.

The coefficients for the Adams–Bashforth methods are given in Table 4.7. As might be expected, the first-order method is the Euler explicit method, but the higher-order methods do not correspond to any of the methods presented earlier.

Similarly, the coefficients for the Adams–Moulton methods are given in Table 4.8. To keep the order of the method equal to the index k, we have terminated the series at $k - 1$. The first-order Adams–Moulton method is backward Euler, and the second order method is the trapezoid rule, but the higher-order methods are new.

The stability of these methods can be analyzed in the manner described earlier and is obviously a chore for all but the lowest order methods. In Table 4.9 we give the stability limits for real values of $hy'/y = \alpha h$; it clearly

TABLE 4.7. Coefficients of the Adams–Bashforth Methods

	$m = 1$	$m = 2$	$m = 3$	$m = 4$	$m = 5$	$m = 6$
β_{1m}	1					
$2\beta_{2m}$	3	−1				
$12\beta_{3m}$	23	−16	5			
$24\beta_{4m}$	55	−59	37	−9		
$720\beta_{5m}$	1901	−2774	2616	−1274	251	
$1440\beta_{6m}$	4277	−7923	9982	−7298	2877	−475

TABLE 4.8. Coefficients of the Adams–Moulton Methods

	$m = 0$	$m = 1$	$m = 2$	$m = 3$	$m = 4$	$m = 5$
β^*_{1m}	1					
$2\beta^*_{2m}$	1	1				
$12\beta^*_{3m}$	5	8	−1			
$24\beta^*_{4m}$	9	19	−5	1		
$720\beta^*_{5m}$	251	646	−264	106	−19	
$1440\beta^*_{6m}$	475	1427	−798	482	−173	27

TABLE 4.9. Stability Limits of the Adams Methods

Order	Adams–Bashforth	Adams–Moulton
1	2	∞
2	1	∞
3	0.5	6
4	0.3	3
5	0.2	1.9

shows that the implicit methods are always considerably more stable than the corresponding explicit methods, as expected.

Many variations on these methods are possible and many have been developed by various authors. Since it is not our purpose here to give an exhaustive survey of methods, and, as the Adams methods are among the best ones, additional methods will not be presented.

Adams methods have properties similar to those of the methods discussed earlier. The explicit ones are easy to use but not as stable as the more difficult to use implicit methods. An obvious idea is to construct a predictor–corrector method based on the Adams methods. Indeed these methods are among the most commonly used differential equation solvers when high accuracy is required. Adams methods allow easy change of the order of the method, and it is relatively simple to construct computer routines that provide for automatic error control. Generally, these methods consist of an Adams–Bashforth predictor followed by one or more corrector steps of the Adams–Moulton type. In a well-designed method, the overall accuracy is that of the last corrector step. One of the simplest and most popular of these uses the nth-order Adams–Bashforth method as the predictor and the $(n + 1)$st-order Adams–Moulton method as the corrector. The overall accuracy is $(n + 1)$st-order and the stability limit lies between that of the predictor and that of the corrector. We elaborate on this in the next section.

4.10. CHOICE OF METHOD: AUTOMATIC ERROR CONTROL

The preceding sections demonstrate that there are many methods for the numerical solution of initial value problems for ordinary differential equations. The number of methods is so large as to become confusing. For most problems, any of a number of methods is capable of doing the job well at a reasonable cost. There are a few points, however, that should be considered in choosing a method in order to avoid a really poor choice.

Except for very crude calculations, first-order methods should not be used; they are not accurate enough unless an extremely fine step size is used. Second-order methods find considerable service as both ordinary and partial differential equation solvers; in the latter application they are often chosen for their low memory requirements. For the general run of engineering work, in which accuracy of 10^{-2} to 10^{-4} is required, fourth-order methods are usually more efficient than second-order methods and are recommended. They provide higher accuracy at modest cost. For very accurate work in which errors as small as 10^{-10} may be required, methods of very high order (sixteenth-order methods have been used in space applications) are necessary.

Many modern computer subroutines for differential equation solving have built-in devices for the control of order and step size. The user need only state the accuracy with which the solution is required and, of course, the information necessary to define the differential equation (or set of equations). The program selects the optimum order and/or step size to meet the user's demand at lowest cost. A number of such programs are available, based on various algorithms; they may be obtained commercially or through the Internet. Rather than looking deeply into these algorithms, a short overview of what is contained in some of them is given.

One popular scheme developed by Bulirsch and Stoer uses a second-order method (usually leapfrog) and extrapolation (which may be Richardson but more commonly is polynomial or rational extrapolation) to achieve high-order accuracy. In order to provide the requested accuracy, the method contains a built-in error estimator. If the error is not within the allowable limit after the maximum amount of extrapolation allowed has been done, step size is reduced and the calculation repeated.

Another common approach uses Adams methods. Since high-order Adams methods do not allow easy adjustment of the step size, this parameter is not changed very often. Rather, the order of the method is selected to provide the desired error. Only when the order has been increased to the maximum available within the program is the step size adjusted. Naturally, routines that use this approach have special starting algorithms built in.

Subroutines based on these methods generally contain many logic and control statements so their structure cannot be described meaningfully in a short space. A description of the details of one such code can be found in the book of Shampine and Gordon (1975).

The author tested some of these routines and found that for simple problems, the Runge–Kutta methods were as good as the multistep methods. The multistep methods were better for more difficult problems such as ones with variable coefficients.

4.11. SYSTEMS OF EQUATIONS—STIFFNESS

4.11.1. Treatment of Systems of Ordinary Differential Equations

We now turn to systems of ordinary differential equations. As high-order equations can almost always be reduced to systems of first-order equations, they are included as well. Systems of equations are most easily treated by reducing them to a form to which we can apply the methods developed for single first-order equations. This allows almost all of the methods already described to be adopted with little modification. The most important problem is stiffness, to which we devote much of this section.

The fundamental problem is to solve the system of first-order ordinary differential equations:

$$y_i' = f_i(x, y_1, y_2, \ldots, y_m) \qquad i = 1, 2, \ldots, m \tag{4.110}$$

subject to the initial conditions

$$y_i(0) = c_i \qquad 1 = 1, 2, \ldots, m \tag{4.111}$$

Any of the methods described earlier in this chapter can be applied to this system of equations, and the subroutines found on the Internet site described in Appendix A have this capability. For example, the Euler method would give

$$y_{i,n+1} = y_{i,n} + hf_i(x_n, y_{1,n}, y_{2,n}, \ldots, y_{m,n}) \qquad i = 1, 2, \ldots, n \tag{4.112}$$

From here on, we shall use the notation $\mathbf{y}_n = (y_{1,n}, y_{2,n}, \ldots, y_{m,n})$ as a shorthand. In this notation, Eq. (4.112) can be written

$$\mathbf{y}_{n+1} = \mathbf{y}_n + h\mathbf{f}(x_n, \mathbf{y}_n) \tag{4.113}$$

where $\mathbf{f} = (f_1, f_2, \ldots, f_m)$ and the boldface indicates that the quantitiy is a vector (in the linear algebra sense).

Any of the other methods described earlier, including the implicit ones, can be applied to this problem. If implicit methods are used, a system of algebraic equations must be solved at each step. If the ODEs are linear, then this system of equations is linear; otherwise, it is nonlinear. In the former case, Gauss elimination or *LU* decomposition (see Chapter 1) can be used. For the nonlinear case, we need to use an iterative method for solving systems of nonlinear equations such as the Newton–Raphson method or one of the secant methods; the best initial guess of the solution at the new step is usually the converged solution at the preceding step. There is no point in converging the iterative method used for the nonlinear equations to a smaller tolerance than the accuracy with which the equations are approximated; usually, a few iterations are sufficient to accomplish this.

4.11.2. Stiffness

To study the difficulties that arise in solving systems of ordinary differential equations, we shall use the strategy used for a single equation—we first consider the simple but important case of a homogeneous system of linear equations with constant coefficients:

$$y_i = \sum_{j=1}^{m} a_{ij} y_j \tag{4.114}$$

which can be written in matrix notation:

$$\mathbf{y}' = A\mathbf{y} \tag{4.115}$$

If we linearize Eq. (4.110) locally, we obtain a system of equations of the form (4.114) with an inhomogeneous or forcing term. The inhomogeneous term must be correctly accounted for to obtain the correct solution but is less important for the analysis of accuracy and stability. Hence it is sufficient to deal with Eq. (4.114) or Eq. (4.115).

We assume that the numbers of equations and unknowns are equal, so A is a square $m \times m$ matrix; its eigenvalues and eigenvectors play an important role in the determining the behavior of numerical methods for solving the system of equations (4.115). Linear algebra (see Chapter 1) tells us that a matrix possesses a set of eigenvalues and eigenvectors defined by

$$A\phi_k = \lambda_k \phi_k \qquad k = 1,2,\ldots,m \tag{4.116}$$

The number of eigenvectors may be less than or equal to m, but in nearly every case arising from a physical problem, the number of eigenvectors is equal to m, so we shall assume this to be the case. When there are m linearly independent eigenvectors, one can construct a matrix S, whose columns are the eigenvectors of A, such that

$$S^{-1}AS = \Lambda \tag{4.117}$$

where Λ is a diagonal matrix whose elements are the eigenvalues of A.

We may use the eigenvalues and eigenvectors to simplify the system of differential equations with constant coefficients (4.115). The trick is to premultiply Eq. (4.115) by S^{-1}. Then, defining $\mathbf{z} = S^{-1}\mathbf{y}$ and inserting SS^{-1} between A and $\mathbf{y}$, we have

$$S^{-1}\mathbf{y}' = \mathbf{z}' = (S^{-1}AS)(S^{-1}\mathbf{y}) = \Lambda\mathbf{z} \tag{4.118}$$

which, in component form, is the uncoupled set of linear differential equations

$$z_k' = \lambda_k z_k \qquad k = 1,2,\ldots,m \tag{4.119}$$

Each member of this set of equations is of precisely the form (4.39), the equation we used for testing methods for single first-order differential equations, with α replaced by λ_k. It follows that everything we discovered about methods for a single first-order differential equation applies with equal validity to systems of equations or higher-order equations.

There is an important and essential difficulty, however. Except for linear systems of differential equations with constant coefficients, the use of diagonalization to produce a decoupled system of equations is impossible. Thus diagonalization is useful for analyzing and understanding the behavior of methods but, in the general case, the system must be solved as it is written; that is, we must solve the system (4.114) rather than (4.119). Solving Eq. (4.114) by any method produces the same result as solving each of Eqs. (4.119) by the same method with the same step size and then computing $\mathbf{y} = S\mathbf{z}$.

For conditionally stable methods, there is a largest step size h_k that yields a stable solution for the kth equation of the set (4.119); in particular, we must have $h_k < c|\lambda_k|^{-1}$ where c is a constant that depends on the method chosen (e.g., $c = 2$ for the explicit Euler method). The most restrictive condition is the one for the equation containing the eigenvalue that is largest in magnitude (we assume that they all have a negative real part). This step size must be applied to all of the equations in the set. In many cases this causes no problem. In others, one or more of the eigenvalues is so much larger than the others that an undesirably small step size (and thus a large number of steps) is required to solve the equation. This occurs when magnitudes of the eigenvalues cover a wide range.

A system of differential equations in which the ratio of the largest and smallest eigenvalues is very large is called *stiff*. In fact this ratio is called the *stiffness ratio*; it is also one definition of the condition number of a matrix. Thus, a set of differential equations in which the coefficient matrix is ill-conditioned is stiff. Physically, the eigenvalues are usually length or time scales, so a stiff system is one in which there is a large range of scales; many engineering problems have this property. Some examples of problems in which stiffness might be a factor follow:

- Any problem that displays boundary layer type behavior. A system has boundary layer behavior when there is a small length scale that is important in a region over which the solution changes rapidly and a longer scale that is important in the remainder of the solution domain. Problems of this kind are common in fluid mechanics, heat transfer, and the kinetic theory of gases.
- Problems in which there are a great variety of rate constants. A common source of problems of this type is a system in which chemical kinetics plays a role. Chemical rate constants vary over many orders of magnitude and many problems in chemical kinetics are stiff.
- Problems that are governed by parabolic partial differential equations. When discretized, these turn out to be equivalent to solving a stiff system of ordinary differential equations. We examine this in Chapter 6.

Stiffness is one of the most difficult problems to handle in the solution of differential equations. There are a number of methods of dealing with stiffness, but none is completely satisfactory, and the development of better methods for stiff systems is a topic of current research interest. A few of the better methods in current use are described in the following section.

4.11.3. Numerical Methods for Stiff Problems

In a majority of problems involving stiff equations, the solution is relatively smooth in most of the solution domain. That part of the solution is determined almost entirely by the small eigenvalues; the large eigenvalues play a small role in these regions. If a conditionally stable method is used in such a region, the step size allowed by the stability criterion is almost always much smaller than the step size that accuracy would require, making the calculation much more expensive than common sense would suggest it should be. On the other hand, an unconditionally stable implicit method with a large step size that treats the slowly varying part of the solution accurately treats the part of the solution due to the large eigenvalues very inaccurately; this is not important because the large eigenvalues play a minor role in regions where the solution is varying slowly. The stability of implicit methods prevents the solution from blowing up, making them good candidates for solving stiff problems.

Ideally we would like to have a method that, for a single equation, is accurate for a fairly large range of αh and is stable for relatively large values of this parameter. The former is required to capture the slowly varying part of the solution well and the latter is necessary to prevent the large eigenvalues from

TABLE 4.10. Coefficients of the Gear Methods

Coefficient	$k = 2$	$k = 3$	$k = 4$	$k = 5$	$k = 6$
β_0	2/3	6/11	12/25	60/137	60/147
α_1	4/3	18/11	48/25	300/137	360/147
α_2	−1/3	−9/11	−36/25	300/137	450/147
α_3		2/11	16/25	200/137	400/147
α_4			−3/25	−75/137	−225/147
α_5				12/137	72/147
α_6					−10/147

causing trouble. A class of methods of this type was found by Gear (1971). For a single equation they are given by:

$$y_{n+1} = \sum_{i=1}^{k} \alpha_i y_{n+1-i} + h\beta_0 f(x_n, y_n) \tag{4.120}$$

The coefficients for this method are given in Table 4.10. For details of how they were chosen, see Gear (1971).

A number of other approaches to the solution of stiff problems have been suggested. Many of these use explicit methods that sacrifice some accuracy in order to achieve improved stability, but none of them seems to be fully satisfactory. Gear's method is probably the most popular one in use today and variations on it have been developed by various authors. As is shown below, it is far from ideal.

We now illustrate some of these points with an example.

Example 4.16: Solution of a Stiff System A simple stiff system is the pair of equations

$$y_1' = -y_1 + 0.999y_2 \qquad y_2' = -0.001y_2 \tag{4.121}$$

The matrix associated with these equations has eigenvalues $\lambda_1 = -1$ and $\lambda_2 = -0.001$. It can be diagonalized but there is no reason to do so. It is not difficult to verify that the general solution of Eqs. (4.121) is

$$y_1 = c_1 e^{-x} + c_2 e^{-0.001x} \qquad y_2 = c_2 e^{-0.001x} \tag{4.122}$$

The constants are determined by providing initial conditions. This solution shows that y_2 varies very slowly while y_1 has both slowly and rapidly varying components. The stiffness ratio is 1000.

To see the problem that stiffness causes for numerical methods, let us solve the system of equations (4.121) with the initial conditions $y_1 = y_2 = 1$. The

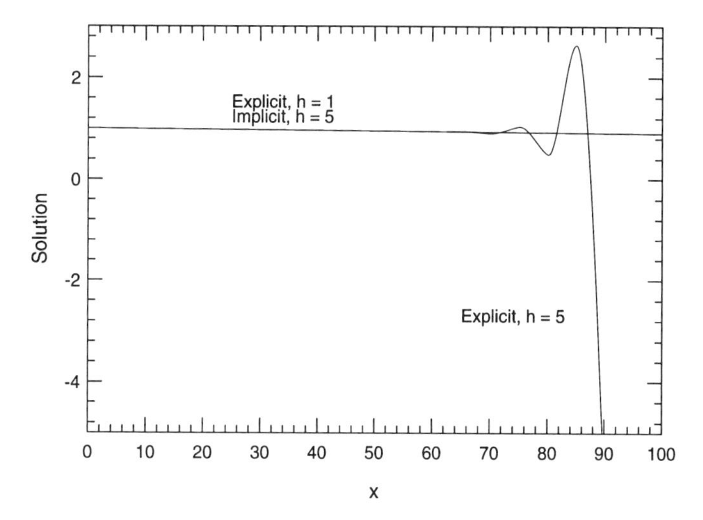

FIGURE 4.14. Solution of a stiff problem with the Euler methods.

constants are then $c_1 = 0$ and $c_2 = 1$, so the exact solution (4.122) is $y_1 = y_2 = \exp(-0.001x)$, which is slowly varying and well behaved. This solution satisfies the simpler system of equations:

$$y_1' = -0.001y_1 \qquad y_2' = -0.001y_2 \tag{4.123}$$

which could be solved accurately by almost any method as long as the step size is less than about 100.

Let us solve Eqs. (4.121) with the initial conditions $y_1 = y_2 = 1$ using the Euler method. Figure 4.14 shows the solution obtained with $h = 1$; the figure also contains two other solutions that will be referred to below. For this case, the product of the step and the large eigenvalue is 1. The Euler method is not unstable and the solution obtained is almost exactly what we would have gotten if we had solved $y' = -0.001y$ with the same method, that is, it is quite accurate and shows no signs of instability.

Now let us solve the same problem using the Euler method with $h = 5$; this step size is small enough to solve Eqs. (4.123) accurately. The result is shown in Figure 4.14. The solution behaves quite well until about $x = 50$. Up to this point, the error is increasing but is not obvious on the figure. In fact, if we did not look at it closely, we might accept the solution up to $x = 70$. After $x = 70$, it is clear that the solution has become unstable, a fact that cannot be overlooked.

The point is that the solution is very smooth, and one would expect to be able to compute it accurately with the step size used. However, the short time scale is lurking in the background waiting to be activated and, as soon as it is brought into play by round-off error, the instability takes over. As a result, we are forced to use a step size much smaller than necessary to capture the smooth solution and pay much more for the solution than seems reasonable. In a simple problem like this one, the problem is not severe; but, in a large system that needs to be integrated for a long range of the independent variable, it can be extremely costly.

In this case, because there was no initial error, it took a while for round-off error to enter and be amplified. Had there been significant error in the initial condition, the instability would have become obvious much sooner. When the error becomes significant, it is multiplied by a factor of approximately -4 at each step and the results deteriorate rapidly.

In this problem, the instability is obvious and easily recognized. There is a danger that, in other problems, the difficulty might not be so obvious and a bad solution may be accepted as accurate. It is important to note that the results depend on the arithmetic accuracy of the machine used and on whether single or double precision is used. A more accurate machine (or double-precision arithmetic) would delay the onset of trouble.

The solution of the same problem using the backward Euler method with $h = 5$ is also shown in Figure 4.14; it is reasonably accurate and is about what we would get if we solve $y_1' = 0.001y_1$ with the same method and step size. The relative error increases linearly and there is no indication of any problem arising from the stiffness. Thus implicit methods can compute the slowly varying portion of the solution of a stiff problem quite effectively with a time step adjusted to the long time scale. Indeed, the solution to this problem could be computed to reasonable accuracy with the implicit Euler method with a step size as large as 100.

The problem is a little different if the initial conditions are such that the rapidly varying component plays a role. If the initial conditions are $y_1(0) = 2$ and $y_2(0) = 1$, then, in Eqs. (4.122), $c_1 = c_2 = 1$, and the solution becomes

$$y_1 = e^{-x} + e^{-0.001x} \qquad y_2 = e^{-0.001x} \tag{4.124}$$

In this solution, y_1 decreases from its initial value of 2 to a value close to 1 in a distance $x < 5$. In this case, in order to obtain the correct long-term solution, we must compute the fast part of the solution fairly accurately. To do this with a fixed step length method requires a very small step size. For the Euler method, a step size of order $h \approx 0.1$ would be needed to yield a 1% error. Higher-order methods could do the job with a much larger step but would still require many steps to reach $x = 10$.

The solution for y_1 is shown in Figure 4.15. After the initial transient, the solution is smooth and resembles the one shown in Figure 4.14. It is reasonable to expect to be able to compute this solution with step size $\Delta x = 0.1$ up to, say, $x = 5$ and a step size of $\Delta x = 5$ (or even larger) thereafter. As Figure 4.15 shows, the explicit method becomes unstable soon after the change in step size while the implicit method has no trouble producing an accurate solution.

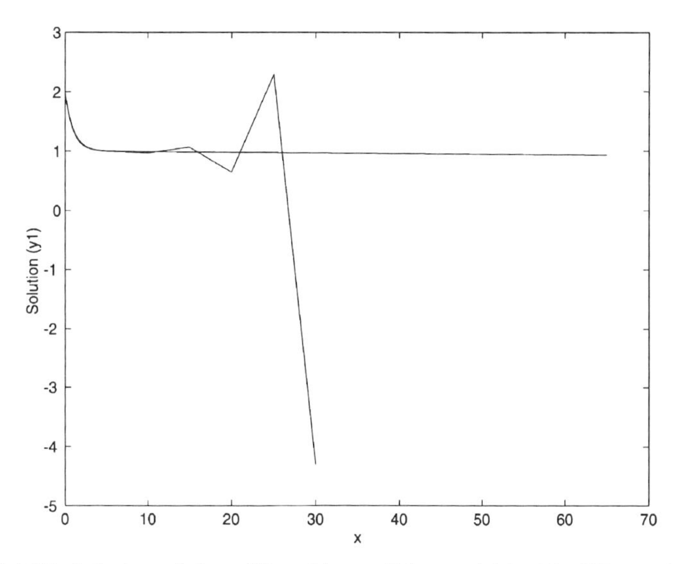

FIGURE 4.15. Solution of the stiff problem of Figure 4.14 with different initial conditions.

4.11.4. Splitting Methods

Methods have been developed for solving partial differential equations that contain difficulties analogous to stiffness but which work only in certain cases. An introduction to some of these ideas is given here because they are even simpler when applied to ordinary differential equations. These methods can accurately compute either the entire solution or just the long-term part of the solution. The idea is to treat the terms that cause the problem differently from the well-behaved terms. As a class, these are called splitting methods. They cannot, however, be applied to every case. The method is best illustrated by an example.

Example 4.17: Solution of a Stiff System with the Split Method Suppose we are interested in solving Eqs. (4.121) with the initial conditions $y_1(0) = 2$ and $y_2(0) = 1$. As we saw above, the solution contains both rapidly varying and slowly varying components. We shall now try to develop a method that takes explicit note of this behavior. Recall that the eigenvalues of these equations are $\lambda_1 = -1$ (the "fast" eigenvalue) and $\lambda_2 = -0.001$ (the "slow" eigenvalue). We make this explicit by writing Eqs. (4.121) as:

$$y_1' = \lambda_1 y_1 + (\lambda_2 - \lambda_1) y_2 \qquad y_2' = \lambda_2 y_2 \tag{4.125}$$

What we would like to do is to treat the terms involving λ_1 with a small time step while reserving a longer time step for the terms containing λ_2. We can do this as follows. Given the initial values $y_{1,0}$ and $y_{2,0}$, we compute

$$\tilde{y}_{1,n+1} = \tilde{y}_{1,n} + \lambda_1 h(\tilde{y}_{1,n} - y_{2,0}) \tag{4.126}$$

that is, we advance the solution using only the terms that contain λ_1.

We used the explicit Euler method to keep the discussion as simple as possible; the advantages of using a more accurate method are obvious. Note that the initial value of y_2 is used at every step in this part of the method. Since y_2 is not updated in this part of the calculation, it is the only value we can use. The step size h must be small enough to satisfy the stability condition $|\lambda_1 h| < 2$ but, for accuracy, it needs to be even smaller. We continue to use Eq. (4.126) until we have computed y_N where N is approximately the stiffness ratio (1000 in this example). When this part of the computation is complete, we update the solution on the long (slow) time scale:

$$\begin{aligned} y_{1,N} &= \tilde{y}_{1,N} + \lambda_2 N h y_{2,0} \\ y_{2,N} &= y_{2,0} + \lambda_2 N h y_{2,0} \end{aligned} \tag{4.127}$$

Here, we applied the Euler method to the parts of the system of equations that involve the slow eigenvalue, λ_2, using a step of Nh. This method is stable because if $|\lambda_1 h| < 2$ then $|\lambda_2 N h| < 2$. The values $y_{1,N}$ and $y_{2,N}$ are then taken as new initial values, and the cycle is repeated as many times as necessary to generate the desired solution.

The solutions produced by this method are not shown because, if the step size h is small enough, they cannot be distinguished from the exact solution.

After the initial transient is complete, one can omit the part of the method that advances the part of the solution arising from the fast eigenvalue, that is, Eqs. (4.126), and use the second part of the method [Eqs. (4.127)] exclusively. Since the first step needs to be done many more times and therefore is responsible for most of the cost, the savings can be enormous. This method is essentially a numerical version of a method used for singular perturbation problems (see, e.g., van Dyke, 1975).

The method just described can, in principle, be applied to any stiff problem. It is designed to produce a solution that is accurate on both the short and long scales, that is, an accurate solution to the entire problem. Should accuracy be necessary on only the long-term scale, it is best to use an implicit method with a large step size. The splitting method is effective when the equations solved on the short scale are much simpler than the full set of equations. In a number of important applications involving partial differential equations, this is the case.

The most difficult part of this method is usually the creation of a good splitting of the system of equations. In many cases, there is more than one way to do it. A method that is frequently used is to carry out the diagonalization described earlier in this section (assuming there are not too many equations) and to set the small eigenvalues equal to zero. When the diagonalization is inverted, the equations that result are the ones to be solved on the fast scale; the equation for the slowly varying solution can then be found by taking the difference between the original equation and the one on the fast scale. For nonlinear problems or ones with variable coefficients, it may be necessary to perform the factorization frequently, increasing the cost considerably. It is also possible to construct methods that use more than two scales.

4.11.5. Variable-Step-Size Methods

Variable-step-size and variable-order methods really show their worth for stiff problems. They are able to use small steps in the initial stages where the rapidly decaying component demands it, but can take larger steps in the later stages where the slowly varying part of the solution dominates. The number of function evaluations for a given error is considerably greater than for nonstiff problems, but it is also very much smaller than the number required by any fixed-step length explicit method.

4.12. INHERENT INSTABILITY

A difficulty that is sometimes overlooked is that higher-order equations have many linearly independent solutions. It often happens that the desired solution is well behaved, but the equation has other solutions that increase without bound in the direction of integration. Consider the following equation that arises in heat transfer and other applications:

$$y'' - k^2 y = 0 \tag{4.128}$$

with boundary conditions $y(0) = y_0$ and $y'(0) = -ky_0$. The exact solution is $y = y_0 e^{-kx}$. The problem is that the differential equation has a second solution e^{+kx}, which is suppressed by the initial conditions. If this problem is solved numerically, truncation and round-off errors will almost certainly introduce a small component of the undesired solution. Once this happens, that component, which acts as an error in this problem, will be amplified exponentially and the result soon becomes meaningless. Such a problem is called *inherently unstable*; none of the numerical methods discussed so far will produce the correct solution to such a problem. When the equation is more complicated, this problem may not be easy to recognize; sometimes it is recognized only after an exhaustive search for a method fails to produce a reasonable solution. For cases of this kind, there are some tricks that sometimes work:

- Integrate backward. If it is feasible to reverse the direction of integration, the growing solution becomes a decaying one and vice versa, and the correct solution is readily computed without interference from the other one. This method will work if it is possible to find reasonable starting conditions for the backward integration.

- If the equations are linear, start the computation with arbitrary initial conditions; the result will almost certainly be a growing solution. Let us call this solution $y_1(x)$. Then we use the method of reduction of order. We assume that the desired solution of the differential equation has the form $y(x) = y_1(x)g(x)$ and find a lower-order system of equations for $g(x)$. The resulting solution may be bounded; if it is not, this is an indication that the original system had more than one growing solution and the technique must be repeated. This method is called *filtering*. For nonlinear equations, the equations can be linearized locally and the method applied over small intervals.

4.13. GROWING SOLUTIONS

The concept of stability is defined only for problems with solutions that remain bounded for all time. A large majority of problems solved by engineers fit into this category. There are, however, exceptions and it is important to know how to deal with them.

In particular, one sometimes needs to compute solutions that are unstable in the physical sense. In cases of this type, an important component of the solution grows with time and/or location and is an essential part of the computation. In many unstable systems, the perturbation grows exponentially on a relatively short scale. A reasonable model for such a system is the differential equation (4.39) in which the coefficient α has positive real part. We need to compute the exponentially growing solution with reasonable accuracy.

For problems of this type, implicit methods may be a disaster. The reason can be seen by considering the implicit Euler method. Recall that use of this method is equivalent to replacing the exact solution $e^{\alpha h}$ by $(1-\alpha h)^{-1}$. When αh is positive, this is a very poor approximation to the exponential. In fact, when $\alpha h \to 1$, the solution diverges in one step! For larger values of αh, it decays. Thus it is best to use explicit methods with carefully chosen time steps for these problems.

The difficulty is that many people, when they encounter such systems, are misled into believing that the system is stiff (which is generally not the case), and they therefore choose implicit methods. Needless to say, this can lead to significant errors in the worst case and much too high an expense in any case.

PROBLEMS

1. The numerical method

$$y_{n+1} = y_n + \Delta x[\beta f(x_n, y_n) + (1-\beta) f(x_{n+1}, y_{n+1})]$$

contains as special cases explicit Euler ($\beta = 1$), backward Euler ($\beta = 0$), and trapezoid rule ($\beta = \frac{1}{2}$). Apply the method to the test equation $y' = \alpha y$.

a. For what values of β is this method unconditionally stable?

b. Find the truncation error as a function of β. For which value of β is the error smallest?

2. The following method was suggested as a means of solving ordinary differential equations:

$$y^* = y_n + \tfrac{1}{3}hf(x_n, y_n)$$

$$y_{n+1} = y_n + \tfrac{1}{2}h\left[3f\left(x_n + \frac{h}{3}, y^*\right) - f(x_n, y_n)\right]$$

a. Classify this method in as many ways as you can, for example, accuracy (order), explicit/implicit, multipoint.

b. Determine its order of accuracy.

c. Discuss the stability of this method using $y' = \alpha y$ *for real* α *only. Hint:* Graph the function that determines the stability.

d. Would you recommend use of this method?

3. The method based on Simpson's rule:

$$y_{n+1} = y_{n-1} + \tfrac{1}{3}h[f(x_{n-1}, y_{n-1}) + 4f(x_n, y_n) + f(x_{n+1}, y_{n+1})]$$

is used as the final corrector in some Runge–Kutta methods given in this chapter.

a. Classify this method in as many ways as you can, for example, accuracy (order), explicit/implicit, multipoint.

b. Apply this method to $y' = \alpha y$ and find the roots. Give the Taylor expansion of the roots for small αh.

c. Investigate the stability of this method for

- α pure real
- α pure imaginary

c. Would you recommend using this method and, if so, for what purposes?

4. Instead of using a differential equation solver (such as the Euler method) as the predictor in a predictor–corrector scheme, it has been suggested that we could find a predicted value by ordinary extrapolation (e.g., Lagrange) and then correct it using a standard implicit scheme. One of the simplest methods of this type would use the values of the solution at x_{n-1} and x_n to "predict" the value at x_{n+1} via linear extrapolation. We could then correct using the trapezoid rule.

a. Write the equations describing this method. Assume equal intervals ($h = x_{n+1} - x_n = x_n - x_{n-1}$) here and throughout this problem.

b. How accurate is this method (i.e., what is its order)?

c. How stable is it? (You may guess or rely on the results in the text. If you work it out, do so only for the case of $y' = \alpha y$ and α real only).

d. What is your opinion of this method?

5. Wile E. Coyote (WEC) is chasing Road Runner (RR). WEC runs so that he is always traveling directly toward RR. If RR runs at a speed of 15 m/s and stays in a straight line (this problem is not very realistic) and WEC runs at 20 m/s, how long does it take WEC to catch RR if they are initially 150 m apart in the direction perpendicular to the rabbit's path?

Derive the differential equation for the WEC's path and solve it numerically. The equations are singular at the instant of catch (so is what happens to RR), so you will need to be careful in this region. It is recommended that you compute the distance between the two at each time step and extrapolate it to the estimate its value at the next time step. Continue the calculation until the extrapolated value is negative; this indicates that capture will occur within the following time step. Then use extrapolation to find the time of capture.

Note: An analytical solution to this problem exists. It can be found in the book by Davis (1962).

6. A mother goose is flying home to her young ones in a heavy wind. If she always flies directly toward the nest, how long will it take her to cover the 500 m to the squawking chicks? Assume that she flies at 10 m/s and the wind is 5 m/s in the direction perpendicular to the line connecting her initial location and the nest.

7. A simple predator–prey model is often used to simulate biological populations. One of the simplest such models is

$$\frac{dx}{dt} = \alpha x - \beta xy \qquad \frac{dy}{dt} = -\gamma y + \eta xy$$

where x represents the prey population and y is the predator population. The behavior of the solution to this system of equations depends on the set of constants chosen and the initial conditions. Assume $\alpha = 1$, $\beta = 0.01$, $\gamma = 1$ and $\eta = 0.001$. There are lots of interesting solutions to this system. Solve the system with the following sets of initial conditions and discuss the behavior found:

a. $x = 100$, $y = 1$

b. $x = 1100$, $y = 120$

c. $x = 20$, $y = 150$

8. Consider the system of ordinary differential equations:

$$\frac{dx}{dt} = 998x + 1998y \qquad \frac{dy}{dt} = -999x - 1999y$$

a. Find the analytical solution of this system of equations with the initial conditions $x(0) = y(0) = 1$.

b. Would you anticipate any difficulty in finding the solution of this problem numerically?

c. Suggest a method (including a recommended value of Δt) that you would use to solve this problem numerically.

d. Advance the solution from $t = 0$ to $t = 2\,\Delta t$ using the method you suggested in part (c) to show that your method works.

9. The Doppler broadened line shape occurs in problems in nuclear reactors, plasma physics, and astrophysics, among others. It can be represented by an integral, but it is simpler to find it by solving the ordinary differential equation:

$$\frac{d^2\psi}{dx^2} + \xi^2 x \frac{d\psi}{dx} + \left[\frac{\xi^4}{4}(1 + x^2) + \frac{\xi^2}{2}\right]\psi = -\frac{\xi^4}{4}$$

Appropriate initial conditions are

$$\psi(\xi,0) = \sqrt{\pi}(\tfrac{1}{2}\xi)\exp(\tfrac{1}{4}\xi^2)\mathrm{erfc}(\tfrac{1}{2}\xi)$$

$$\frac{\partial\psi}{\partial x}(\xi,0) = 0$$

where erfc is the complementary error function (1-erf) and erf is the standard error function. Use this data to generate the value of $\psi(1,x)$.

10. The differential equation

$$\epsilon y'' + y' + y = 0$$

is stiff if ϵ is small. Solve it numerically with the initial conditions $y(0) = 1$ and $y'(0) = 0$ up to $x = 2$ with 1% accuracy for $\epsilon = 0.001$.

CHAPTER 5

ORDINARY DIFFERENTIAL EQUATIONS: II. BOUNDARY VALUE PROBLEMS

For first-order ordinary differential equations, only one additional datum is required to make the solution unique. In solving first-order equations numerically, one may start where that value is given and "march" the solution in the direction of increasing (or decreasing) independent variable. That is why they are called initial value problems. Methods for solving this type of problem were given in the preceding chapter.

For higher-order equations or systems of ordinary differential equations, the number of data required to make the solution unique is equal to the order of the equation or system. If all of the data are given at the same value of the independent variable, we again have an *initial value problem*. As demonstrated in the preceding chapter, methods designed for single equations can be used to solve initial value problems for several equations.

If the data are given at more than one point, we have a *boundary value problem*. For a second-order equation, data can be given at only two values of the independent variable. In principle, for systems of order higher than 2, data can be given at more than two values of the independent variable, but this is rare in practice; data are never given at more than two points. We therefore assume we are dealing with two-point boundary value problems.

Many applications produce boundary value problems. When the method of separation of variables is applied to partial differential equations, one often has a boundary value problem to solve. Finally, many partial differential equations behave in a manner similar to boundary value problems for ordinary differential equations. The methods of this chapter therefore play an important role in Chapters 6 to 8.

There are two major approaches to the numerical solution of boundary value problems; each has advantages and disadvantages. The approach we shall consider first, because it relies strongly on the methods of the preceding chapter, is called *shooting*, a name derived from the original application of this method—the adjustment of artillery settings. The second approach treats the problem more directly by applying a finite difference approximation to the differential equations and boundary conditions, thereby reducing the problem to a system of algebraic equations that is solved with algebraic equation solvers of the kind described in Chapter 1. There are many methods of this kind and they occupy the largest part of this chapter.

5.1. SHOOTING

As we have noted, a boundary value problem cannot arise from a first-order equation, so second-order equations or systems provide the simplest problems of this type. A second-order quasi-linear boundary value problem with the values of the solution given at both boundaries has the form:

$$y'' = f(x, y, y') \qquad y(0) = A \qquad y(1) = B \tag{5.1}$$

The simplest boundary conditions have been chosen for purposes of illustration, but, as we shall see, the technique applies equally well when the boundary conditions contain the first derivative of the dependent variable.

We assume that the problem is well posed so there is a unique exact solution to the problem. If we knew that solution, we would have not only its value at the initial point but also its first derivative there, $y'(0)$. The solution could then be computed using the methods of the preceding chapter. At the final point [$x = 1$ for problem (5.1)], the solution would match the boundary condition $y(1) = B$ to within the accuracy of the method used. Thus everything hinges on being able to find the first derivative of the solution at the initial point. Of course, $y'(0)$ is not known and can be found only as part of the solution process. Thus one approach is to use an iterative method to find the unknown initial value $y'(0)$. A simple iterative scheme for solving the problem (5.1) is the following. We begin by choosing a value of $y'(0)$; the choice may be based on an approximate solution to the problem or, if none is available, it may be a pure guess. We then solve the initial value problem defined by the differential equation, the prescribed initial condition $y(0) = A$ and the guessed value of $y'(0)$ using any of the methods of the preceding chapter. The solution is calculated until $x = 1$, where the second boundary condition is given. Unless we are extremely lucky, the value of $y(1)$ obtained will not be the value (B) that the boundary condition demands. Consequently, we need to adjust the initial slope and try again. The procedure is repeated until the computed value at the final point matches the boundary condition. This should be done systematically, but, to introduce the method, we illustrate the process with an example based on a hit-or-miss method.

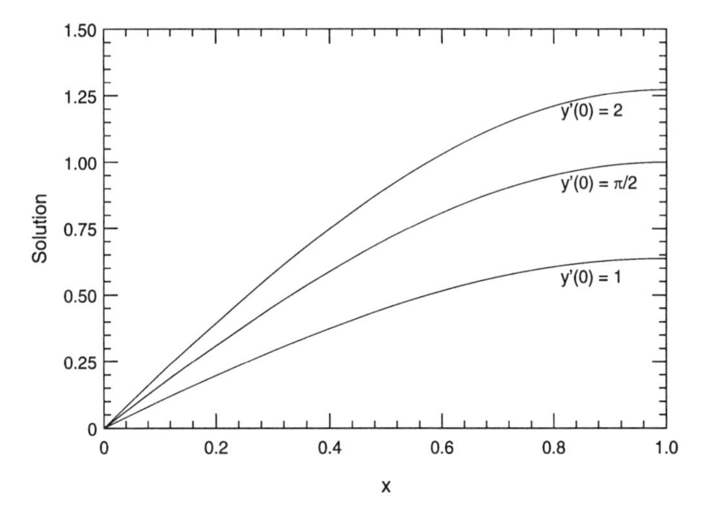

FIGURE 5.1. Illustration of the solution of a boundary value problem by the shooting method. The top curve is the solution found in the first trial, the bottom curve represents the second trial, and the middle curve is the correct solution obtained after iterating several times.

Example 5.1: Solution of a Boundary Value Problem by Shooting Let us try to find the solution of the differential equation:

$$y'' + \tfrac{1}{4}\pi^2 y = 0$$

with the boundary conditions:

$$y(0) = 0 \qquad y(1) = 1$$

The exact solution is $y(x) = \sin \pi x/2$, whose initial slope is $\pi/2$, but, for purposes of this example, we assume that this is not known. The process is illustrated in Figure 5.1. In this illustration, we use analytical rather than numerical solutions; we will discuss what happens when numerical methods are used later. The first choice for the initial slope, for which we take $y_1'(0) = 2$, is too large and the resulting $y_1(1) = 1.27$ is larger than the boundary value $y(1) = 1$. Reducing the initial slope to $y_2'(0) = 1$ results in a value $y_2(1) = 0.64$ that is too low. By trial and error, we eventually find the correct value y' is 1.57 ($\approx \pi/2$), which produces $y(1) = 1$, as required.

In the shooting method, each time we construct a trial solution with a new initial slope, ($y'(0)$), we obtain a new value of $y(1)$. These results can be represented in graphical form by plotting $y(1)$ against $y'(0)$. Each "shot" gives a point in this two-dimensional space. The collection of points resulting from a large number of trials can be connected to produce a smooth curve.

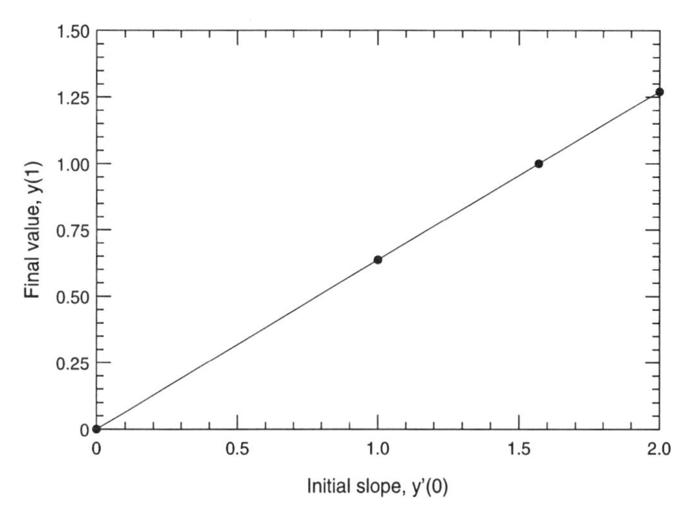

FIGURE 5.2. Functional relationship between the initial slope and final value of the solution in the shooting process illustrated in Figure 5.1. The dots represent the actual trials; the curve (actually a straight line) represents the values that would be obtained if a large number of cases were run.

Thus there is a functional relationship between the final value of the solution, $y(1)$, and its initial slope, $y'(0)$; let us call it $y(1) = f(y'(0))$. This relationship for this example is shown in Figure 5.2; the dots represent the trials shown in Figure 5.1 plus an additional point at $(0,0)$. Looked at in this way, the problem becomes one of finding the root of the equation defined by setting the function $y(1) = f(y'(0))$ equal to the prescribed boundary value, B $(= 1)$ and the differential equation solver is just a complicated way of evaluating the function.

For linear differential equations, shooting is especially simple because there is a superposition principle that states that any linear combination of solutions of the differential equation is also a solution of the equation. As a result, the functional relationship between $y'(0)$ and $y(1)$ is linear (as shown in Figure 5.2) and the correct solution to the problem can be found with only two trials.

Suppose we compute any two solutions to the differential equation $y_1(x)$ and $y_2(x)$, both of which have the correct initial value, that is, $y_1(0) = y_2(0) = A$, but different initial slopes $y_1'(0)$ and $y_2'(0)$. Then, provided that

$$c_1 + c_2 = 1 \tag{5.2}$$

the linear combination

$$y(x) = c_1 y_1(x) + c_2 y_2(x) \tag{5.3}$$

is also a solution of the differential equation that satisfies the given initial condition. Its value at the final point $x = 1$ is

$$y(1) = c_1 y_1(1) + c_2 y_2(1) \tag{5.4}$$

Setting $y(1) = B$, we have two linear algebraic equations (5.2) and (5.4) for the constants c_1 and c_2. Solving these equations, we have

$$c_1 = \frac{B - y_2(1)}{y_1(1) - y_2(1)} \qquad c_2 = 1 - c_1 \tag{5.5}$$

With these values of the constants, Eq. (5.3) provides the required solution to the problem and there is no need to compute a third solution.

Dealing with other boundary conditions is not difficult. For example, if $y'(0)$ is specified rather than $y(0)$, we could guess $y(0)$ and iterate on it until the correct solution is obtained; in the linear case, iteration is not required. If the boundary condition at $x = 1$ involves $y'(1)$, that is also easily handled. We compute whatever quantity appears in the boundary condition and adjust the guessed initial condition until the required value is obtained.

The approach just given can be extended to higher-order linear systems. For example, for a linear fourth-order problem in which two conditions are given at each end, it is necessary to guess two additional data at the initial point; these might be the second and third derivatives if the function and its first derivative are given. The solution is computed up to the final point, and the data specified by the boundary conditions are evaluated. These two quantities may be regarded as functions of the two guessed initial data. Since two initial values need to be found, it is necessary to compute three solutions. The correct solution is a linear combination of these, the coefficients being obtained by solving a set of three linear equations. In the general case, if n conditions are specified at the final point, $n + 1$ solutions must be generated and a system of $n + 1$ algebraic equations solved. If the numbers of conditions at the two endpoints are unequal, it is best to start at the side at which the larger number of conditions is given.

The major advantage of the shooting method is that the accuracy can be as high as desired. By using an automatic error control integrator, it is possible to obtain very high and guaranteed accuracy. We shall see later that, although direct methods are capable of high accuracy, they have more difficulty delivering it.

On the other hand, there are problems for which the shooting method is not well behaved; the iterative procedure may be unstable or slowly convergent. Situations of this kind arise in problems in which the final value of the function is sensitive to the initial slope, usually because the differential equation is stiff. There is also the possibility that the solution contains a boundary layer, an internal layer, or a critical layer, all of which can produce this kind of sensitivity. We illustrate these situations with examples.

Example 5.2: Shooting for a Stiff Problem In many problems in mechanics the mathematical difficulty is due to the coefficient of the highest order derivative being very small. In the limit of the coefficient tending to zero, the order of the equation is reduced, and it is impossible to satisfy all of the boundary conditions. The solution may develop a discontinuity in this limit. This is called a singular perturbation problem. When the coefficient is small but finite, the solution may have a thin layer, known as a boundary layer (because it is usually found near the boundary). This is a classic case of a problem with disparate scales, which means that the equation is very stiff.

To make this definite, let us consider the differential equation

$$\epsilon y'' + y' = 0$$

with the boundary conditions

$$y(0) = 0 \qquad y(x) \to 1 \quad \text{as} \quad x \to \infty$$

This problem is one of the simplest that displays boundary-layer-type behavior. The exact solution,

$$y(x) = 1 - e^{-x/\epsilon}$$

rises from $y = 0$ at $x = 0$ to $y = 1$ in a distance of order ϵ and remains there. As $\epsilon \to 0$, the solution approaches a step function. When $\epsilon = 0$, the differential equation becomes $y' = 0$ whose solution, $y =$ constant, is capable of matching the boundary condition at infinity or the one at $x = 0$ but not both.

This behavior is a consequence of the differential equation being only of first order when ϵ is exactly zero. The reduced equation has only one solution, so only one boundary condition can be satisfied. Problems that have this property are called singular perturbation problems and have been extensively investigated from an analytical point of view (see van Dyke, 1964). As we have already noted, the equations are stiff and the boundary value problem is difficult to solve. We now solve this problem by shooting. (We will revisit it later.)

Since the differential equation has an analytical solution, we can solve this problem without a computer. If we treat it as an initial value problem with $y(0) = 0$ and $y'(0) = A$, the solution is

$$y(x) = A\epsilon(1 - e^{-x/\epsilon})$$

The asymptotic value of y is $A\epsilon$ and the correct initial slope, $A = 1/\epsilon$, is very large.

Because this problem is linear, the method introduced above in which we find two solutions and create the exact solution by taking a linear combination of them can be applied to this problem. However, if ϵ is very small, the two computed values of $y(\infty)$ may have very different magnitudes. If the ratio of these magnitudes becomes as large as the inverse of the accuracy of the computer ($\sim 10^7$), then round-off error will make the solution inaccurate. In other words, the system of algebraic equations is ill-conditioned. The effects of numerical error are investigated further below.

Other kinds of behavior that are difficult to treat numerically are associated with *critical points*. Turning points are locations at which at least one of the coefficients in the differential equation (but not the coefficient of the highest derivative) changes sign. At such a point, the nature of the solution may change (e.g., from exponential-like to oscillatory) or there may be an *internal layer*, a region similar to a boundary layer within the solution domain where the solution undergoes rapid change. At a critical point, the coefficient of the highest derivative passes through zero and the problem is even more serious. Some examples containing these kinds of points are found in the examples that follow.

It is also important to remember that methods for integrating initial value problems in differential equations contain numerical errors. The result of a shooting procedure applied to a boundary value problem cannot be more accurate than the integration scheme used to generate the trial solutions. For this reason, it does not make sense to reduce the error due to incomplete convergence of the iteration scheme to a level much below that of the accuracy of the integration scheme, that is, the iteration process should be stopped short of complete convergence.

Example 5.3: An Internal Layer As an example of a problem with an internal layer, consider the equation:

$$\epsilon^2 y'' + 2xy' = 0$$

The general solution to this equations is

$$y(x) = c_1 + c_2 \text{erf}\left(\frac{x}{\epsilon}\right)$$

where erf is the well-known error function. If ϵ is small, the solution grows very rapidly near $x = 0$.

First, let us solve the initial value problem with $y(-1) = 1$ and $y'(-1) = 1 \times 10^{-30}$ and compute the solution up to $x = 1$. Since the purpose of this example is to illustrate the effect of numerical error, Euler's method with double-precision arithmetic was used; this was done to eliminate the round-off error resulting from the large number of steps required. The value of the computed $y(1)$ is given in Figure 5.3 as a function of the step size h; the extreme sensitivity and nonlinearity of $y(1)$ as a function of the step size are

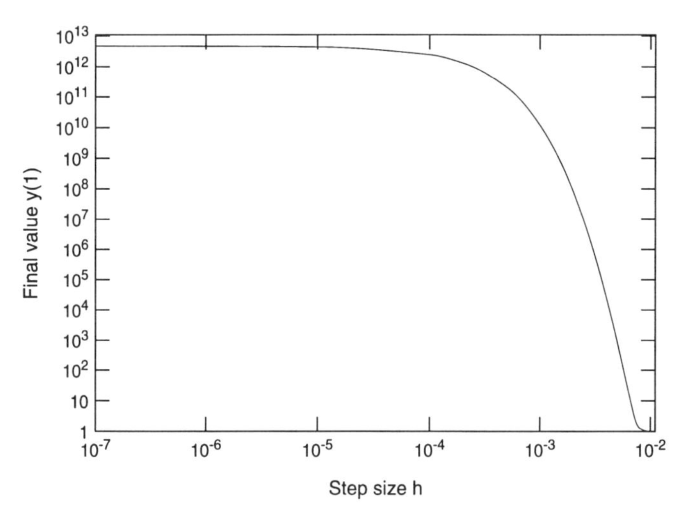

FIGURE 5.3. Relationship between $y(1)$ and $y'(0)$ for a problem with an internal layer. For details, see Example 5.3.

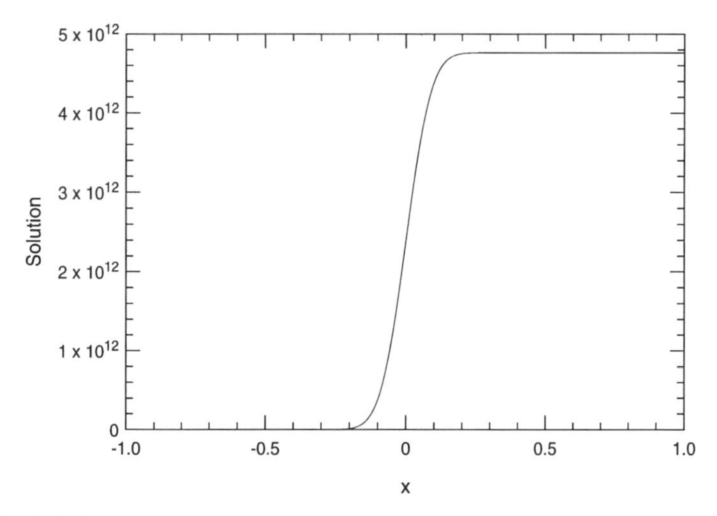

FIGURE 5.4. Solution of a problem with an internal layer. For details, see Example 5.3.

obvious. The result is also very sensitive to the initial slope but that dependence is linear.

The solution is shown in Figure 5.4 and is the well-known S-shaped curve. Problems similar to this often occur as models for the behavior of some quantity that initially has exponential (in this case e^{x^2}) growth followed by saturation due to nonlinearity or some other factor. This type of model is often used to represent the growth of industries (or whole economies), epidemics, or catastrophes. It is very difficult to predict the final value of the quantity given only data for the early stages of growth. One must be very cautious about accepting predictions based on models of this kind. Unfortunately, examples like this one are often found in the popular press and are taken far too seriously. It is also important to note that the rapid growth is sometimes taken as an indicator of "stiffness," but this problem is not stiff in the sense defined in the preceding chapter; it simply has an exponentially growing solution. Use of an implicit method (based on the belief that the problem is stiff) will produce larger errors than an explicit method such as the one used in this example and should be rigorously avoided. Using a higher-order method or one with built-in error control is a good idea.

If the differential equation of this example were to be solved with the boundary condition $y(1) = 10$ by shooting, and the first two guesses of the initial slope were of the order of $y'(0) \approx 1$, it could be very difficult to find the correct answer, which is $\approx 4.2 \times 10^{-43}$, because round-off error might destroy the needed information. This result could only be found with a good initial estimate, which would probably need to be found by a direct search. Thus stiffness can make solution of boundary value problems with the shooting method very difficult. The difficulty may be even more severe when the problem is nonlinear.

Example 5.4: A Critical Layer As noted earlier, the solution of a differential equation may contain a critical layer if the coefficient of the highest derivative vanishes somewhere within the solution domain. Such a point is a singular point of the differential equation; we will assume that it is a regular singular point, as this is the case in essentially every physical application. At such a point, one of the solutions of the differential equation is singular, and the requirement that the solution be well behaved acts as a constraint. Problems of this kind are very difficult to deal with both analytically and numerically.

Here we consider a problem that illustrates the major difficulties. Most of problems that arise in applications are sufficiently complicated that they are beyond the scope of this book. We shall therefore treat the differential equation:

$$x^2y'' + 2xy' - x^2y = 0$$

which has a singularity at $x = 0$. This equation usually arises from problems in a spherical coordinate system; in that case, x is the radial coordinate, and we are interested only in the region $x > 0$. Since the purpose here is to illustrate a point, we shall adopt the unphysical boundary conditions:

$$y(-1) = 1 \qquad y(1) = 1$$

The differential equation has the general solution:

$$y(x) = c_1 \frac{\sinh x}{x} + c_2 \frac{\cosh x}{x}$$

as can be verified by direct substitution. The second solution is singular, that is, it becomes infinite at $x = 0$ so the coefficient c_2 must be zero in any physically meaningful solution. This means that only one boundary condition can be provided. However, the boundary conditions given above are consistent (if redundant) and the problem does have a unique solution. The solution of the problem that satisfies the boundary conditions is

$$y(x) = \frac{\sinh x}{x \sinh 1}$$

By dividing the differential equation by x and taking the limit as $x \to 0$, we see that any solution of this equation that is bounded and has bounded derivatives must have $y'(0) = 0$. The solution given above meets this criterion, but the solution that was discarded does not. If one attempts to solve this problem by the shooting method starting at $x = -1$, with a guessed value of $y'(-1)$, it is almost certain that, when $x = 0$ is reached, y' will not be zero. If $y' \neq 0$ the solution procedure will stop due to an attempted division by zero. [Even if $y'(0)$ were zero, the computer would object unless it is carefully programmed.] It is more likely that a small round-off error or the choice of the step size will prevent x from ever being exactly zero. At the point closest to $x = 0$, there will be a division by a small but nonzero number. The result will be a very large but unpredictable number and the computation will become meaningless. No amount of readjusting the initial condition is likely to improve the situation.

In a case of this kind, a method that sometimes works is to shoot from both sides toward the center. For this case, one guesses y' at both boundaries, solves two initial value problems (one starting at $x = -1$ computed in the positive x direction and another starting at $x = 1$ computed in the direction of decreasing x), and tries to match both the function and first derivative at $x = 0$. In more difficult cases, the point at which the singularity occurs is not known in advance (it may be one of the important parameters to be determined), and it may be necessary to iterate to find the proper matching point. In other cases, the singularity can be treated by providing "jump conditions" across it. We shall not give details of how these problems are solved.

When the differential equation and/or the boundary conditions are nonlinear, the relationship between the guessed data at the initial point and the given data at the final point also becomes nonlinear. The straight line of Figure 5.2 is replaced by a curve, and the problem cannot be solved exactly after two trials—iterative solution is required. However, because data for only a few values of the guessed condition(s) are available, it is not possible to calculate the derivative of the function relating the initial and final data exactly, so use of the Newton–Raphson method is impossible. The secant method can be used and is recommended. How this can be done is demonstrated in the following example.

Example 5.5: A Nonlinear Shooting Problem As was stated above, one effect of nonlinearity is to cause the relationship between the guessed datum and the boundary condition to be fit at the final point to become nonlinear, making an iterative solution method necessary. To illustrate how this can be done, we take a third-order problem that is well known in fluid mechanics, the Blasius problem. The problem is to solve the differential equation:

$$f''' + \tfrac{1}{2} f f'' = 0$$

with the following boundary conditions:

$$f(0) = 0 \qquad f'(0) = 0 \qquad f'(y) \to 1 \quad \text{as} \quad y \to \infty$$

The derivative of the solution is the fluid velocity parallel to a wall in the viscous boundary layer near a solid surface. The boundary conditions at $y = 0$ represent the conditions of impermeabilty (no flow through the surface) and no-slip (zero velocity parallel to the wall as a result of friction). The boundary condition far from the surface is the statement that the velocity becomes constant far from the wall.

Following the method outlined at the beginning of the chapter, we let $y_1 = f$ and rewrite the differential equation as three first-order equations:

$$y_1' = y_2 \qquad y_2' = y_3 \qquad y_3' = -\tfrac{1}{2} y_1 y_3$$

In terms of these variables, the boundary conditions can be written:

$$y_1(0) = y_2(0) = 0 \qquad y_2(\infty) = 1$$

Of course, the boundary condition at infinity is a problem. Because the solution to this problem is available in the literature (Schlichting, 1987), we know that the error incurred by enforcing the boundary condition at $y = 10$ is very small. If we did not know this, we could enforce this condition at $y = 10$ and then repeat the problem with the condition enforced at, say, $y = 20$. If the two solutions are in adequate agreement for $y < 10$, the solution may be accepted; if not, we can repeat the process with the condition enforced at a still larger value of y. With the condition enforced at $y = 10$ the results are accurate to the number of significant figures, presented below.

To start the shooting procedure, we guess two values of $y_3(0)$ and solve the resulting initial value problems. The fourth-order Runge–Kutta method with a fixed step size $h = 0.1$ and double-precision arithmetic was used. For each

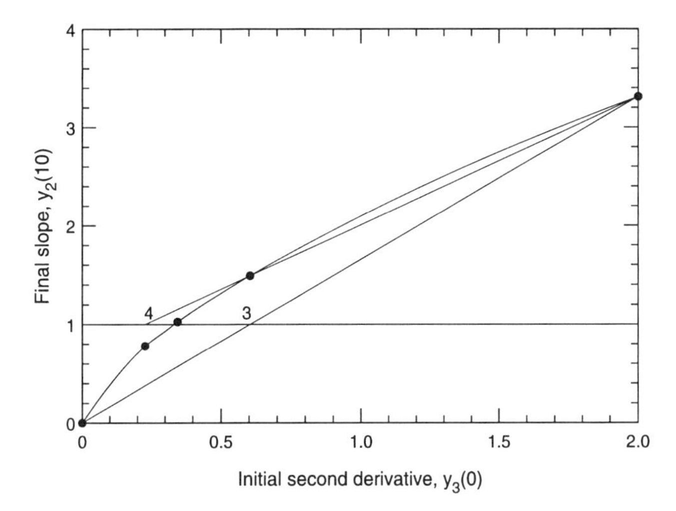

FIGURE 5.5. Relationship between the final slope and the initial second derivative in the nonlinear shooting example of Example 5.5.

guessed value of $y_3(0)$, we find a value of $y_2(10)$. For the first guess, we chose $y_3(0) = 0$, which gives $y_2(x) = 0$ everywhere and, in particular, $y_2(10) = 0$. A second try with $y_3(0) = 2$ produces $y_2(10) = 3.31$. The complete functional relationship between $y_3(0)$ and $y_2(10)$ is shown in Figure 5.5; the points indicate the values obtained in the actual trials.

Having the first two trial solutions, we connect them with a straight line and determine the value of $y_3(0)$ at which this line crosses the horizontal line $y_2(10) = 1$. This produces the value of $y_3(0)$ marked 3 in the figure, which is used as the starting value for the next iteration; the resulting value of $y_2(10)$ is indicated by the black dot above the 3. A straight line is then drawn between the second and third results and produces the result marked 4. This procedure is the well-known secant method for finding the roots of equations. To converge to 10^{-6} accuracy requires eight solutions of the differential equation; the final result is 0.332057, a result that is quite important as the skin friction (and heat transfer) at the surface are proportional to it. This result is, of course, not perfectly accurate due to the error produced by the Runge–Kutta method. However, repeating the calculation with $h = 0.05$ gives the same result to six significant figures, so we conclude that the value cited is correct to the number of decimal places given.

The program BLASIUS, used to solve this problem, can be found on the Internet site described in Appendix A.

This example posed no difficulty other than nonlinearity. The convergence rate is typical of what one finds for nonlinear problems. Had the equation been stiff, the target data would be much more sensitive to the initial conditions and convergence would probably be much slower or even impossible to obtain.

Extension of the method used in the example to higher-order systems in which several data elements are given at each endpoint is straightforward in principle but may be complicated in practice. All of the data at the final point are regarded as functions of the missing initial data. In essence, we have to solve a system of nonlinear equations and this must be done iteratively. This can be accomplished by one of the generalizations of the secant method such as Broyden's method (Gill et al., 1981) for which the convergence is as fast as the secant method for a single equation.

The ability to use a method of any accuracy to generate the solution makes shooting the best method when high accuracy is required. Its disadvantages are that it usually finds only a single solution in problems that have multiple solutions (such as eigenvalue problems considered later) and the difficulties that it has with stiff systems.

We shall now go on to the other broad class of methods for boundary value problems called direct methods.

5.2. DIRECT METHODS: INTRODUCTION

The alternative to shooting is a kind of "sledge hammer" approach. *Direct methods* are based on finite difference approximations (or equivalents to be introduced below) to the differential equation at a predetermined set of grid points. The boundary conditions are incorporated directly into the discretized equations. The result is a set of algebraic equations for the values of the solution at the mesh points. These equations are solved by a method appropriate to their nature; they may be linear or nonlinear, homogeneous or inhomogeneous, according to the nature of the original differential equations. This may seem an expensive approach, but the algebraic equations that result from approximating differential equations almost always have simple structures (e.g., they may be tridiagonal), so computing a solution is relatively inexpensive, making this approach competitive with the shooting method.

The method is illustrated by taking Eq. (5.1) as an example. Since, at this point, there is no reason to do otherwise, we approximate this equation on a uniform grid. The solution interval $(0, 1)$ is broken into N equally spaced subintervals each of size h so that $Nh = 1$. The grid points are thus $x_i = ih$, $i = 0, 1, \ldots, N$. Again because there is no reason to do otherwise, we apply central difference approximations for the approximation of all derivatives. Thus y' at the ith grid point is approximated by

$$y_i' = \frac{y_{i+1} - y_{i-1}}{2h} \tag{5.6}$$

and the second derivative is approximated by

$$y_i'' = \frac{y_{i+1} - 2y_i + y_{i-1}}{h^2} \tag{5.7}$$

When these approximations are employed, the differential equation (5.1) at the ith mesh point is approximated by

$$y_{i+1} - 2y_i + y_{i-1} = h^2 f(x_i, y_i, (y_{i+1} - y_{i-1})/2h) \tag{5.8}$$

This approximation is enforced at every interior grid point, x_i, $i = 1, 2, \ldots, N - 1$. In the equation with $i = 1$, which is an approximation to the differential equation at the first interior point, the value of y at the boundary, $y(0) = y_0$, occurs in the equation; it is replaced by the known boundary value A. In the same manner, we can substitute B for y_N where it occurs in the equation at mesh point $N - 1$. The result is then a system of $N - 1$ algebraic equations in $N - 1$ unknowns. It is important that finite difference equations inherit the character of the original differential equation. If the latter is linear, so are the finite difference equations; if the differential equation and the boundary conditions are homogeneous, the difference equations are also homogeneous.

The set of finite difference equations (5.8) produced by this method must be solved. We deal with the linear case first; in this case, not only are Eqs. (5.8) linear, they are tridiagonal. That is, the ith equation contains only the variables y_{i-1}, y_i, and y_{i+1}. A system of linear equations of this form is extremely easy to solve numerically; an efficient algorithm for doing so was described in Chapter 1 and a subroutine for the task (TRDIAG) is given on the Internet site described in Appendix A. In fact, unless very high accuracy is required, this approach is usually preferred to shooting because it is usually computationally more efficient, easier to program, and the error tends to be distributed more evenly over the domain.

In the nonlinear case, Eqs. (5.8) are a coupled system of nonlinear algebraic equations; in most cases, it is relatively easy to solve this system. All standard methods for solving nonlinear systems use some kind of linearization; for the system (5.8), the linearized equations are tridiagonal and thus easy to solve.

We also need methods for dealing with other types of boundary conditions. The boundary conditions in the problem (5.1), that is, ones that give the value of the solution at the boundary, are often called Dirichlet boundary conditions. They are the easiest ones to handle, which is why they were treated first. More generally, either boundary condition may contain the first derivative as well as the function. If it does, the value of the dependent variable at the boundary is unknown and must be computed as part of the solution. It is then necessary to solve the differential equation at the boundary itself; this is done by applying the finite difference approximation at the boundary point; for the approximations being used here, the equation at the boundary is identical to those at the interior points. The unusual feature is that the finite difference equation at the boundary references a point located a distance h outside the boundary; this is called a *phantom* or ghost point. At the left boundary, the phantom point is at $x_{-1} = -h$ and y_{-1} occurs in the finite difference equation at the boundary. The system of equations then contains more unknowns than equations; this situation is remedied by applying the boundary condition. For example, the boundary condition:

$$ay'(0) + by(0) = c \tag{5.9}$$

can be approximated by the central difference approximation:

$$a\frac{y_1 - y_{-1}}{2h} + by_0 = c \tag{5.10}$$

Equation (5.10) can be solved for y_{-1} and substituted into the finite difference equation at $x = x_0$. Carrying out a similar procedure at $x = x_N$, the point on the right boundary, we obtain a set of $N + 1$ equations containing $N + 1$ unknowns. Alternatively, one could regard y_{-1} and y_{N+1} as additional variables, include Eq. (5.10) and its counterpart at the right boundary of the equation set, and solve a set of $N + 3$ equations in $N + 3$ unknowns. This set of equations obtained in this way is not quite tridiagonal, a disadvantage that favors use of the first approach.

There are many other methods of forming difference approximations; any of the formulas introduced in Chapter 4 can be used for this purpose; we will consider some of these later in this chapter. One method reduces the second-order differential equation to a system of two first-order equations and finite differences those. A method of this kind designed for parabolic partial differential equations is given in Chapter 6 and could be used here as well.

Example 5.6: Direct Solution of a Second-Order BVP As the first example of the use of direct methods, we shall solve a boundary value problem (BVP) that presents few difficulties. The neutron distribution in a material obeys the following equation:

$$\frac{d^2\phi}{dx^2} - \kappa^2\phi = -\frac{S(x)}{D}$$

where ϕ is the (poorly named) neutron flux (actually the neutron number density multiplied by a scalar speed), κ is a nuclear property (inverse diffusion length), D the diffusion coefficient, and $S(x)$ represents a neutron source. For simplicity we take S to be constant, although it would not be constant is a more realistic situation. The boundary conditions are

$$-D\frac{d\phi}{dx} = \pm\lambda\phi \quad \text{at} \quad x = \pm\frac{t}{2}$$

where t is the thickness of the slab and λ is a given constant.

The number of parameters can be reduced by introducing the dimensionless variables

$$\alpha^2 = \kappa^2 t^2 \qquad \beta = \frac{\lambda t}{D} \qquad \theta = \frac{\phi}{\phi_0} \qquad \xi = \frac{x}{t} \qquad Q = \frac{St^2\phi_0}{D}$$

where ϕ_0 is an arbitrary flux level, to obtain

$$\frac{d^2\theta}{d\xi^2} - \alpha^2\theta = -Q$$

$$\frac{d\theta}{d\xi} = \mp\beta\theta \quad \text{at} \quad \xi = \pm\frac{1}{2}$$

This problem can be solved analytically, allowing exact determination of the error in the numerical solution. We leave this as an exercise for the reader.

Applying the second-order central difference approximation to the differential equation gives

$$\theta_{j-1} - (2 + \alpha^2 \Delta\xi^2)\theta_j + \theta_{j+1} = -\Delta\xi^2 Q \qquad j = 0, 1, \ldots, N$$

where $\xi_j = jh$. At the boundaries $j = 0$ and $j = N$, the boundary conditions give

$$\theta_1 - \theta_{-1} = 2\beta\,\Delta\xi\theta_0$$

$$\theta_{N+1} - \theta_{N-1} = -2\beta\,\Delta\xi\theta_N$$

These can be substituted into the difference equations for $j = 0$ and $j = N$, respectively, to yield the tridiagonal system of algebraic equations that needs to be solved.

The program NEUDIFF found on the Internet site described in Appendix A illustrates how the solution of this problem may be obtained through the use of the subroutine for the solution of tridiagonal systems, TRDIAG.

Because the solution to this problem is very smooth, the results obtained are sufficiently close to the exact solution that plotting them does not show very much. Instead, in Table 5.1, we give the error at $\xi = 0$, the midpoint of the domain, for various grid sizes with the parameter values $\alpha = \beta = 1$. The exact solution at $\xi = 0$ is $\theta = 0.393$. The table shows that using just two intervals produces an accuracy of better than 1%. Finally, in the last column in Table 5.1 the error divided by the square of the gird size, which should be constant for small enough grid size, is given. The ratio is indeed nearly constant, which clearly shows the second-order nature of the method. As stated earlier, this is a relatively easy problem.

We next take a more difficult example.

Example 5.7: Direct Solution of a Stiff Problem For this example, we take the stiff problem used in Example 5.2:

$$\epsilon y'' + y' = 0 \qquad y(0) = 0 \qquad y(x) \to 1 \quad \text{as} \quad x \to \infty$$

TABLE 5.1. Solution of a Boundary Value Problem by a Direct Method

N	$\Delta\xi$	Error	Error/$\Delta\xi^2$
2	0.5	2.76×10^{-3}	0.0110
4	0.25	7.03×10^{-4}	0.0112
6	0.167	3.14×10^{-4}	0.0113
8	0.125	1.77×10^{-4}	0.0113
10	0.100	1.13×10^{-4}	0.0113
16	0.0625	4.41×10^{-5}	0.0113
20	0.0500	2.82×10^{-5}	0.0113

TABLE 5.2. Direct Solution of a Stiff Boundary Value Problem

x	y
1	50.9
2	0.2
3	50.7
4	0.4
5	50.5
6	0.6
7	50.3
8	0.8
9	50.1
10	1.0

An added difficulty is that one of the boundary conditions is imposed at infinity, a situation encountered earlier in Example 5.3. As we did there, we shall get around the difficulty by setting $y = 1$ at some large but finite value of x, say $x = b$. As long as $b \gg 1/\epsilon$, this causes no problem. If the behavior of the solution were not known, we could apply the condition at some x, compute the solution, repeat the calculation with a larger final value of x, and see whether the solution changes by more than an acceptable amount.

The finite difference approximations we will use are the ones used in the preceding problem. Specifically, the difference equations are

$$(2\epsilon + \Delta x)y_{i+1} - 4\epsilon y_i + (2\epsilon - \Delta x)y_{i-1} = 0 \qquad i = 1,2,\ldots,N-1$$

with $y_0 = 0$ and $y_N = 1$.

Since the rapid rise of the solution occurs within a distance of order ϵ, the key parameter in determining the accuracy of the solution is $\Delta x/\epsilon$; it must be small to obtain an accurate solution.

An example of how badly things can go is shown in Table 5.2. The boundary condition $y = 1$ was applied at $x = 10$ and the parameters were $\Delta x = 1$ and

$\epsilon = 0.001$. It is obvious from Table 5.2 that the solution bears no relation to the correct solution. To explain this, we note, that when $\Delta x/\epsilon$ is large, the diagonal elements of the matrix corresponding to the system of equations (5.2) are much smaller than the off-diagonal elements. Consequently, the odd-index variables, $y_1, y_3, \ldots$, are closely coupled as are the even-index variables. The two sets of variables are, however, almost uncoupled from each other. When an odd number of points is used, only the even-index variables are tightly coupled to the boundary conditions, and they represent an approximation to the actual solution that is smooth but not accurate. The odd-index variables are not tied to anything definite and simply reflect the fact that the matrix is almost singular; in fact, for odd N, the matrix does become singular in the linear algebra sense as $\epsilon \to 0$.

When the number of points is even, the matrix does not become singular as $\epsilon \to 0$, but the results are still not good. The even-numbered points are tightly coupled to the boundary condition at $x = 0$ and, as $\epsilon \to 0$, $y_{2j} \to y_0$. The odd points are coupled to the condition at x_N and $y_{2j-1} \to y_N$ in this limit.

To obtain an accurate solution, we need to use a grid size $\Delta x < 1/\epsilon$, which, for the parameters given above, means that we will need a grid of 10,000 points! This is clearly not a efficient method of solving this problem. The solution can be obtained more economically with variable mesh spacing. What is needed is close spacing of the points near $x = 0$ where the solution is growing rapidly; wider spacing can be used when $x > 1$ where the solution is changing more slowly. An approach of this kind is described in Section 5.5.

5.3. HIGHER-ORDER DIRECT METHODS

For initial value problems, we found that high accuracy is more readily obtained with high-order methods than small step sizes. The same should be true for boundary value problems. It should be recalled, however, that some higher-order methods for initial value problems are difficult to start; that is to say, special methods are required for the first few steps. For direct methods applied to boundary value problems, the analogous problem is the difficulty of maintaining accuracy at the boundaries.

An example serves to illustrate the problem. Suppose that we try to solve the ordinary differential equation

$$y'' = f \tag{5.11}$$

with the boundary conditions

$$y(0) = A \qquad y(1) = B \tag{5.12}$$

using the fourth-order central difference scheme given in Table 4.2. The resulting finite difference equations would be

$$-y_{i-2} + 16y_{i-1} - 30y_i + 16y_{i+1} - y_{i+2} = 12h^2 f_i \tag{5.13}$$

(Uniform spacing is used to avoid introducing too many complications at once, and the issue of how f_i is evaluated is ignored for now.) Consider what happens near the left boundary. The value $y(0) = y_0 = A$ is known and can be used wherever it occurs in the equation set. This quantity occurs in Eq. (5.13) with $i = 2$. The given value can be substituted into the equation and, when the term containing it is moved to the right-hand side of the equation, there is no difficulty. In the equation with $i = 1$, however, there is a problem that is not so easily dealt with. The equation contains y_0, which can be treated in the same way as it was in the $i = 2$ equation, but it also contains y_{-1}, which is the value of the solution at a point not in the solution domain, a phantom point. Its value can only be obtained by making a further approximation.

There are a number of methods of dealing with this difficulty. In one approach, we construct an approximation to the function at the phantom point. An alternative is to use a different difference formula near the boundary than in the interior.

If the latter approach is adopted, and fourth-order accuracy is to be maintained, we need to construct a fourth-order approximation to the second derivative at x_1 that does not reference the value y_{-1}. Such a formula would have to include y_0, y_1, y_2, y_3, y_4, and y_5, while the approximation (5.13) at $i = 2$ requires only y_0, y_1, y_2, y_3, and y_4. [The difference approximation (5.13) is fourth-order accurate in part because it is a central difference approximation on a uniform grid; symmetry causes the third-order error terms to cancel out.] We state without proof that using a difference approximation at x_1 that contains points not included in the approximation at x_2 causes the results to become unreliable. It also violates our sense that the solution at x_1 should be less affected by what happens at x_5 than is the solution at x_2. We conclude that this is probably not a good approach.

A better method, and the recommended one, is to use a difference approximation at x_1 that uses only the values y_0, y_1, y_2, y_3, and y_4. The highest order of accuracy that a difference approximation using only these data can have is third order, and the approximation is

$$y''(x_1) \approx \frac{1}{12h^2}(11y_0 - 20y_1 + 6y_2 + 4y_3 - y_4) \tag{5.14}$$

The analogous formula at the other end is

$$y''(x_1) \approx \frac{1}{12h^2}(-y_{N-4} + 4y_{N-3} + 6y_{N-2} - 20y_{N-1} + 11y_N) \tag{5.15}$$

The truncation error in these formulas is $-h^3 f^{v}/12$. However, because the third-order approximation is applied at only one grid point, it actually produces an error proportional to h^4 so the overall method remains fourth-order accurate globally. Thus only a little is lost by using this approximation. In general, one can use an approximation one order less accurate than the approximation used in the interior at a boundary without sacrificing global accuracy. We shall demonstrate this in Example 5.8.

The system of equations (5.13), if we ignore the approximations at the boundaries for the moment, is pentadiagonal—that is, the matrix of the problem has nonzero elements only on the main diagonal and two diagonals on either side of it. As discussed in Chapter 1, the solution of pentadiagonal systems can be achieved by using a generalization of the method used for a tridiagonal systems at a cost that is about three times as great. The fourth-order accurate boundary conditions introduce two extra elements in the first and last rows of the matrix, necessitating the construction of a modified pentadiagonal solver. For the boundary condition approximations (5.14) and (5.15), less modification is required.

Example 5.8: A Fourth-Order Direct Method A problem similar to the one of Example 5.3 is solved using the method recommended above. The important change is that, to avoid undue complication, the boundary conditions are changed from the ones of the earlier problem to ones that require the solution to be zero at the boundary:

$$\phi(0) = \phi(1) = 0$$

In matrix form the problem is

$$\begin{pmatrix} (-20 - 12k^2h^2) & 6 & 4 & -1 & 0 & \cdots \\ 16 & (-30 - 12k^2h^2) & 16 & -1 & 0 & \cdots \\ -1 & 16 & (-30 - 12k^2h^2) & 16 & -1 & \cdots \\ \vdots & \vdots & \vdots & \vdots & \vdots & \ddots \\ \vdots & \vdots & \vdots & \vdots & \vdots & \ddots \end{pmatrix} \begin{pmatrix} y_1 \\ y_2 \\ y_3 \\ \vdots \\ \vdots \end{pmatrix}$$

$$= 12h^2 \begin{pmatrix} Q_1 \\ Q_2 \\ Q_3 \\ \vdots \\ \vdots \end{pmatrix} \tag{5.16}$$

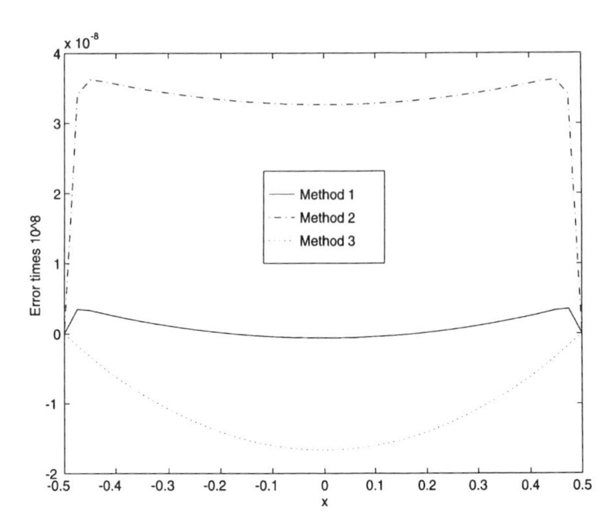

FIGURE 5.6. Error in the solution of a boundary value problem with a fourth-order central difference approximation and three approximations to the second derivative near the boundary. The error for method 2 has been multiplied by 10; the error for method 3 has been multiplied by 100.

These equations are a pentadiagonal system with one extra element each in the first and last rows; the elements of the last row are identical to those in the first one with the order reversed. Thus the matrix is almost pentadiagonal. This method is sufficiently accurate for this problem that double precision is required in order to compute the difference between this solution and the exact one. The error is shown in Figure 5.6, in which this method is designated as method 1.

For comparison, results obtained by the use of another method in which the second derivative at x_1 is approximated by the second-order central difference approximation:

$$y''(x_1) \approx \frac{1}{h^2}(y_2 - 2y_1 + y_0) \tag{5.17}$$

is also tried. This produces a standard pentadiagonal system of equations; this is denoted as method 2 in Figure 5.6.

The error in the solution produced by each of these methods with 40 points is shown in Figure 5.6. Method 1 produces a larger error near the boundary than in the center of the region. The error near the boundaries is positive, that is, the solution tends to be smaller than the true value and is due mainly to the approximations (5.14) and (5.15) used at the boundary.

The use of the second-order approximation (5.17), which was designated as method 2, produces an error that, for this problem, is nearly constant across the region. As Table 5.3 shows, the error behaves in a fourth-order manner but

TABLE 5.3. Errors in Fourth-Order Solutions of a Boundary Value Problem

Number of Intervals N	$\epsilon/\Delta x^4$ Method 1	$\epsilon/\Delta x^4$ Method 2	$\epsilon/\Delta x^4$ Method 3
4	0.00164	0.0773	−0.000426
6	0.00234	0.0797	−0.000426
8	0.00213	0.0806	−0.000427
10	0.00180	0.0813	−0.000427
20	0.00066	0.0828	−0.000427
30	0.00013	0.0834	−0.000427
40	−0.00016	0.0837	−0.000427
50	−0.00034	0.0839	−0.000427

is larger than the error produced by method 1. The last column in Table 5.3 relates to a fourth method described in the next section.

This example shows that multipoint difference formulas are an effective but somewhat difficult way to achieve accuracy in boundary value problems. The difficulty is increased further when still higher-order accuracy is demanded. An alternative approach is to seek methods that use only three points but that are more accurate than the standard formulas. We demonstrate how that can be done in the next section.

5.4. COMPACT METHODS

A method of improving the accuracy of difference approximations without using the values of the function at a large number of grid points in the formula is to use the compact difference method introduced in Section 4.1. We recall that the compact difference formulas can be found by looking at the standard central difference approximation and its error:

$$\frac{y_{i+1} - 2y_i + y_{i-1}}{h^2} = y''(x_i) + \frac{h^2}{12} y^{iv}(x_i) \tag{5.18}$$

We then construct a linear combination of the second derivatives at these three points, which approximates y_i'' with the same error. In fact, by applying Taylor series to y'', one can show that

$$\frac{1}{12}(y_{i+1}'' + 10y_i'' + y_{i-1}'') = y''(x_i) + \frac{h^2}{12} y^{iv}(x_i) \tag{5.19}$$

The right-hand sides of the last two equations are identical. Consequently, if we equate the left-hand sides of these equations, the second-order error terms drop out and the resulting approximation has fourth-order accuracy. We can then compute y_i'' by solving the tridiagonal system of equations:

$$\frac{1}{12}(y_{i+1}'' + 10y_i'' + y_{i-1}'') = \frac{y_{i+1} - 2y_i + y_{i-1}}{h^2} \tag{5.20}$$

To apply this method to the solution of differential equations, it is best to write Eq. (5.20) in matrix form:

$$A\mathbf{y}'' = B\mathbf{y} \tag{5.21}$$

where $\mathbf{y}''$ and $\mathbf{y}$ are vectors (in the linear algebra sense) whose components are the second derivatives and functions at the mesh points, and A and B are tridiagonal matrices whose components may be determined by inspection of Eq. (5.20). Equation (5.21) can be written

$$\mathbf{y}'' = A^{-1}B\mathbf{y} \tag{5.22}$$

and the differential equation (5.11) becomes

$$A^{-1}B\mathbf{y} = \mathbf{f} \tag{5.23}$$

Multiplying by A and writing the result in component form, we have

$$y_{i+1} - 2y_i + y_{i-1} = \frac{h^2}{12}(f_{i+1} + 10f_i + f_{i-1}) \tag{5.24}$$

which is very similar to the standard second-order finite difference approximation. The major difference is that the inhomogeneous term at the point x_i is replaced by a linear combination of the values at three points.

If the differential equation does not contain the first derivative, there is no problem in applying this method. This approach is only slightly more expensive than the second-order method and implementation of the boundary conditions is no problem. It is called the *compact fourth-order method.* This method is limited to application to linear equations on uniform grids but, when it can be used, it is very powerful.

These methods are also called *Padé* methods for their resemblance to the Padé approximations that are sometimes used to compute functions on computers.

Example 5.9: Direct Solution with a Compact Method We now solve the problem of the preceding example using the compact fourth-order scheme. As noted above, this is almost as easy as using a second-order method; the only difference is in the matrix elements and the right-hand side. For this particular problem the right-hand side is in fact unchanged and the equations are

$$\left(1 - \frac{\alpha^2 h^2}{12}\right)\theta_{i-1} - \left(2 + \frac{5\alpha^2 h^2}{6}\right)\theta_i + \left(1 - \frac{\alpha^2 h^2}{12}\right)\theta_{i+1} = Qh^2 \quad (5.25)$$

This problem can be solved by the method used for the second-order method but requires a double-precision tridiagonal solver for accurate calculation of the error. The results are shown as method 3 in Figure 5.6 and Table 5.3. The table shows this method behaves in a very stable fourth-order fashion and is actually considerably more accurate than either of the two methods presented earlier. Furthermore, for this problem, the error is distributed much like the solution itself—that is, the relative error is approximately constant, a very desirable property.

This example shows the efficiency of the compact fourth-order method. Due to their accuracy, compact methods have been the subject of considerable research and have been extended to even higher-order accuracy and nonuniform grids. For a recent example, see Poinsot and Lele (1992). As we have already seen, the major drawback is that they may be difficult to apply to some types of equations.

Another approach to the construction of compact fourth-order methods is based on recognizing that Hermite interpolation is a fourth-order accurate interpolation method. Using Hermite interpolation as the basis for a numerical method for solving second-order differential equations requires that both the dependent variable and its first derivative as well be treated as unknowns. This doubles the number of equations and makes the matrix block structured, but the additional accuracy may be worth the extra cost. Methods based on Hermite interpolation have been proposed but have not achieved the popularity of the methods discussed here. For that reason, we shall say no more about them.

5.5. NONUNIFORM GRIDS

As we demonstrated in the preceding two sections, higher-order methods are highly desirable but difficult to apply to boundary value problems. The difficulty with the boundary conditions grows worse as the order increases. An alternative approach to obtaining high accuracy is clearly desirable. For both quadrature and initial value problems, we saw that high accuracy can be obtained at relatively low cost with low-order methods if the resolution (step size) is allowed to vary within the domain. We found that adaptive quadrature and variable step size integrators for ordinary differential equation initial value problems are efficient methods for solving the problems for which they are designed. It therefore makes sense to look at similar methods for direct

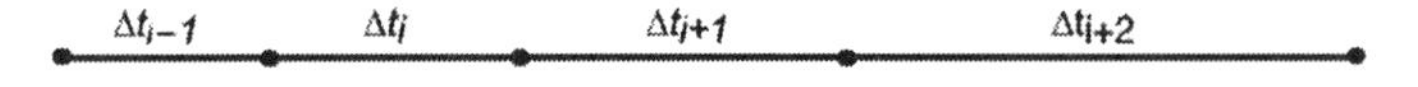

FIGURE 5.7. Section of a nonuniform grid.

boundary value methods. (In the shooting method it is natural to use a variable step integrator, making it easy to achieve any desired level of accuracy.) The equivalent of variable step size is the use of a nonuniform grid or, what is nearly the same thing, a transformation of the independent variable; this approach is discussed in this section. A more flexible and elegant method, the use of adaptive grids, will be taken up in Section 5.7.

The concept behind a nonuniform grid is simple. We note a few facts about finite difference methods. First, all finite difference approximations have a local error proportional to the product of a power of the step size and some derivative of the solution. Second, an optimum computation is one that distributes this error more or less evenly over the solution domain. Finally, we have no control over the derivatives of the solution and, for simplicity, would like to use the same approximations everywhere. The only way to accommodate all of these constraints and desires simultaneously is by allowing the grid size to vary. The grid spacing must be small where the higher derivatives of the function are large; it may be larger where the function is smooth. By allowing grid size variation, we should be able to maximize the accuracy for a given number of grid points or, what is nearly the same thing, a given amount of computation. The major constraint imposed in this section is that the grid be selected prior to the computation. In Section 5.7 we present a method that allows the grid to be developed adaptively as the computation proceeds.

There has been some controversy about the method of choosing the difference approximation. Although our interest is in boundary value problems, for which the differential equations are always of order higher than 1, we can learn something by looking at first derivatives. Specifically, let us consider some possibilities.

5.5.1. Finite Difference Approximations

First, suppose that we use a nonuniform mesh, a portion of which is shown in Figure 5.7. One approximation for the derivative at x_i is the central difference formula

$$y_i' \approx \frac{y_{i+1} - y_{i-1}}{\Delta x_i + \Delta x_{i+1}} \tag{5.26}$$

A formal Taylor series analysis applied to the right-hand side shows that

$$\frac{y_{i+1} - y_{i-1}}{\Delta x_i + \Delta x_{i+1}} \approx y_i' + \left(\frac{\Delta x_i - \Delta x_{i+1}}{2}\right) y_i'' + \left(\frac{\Delta x_i^2 + \Delta x_{i+1}^2}{12}\right) y''' \tag{5.27}$$

and the method is formally only first-order accurate; that is, there is an error term that decreases linearly as Δx decreases. However, the first-order error term is proportional to the difference of two adjacent grid sizes; if the mesh spacing changes slowly, $\Delta x_{i+1} - \Delta x_i$ is small and the error is more like that of a second-order method than a first-order one. In a recent paper (Ferziger and Perić, 1997), it was shown that, for at least two reasonable choices of grids, the error is very nearly of second order.

A formally second-order accurate approximation is

$$\frac{(\Delta x_i/\Delta x_{i+1})(y_{i+1} - y_i) - (\Delta x_{i+1}/\Delta x_i)(y_{i-1} - y_i)}{\Delta x_i + \Delta x_{i+1}} = y_i' + \frac{\Delta x_i \Delta x_{i+1}}{6} y_i''' \tag{5.28}$$

5.5.2. Coordinate Transformations

The use of a nonuniform grid is also equivalent to using a coordinate transformation. Suppose that the independent variable is transformed from x to

$$\eta = g(x) \tag{5.29}$$

before any finite difference approximations are used. Then

$$\frac{d}{dx} = \frac{d\eta}{dx}\frac{d}{d\eta} = g' \frac{d}{d\eta} \tag{5.30}$$

Unless the transformation is linear, a uniform grid in η corresponds to a nonuniform grid in x and vice versa showing that the two approaches are related. The situation is shown in Figure 5.8. Assuming that a uniform grid is used in the η coordinate, dy/dx at x_i can be approximated by

$$y'(x_i) \approx g'(x_i) \left[\frac{y(x_{i+1}) - y(x_{i-1})}{2\Delta\eta} \right] \tag{5.31}$$

The term in brackets is an approximation to $dy/d\eta$ and has an error proportional to $\Delta\eta^2$, and the approximation is thus second-order accurate in terms of η. Since the x and η coordinates system are equally valid, we should regard (5.31) as a second-order approximation. All of this assumes that g' is evaluated exactly. If g' is evaluated numerically, an additional error is introduced and account has to be taken of it; there are published results in which the dominant error arises from the evaluation of the derivatives associated with the coordinate transformation.

With a good choice of transformation, it is possible to make the solution a very smooth function of the transformed variable. To take an extreme example, we note that the equation $dy/dx = f$ becomes $dy/d\eta = 1$ if the transformation is defined by $g' = f$. Of course, it is possible to use this transformation only

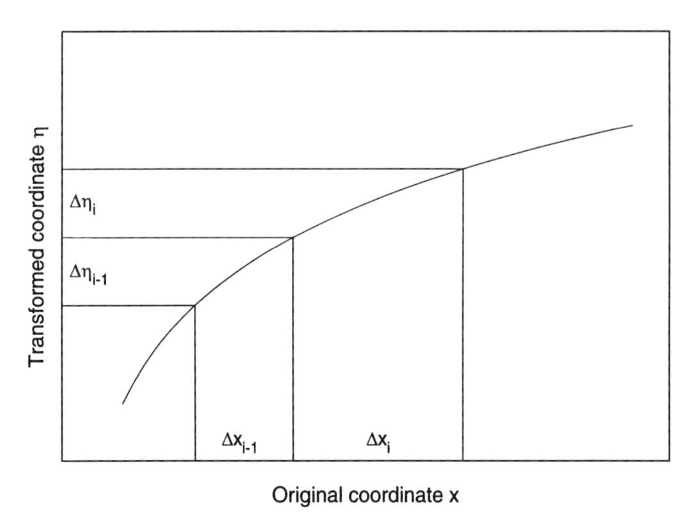

FIGURE 5.8. Relationship between the original and transformed coordinates.

if the solution is already known, so this result is of no practical value, but it illustrates the effect a good choice of transformation can have on the accuracy of the results.

In practice, the choice of a coordinate transformation is based on previous experience or, when available, on an approximate solution of the problem. It is important to note that no one coordinate transformation, or even any one particular method of constructing coordinate transformations, is optimal for all problems.

Example 5.10: Direct Solution on a Nonuniform Grid In this example a variable grid is used to solve the problems considered in earlier examples. We shall use one of the most common and easiest methods of constructing a grid system—we use the *compound interest grid* introduced in Section 2.4. In this grid, each interval is a constant multiple of the preceding one:

$$x_{i+1} - x_i = \alpha(x_i - x_{i-1}) \tag{5.32}$$

With $\alpha > 1$, the grid size grows with x so there are more points at small values of x, while $\alpha < 1$ places more mesh points at large values of x. For the second derivative, we use the finite difference approximation (4.12) while, for the first derivative, Eq. (4.8) is used.

For the problem of Example 5.4, which has a smooth solution, using a variable grid helps, but the error reduction is much smaller than it is for a problem with a solution that is not as smooth. We shall use the grid of Eq. (5.32) for $0 < x < 0.5$, while for $-0.5 < x < 0$ the reflection of this grid is used. We will then vary the constant α in Eq. (5.32) to minimize the root-mean-square (rms) difference between the exact closed-form solution and the one

TABLE 5.4. Nonuniform Grid Solution of the Problem of Example 5.5

N	Optimum α	rms Error	Error (uniform grid)
10	0.825	1.14×10^{-5}	5.50×10^{-5}
20	0.91	2.62×10^{-6}	1.57×10^{-5}
30	0.95	1.56×10^{-6}	6.96×10^{-6}

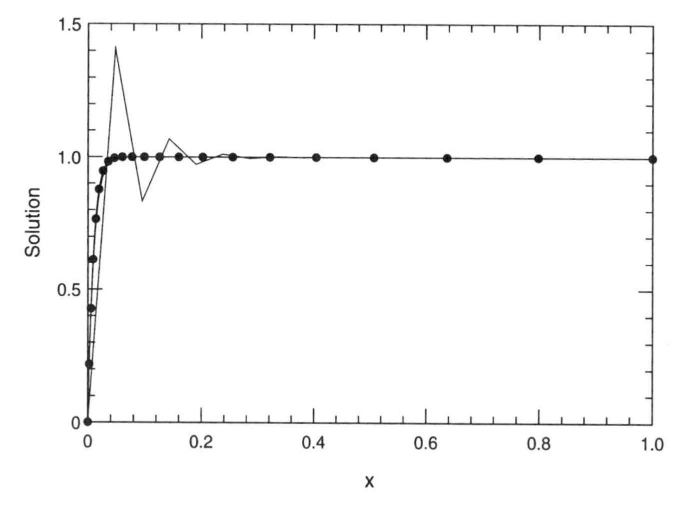

FIGURE 5.9. Solution of $0.01y'' + y' = 0$ on uniform and nonuniform grids.

obtained from the finite difference approximation. The best result is obtained with values of α slightly less than 1. The optimum value of α increases toward unity as the number of mesh points increases; in fact, it appears that $1 - \alpha \sim 1/N$. As Table 5.4 shows, the variable grid reduces the rms error by as much as a factor of 5 or 6. Grids better than the compound interest grid could be constructed for this problem (or any other problem), but it is hardly worth the effort. The only reason for doing this problem is to show that nonuniform grids can yield significant improvement at almost no increase in cost.

A much more dramatic error reduction is found for the problem of Example 5.7. In this case, more resolution is clearly needed at small x, so $\alpha > 1$ is required. Some results for $\epsilon = 0.01$ are shown in Figure 5.9; 20 points were used in these calculations. For $\alpha = 1$ (the uniform grid) the solution displays the kind of oscillations that we saw in Example 5.5, but they are less exaggerated in the present case because the larger value of ϵ makes this problem less stiff than the earlier one. The difficulty is severe enough, however, that $\alpha = 1.25$ is needed to give the best results, producing the highly distorted grid illustrated by the points in Figure 5.9. The errors are too small to be seen clearly in the figure. In fact, the maximum error is approximately 0.03 and occurs at $x = 0.03$ in this case. The error reduction achieved by the nonuniform grid has turned a totally unacceptable solution into one that is probably sufficiently accurate for engineering purposes.

With more grid points, a value of α closer to 1 is required. Using the optimum value of α for each number of grid points, we find that the error reduction is about two orders of magnitude and is almost independent of the number of grid points. If a uniform grid is used, 80 intervals are required to reduce the rms error to what is achieved with a nonuniform mesh of 10 intervals, and 500 intervals are needed to reduce the maximum error to what the nonuniform 20 interval calculation produces.

The compound interest grid is well suited to this problem and the results are spectacular. As already noted, the optimum grid is problem dependent; a little bit of advance knowledge about the solution can be an excellent aid to producing accurate results with relatively little effort. In the present case, the boundary layer has a thickness a few times ϵ (0.01 in the previous example). The grid stretching needs to be selected so that several points lie within the boundary layer. A rule of thumb suggests that a minimum of 10 points must be in the boundary layer to reduce the relative error to 1%.

In the case of partial differential equations (for which the costs are much higher), the use of a well-chosen nonuniform grid may make the difference between a practical tool and a much-too-costly curiosity.

5.6. FINITE ELEMENT METHODS

In Section 4.3, we showed that it is possible to find approximations to differential equations by integrating them; this is an alternative to finite difference approximation. When this technique is applied to ordinary or partial differential equations, it is called the *finite volume* method. Many of the equations encountered in engineering are conservation laws expressed in differential form. The integration over a small volume that is the essence of the finite volume method is equivalent to constructing the conservation law for a small but finite portion of the domain; for an ordinary differential equation, the finite volume method gives the conservation law for a finite range of the independent variable x. By choosing consistent approximations, it is possible to assure that the integrated (global) form of the conservation law is built into the discretized equations; this is a very desirable property.

Finite volume methods may in turn be regarded as the simplest members of a broader class of methods called *finite element* methods. The nomenclature comes from the mechanics of solids, the field in which the method was first developed. One version of the method consists of multiplying the equation by a "weight" function (which may be a known function or the solution itself) and integrating the result over a small part of the solution domain; for an ordinary differential equation, one might integrate over the line segment between two grid points. For ease of integration, the weight function is usually a polynomial on each segment, and the solution is approximated by piecewise polynomial.

The discretized equations of the finite element method are found by minimizing the resulting integral with respect to the variables to obtain a system of algebraic equations for the coefficients of the polynomials, which may, in turn, be expressed in terms of the values of the function at the nodes.

Although this approach may seem rather arbitrary, it takes on true significance for nondissipative linear systems. For such systems one can show that solving the differential equation (ordinary or partial) is equivalent to finding the minimum or maximum value of the integral of a quadratic functional of the solution. In a typical case, the integral might represent the potential energy of the system, and the solution is the function that minimizes this energy. An example of this method is the Lagrangian approach to classical mechanics. A complete development of this version of the theory requires some knowledge of the calculus of variations. For this reason, we shall give only a short introduction to the subject by means of an example. For further details, the reader is referred to specialized books on the subject such as the one by Hughes (1987).

To make all of this concrete, we consider a simple example that illustrates the method. Solving the differential equation

$$\frac{d^2y}{dx^2} - k^2y = f(x) \tag{5.33}$$

together with the boundary conditions

$$y'(0) = y'(1) = 0 \tag{5.34}$$

is equivalent to minimizing the integral

$$F = \int_0^1 (y''^2 + k^2y^2 + 2yf)\,dx \tag{5.35}$$

To see this, substitute $y = \tilde{y} + \delta y$, where $\tilde{y}$ is the function for which F is minimal and δy is a small "variation" from that function, into Eq. (5.35). After neglecting terms quadratic in δy and integrating by parts to eliminate the derivative of δy, one finds that the function that minimizes the functional F is is also a solution to Eqs. (5.33) and (5.34). The reader unfamiliar with the calculus of variations is advised to perform all of the steps in this calculation.

We now attempt to solve the problem by finding the piecewise linear function, that is, the function of the type

$$y(x) = \frac{x - x_i}{x_{i+1} - x_i} y_{i+1} + \frac{x_{i+1} - x}{x_{i+1} - x_i} y_i \qquad x_i < x < x_{i+1} \tag{5.36}$$

that produces the smallest value of F. In the precise sense of the minimization, this is also the piecewise linear function that is the best approximation to the exact solution of the original problem. We also approximate the forcing function, f, by a piecewise linear function. To keep the bookkeeping simple we assume that all of the grid sizes are equal and let $\Delta x_i = h$. Then, computing the integrals, we find

$$F = \frac{1}{h}\sum_{i=0}^{N}(y_{i+1} - y_i)^2 + \frac{k^2 h}{3}\sum_{i=0}^{N}(y_{i+1}^2 + y_i y_{i+1} + y_i^2)$$
$$+ \frac{h}{3}\sum_{i=0}^{N}(2y_{i+1}f_{i+1} + y_i f_{i+1} + y_{i+1}f_i + 2y_i f_i) \tag{5.37}$$

We have evaluated F for an arbitrary piecewise linear function. Just as the exact solution of the differential equation is the function that minimizes the exact F [Eq. (5.35)], it is reasonable to assume that the best piecewise linear approximation is the one that minimizes the approximate F [Eq. (5.37)]. This minimization is accomplished by differentiating Eq. (5.37) with respect to each of the y_i. Through a fair amount of tedious work, we obtain the system of linear equations:

$$(y_{i+1} - 2y_i + y_{i-1}) - \frac{k^2h^2}{6}(y_{i+1} + 4y_i + y_{i-1}) = \frac{h^2}{6}(f_{i+1} + 4f_i + f_{i-1}) \tag{5.38}$$

for $i = 1,2,\ldots,N$. For $i = 0$, the terms involving y_{-1} and f_{-1} are missing, and for $j = N$ the terms involving y_{N+1} and f_{N+1} are also missing. These equations are different from the ones obtained by the finite difference method but are similar to the equations obtained by compact difference methods in Section 5.4. If we regard Eq. (5.38) as a finite difference approximation and apply Taylor series analysis to it, we find that it is second-order accurate. In fact, any finite element approximation can be regarded as a finite difference method but, in some cases, it is stretching things to make the connection. The finite element approach produces new and valuable approximations that would probably not have been discovered by the methods presented earlier.

There are many finite element methods. The principal advantages of the approach are that (1) the locations of the grid points can be allowed to be completely arbitrary with little increase in complication, and (2) the accuracy of the approximation can be increased by using higher-order polynomials with little added difficulty other than an increase in the amount of calculation required; however, the process can be automated. For partial differential equations in two and three dimensions, the first advantage becomes a very powerful one; the computational points need not be arranged on a rectangular grid (triangles are used quite commonly in two dimensions). This allows the method to fit irregular boundaries very accurately and efficiently, while retaining the second advantage. The major disadvantages of finite element methods are that they require a bit more work to derive and the resulting systems of algebraic equations are generally not quite as regular as those produced by finite differences; as a result, the cost of solving the equations is increased considerably.

These advantages have made the finite element method extremely popular. The power of the method shows itself best, however, in systems for which the solution is equivalent to maximizing (or minimizing) a functional, and it is the preferred method for such problems. The method may be used for other systems (e.g., dissipative systems) and is often quite powerful in these applications as well. Whether its advantages are worth the difficulties is a matter of debate and personal taste.

5.7. ADAPTIVE GRIDS

The concept of an adaptive method was introduced in Chapter 3. The idea is that the grid points used in a calculation are chosen while the computation is in progress and depend on the solution produced. Also, recall that, in initial value problems, it is advantageous to change the step size as the computation proceeds. The best strategy is to select the step size so that approximately the same amount of error is introduced at each step; this is easily done in the shooting approach. We would like to have an equivalent method for the direct approach to solving boundary value problems.

The difficulty is that we often do not know what part of the solution domain requires the highest resolution before we begin the solution procedure. If we did, it would not be difficult to construct an effective grid. If that information is not available in advance, the only feasible approach is to make a trial calculation, determine where the region requiring the finest grid is, refine the grid in that area, and make a new calculation. This is the essential idea behind adaptive grid methods for boundary value problems.

Although adaptive grid methods can be applied to ordinary differential equations (ODEs), there are only a few ODE problems for which they are really needed. These are situations in which a long sequence of difficult problems need to be solved with high accuracy. (The author has used them in simulations of transition from laminar to turbulent fluid flow.) It is for partial differential equations (PDEs) that adaptive grid methods really show their worth, but they are easier to illustrate for ODEs.

A large number of adaptive grid methods have been proposed and more are being developed. Most of them fall into two broad categories. The first group consists of *moving grid* or *displacement* methods. In these, the total number of grid points is fixed, but the points are moved so as to minimize the total error. This approach lends itself to sophisticated mathematical treatment, but, especially for partial differential equations, it sometimes produces rather peculiar grids that can make a numerical method unstable. For that reason, their popularity has been decreasing.

The other type of adaptive grid approach adds new points in the regions in which the errors are large; these are called *refinement* methods. There are a number of methods based on this approach, of which we shall discuss two. All refinement methods start by computing solutions on two different grids; a good choice, because the adaptive methods automatically generate an appropriate nonuniform grid, is to use uniform grids with grid sizes in the ratio 2 : 1. From these, the error can be estimated using the Richardson estimator described in Section 4.5. Where the error is greater than the required tolerance, the grid is refined, again a 2 : 1 refinement ratio is usual but others can be used. In the initial stages of this kind of calculation, one often finds that the error is large everywhere, so the entire grid needs to be refined. In this case, we keep only the solution on the latest (and finest) grid and discard the coarser ones, and a new solution is computed on the refined grid. When a level is reached at which the error is small enough that only part of the grid needs to be refined, there are two versions of the refinement method that can be used to continue the calculation.

In one type of method, the grid is refined where necessary and a new calculation is made on the entire domain. In one dimension, this is the simpler approach. The major difficulty with it is that the abrupt change in grid size near the boundary of the refinement region may cause errors if the approximations used to discretize the equations are not carefully chosen; this can be overcome by a judicious choice of approximations used. In two or three dimensions, this method requires somewhat complicated programming to deal with the connections between the grid points.

In the other type of refinement method, a new calculation is made only on the refined part of the domain; the boundary conditions for this calculation are taken from the coarser grid results. In some problems, the solution outside the refinement region can be accepted as accurate and need never be improved further. When the solutions in the refined and unrefined regions interact, this can lead to problems as we shall see in Example 5.11. It is safer to not accept the solution outside the refinement region and modify it using a correction obtained from the fine grid solution in order to guarantee that the solution produced on the latter is a smooth version of the fine grid solution (Caruso et al., 1985). We shall defer consideration of this type of method until Chapter 7.

In either type of method, further refinement of the grid can be made if the error remains too large.

Example 5.11: Solution of a BVP on an Adaptive Grid We again consider the problem treated in Example 5.5:

$$\epsilon y'' + y' = 0$$

with the boundary conditions

$$y(0) = 0 \qquad y(1) = 1$$

and $\epsilon = 0.01$. We require the solution to have accuracy 10^{-4}; the solution at a given point is accepted as final when the Richardson estimate of the error

TABLE 5.5. Adaptive Grid Used in Solving $\epsilon y'' + y' = 0$

Level	Grid Size	First Point (x_{min})	Last Point (x_{max})	Number of Interior Points
1	0.5	0	1	1
2	0.25	0	1	3
3	0.125	0	1	7
4	0.0625	0	1	15
5	0.03125	0	1	31
6	0.015625	0.203125	1	50
7	0.007813	0.820313	1	22
8	0.003906	0.925781	1	18
9	0.001953	0.931641	1	34
10	0.000977	0.950195	1	50
11	0.000488	0.969238	1	62
	Total number of points			293
	Points on uniform fine grid			2048

is less that this. Second-order central difference approximations are applied to each term in the equation. The initial grid has just one interior point, that is, $\Delta x = 0.5$, which is far from fine enough. A second solution is made on a grid with $\Delta x = 0.25$ and the error is estimated. This error is, of course, much larger than the requested error at every grid point so the entire grid is refined again.

We shall not present the details of the solution at every level of refinement. Some properties of the grids used in the calculation are given in Table 5.5. The table shows that five levels of refinement are needed before any part of the solution is accepted. Knowing that this is a difficult problem, we might have started with a grid of approximately this fineness and eliminated the first few levels, but the cost of computing the solutions at these levels is small and it is just as easy to let the method handle the grid refinement automatically. At the fifth level, the solution is accepted only for $x < 0.2$; the solution is essentially zero up to this point, but this much refinement is required to eliminate the oscillations that result from using central difference approximations on coarse grids. After this level, increasingly smaller regions are refined at each succeeding level. The number of levels required to meet the error criterion was 11, the number of points used at each level is shown in the last column of Table 5.5 and the total number of points used on all of the grids was 293. A uniform grid of the size of the finest grid would have 2048 points. Of course, the nonuniform grid of Example 5.10 is also very efficient (we used a different measure of effectiveness there), so this comparison overemphasizes the effectiveness of the adaptive grid. The important advantage of the adaptive grid relative to the nonuniform grid (the two could be combined) is that there is no need to search for an optimum parameter.

Thus the adaptive method requires one seventh as much memory as the uniform grid method. If the cost were proportional to the number of points, about this much saving in computation time would have been obtained. It is true that the adaptive method has overhead costs due to estimating the errors and generating the new grids, but the single-grid calculation also requires calculations on more than one grid to assure that the error is within the allowed limit. Generally, the more difficult the calculation, the greater the advantage of the adaptive method. When the problem was redone with $\epsilon = 0.1$, an easier case, the savings were only about a factor of 2.

The advantage of the adaptive grid method increases in two and three dimensions because the ratio of the size of the refined region to the original domain tends to be smaller in those cases.

The solution is shown in Figures 5.10 and 5.11; the grid used can be seen from the points shown on the x axis of Figure 5.11. The adaptive grid method produces a nonuniform grid that is well suited to the problem in an automated fashion. It should be noted that comparison of the computed solution to the exact one shows that the maximum error is actually approximately 3.3×10^{-4}, which is greater than the requested error. The reason is that the solution is accepted whenever the error estimated by the Richardson method is less than 1.0×10^{-4} at a point. In this problem, the largest error always occurs at the last accepted point. This point provides the boundary condition for the next finer grid so the error is passed down to the finer grids. Thus the error accumulates and is larger at the finer levels. To achieve the demanded error, the criterion used for refinement needs to be tighter than the acceptable error in the final solution.

The code used to perform these calculations, ADAPGRID, is found on the Internet site described in Appendix A.

5.8. EIGENVALUE PROBLEMS

Eigenvalue problems are linear homogeneous boundary value problems that have solutions only when a parameter (the eigenvalue) has certain discrete values. These problems are similar to the ones dealt with in the preceding sections and can be solved using the same methods. We discuss them separately in part because we wish to derive some results to be used in later chapters. Experience shows that solutions of eigenvalue problems are often very sensitive to small changes in the estimated eigenvalue, especially when the estimate is close to the exact eigenvalue. This can make finding accurate solutions difficult. The most commonly used method of solving eigenvalue problems is the direct method, although shooting is often used when high accuracy is required. We shall therefore discuss direct methods first and then consider shooting methods. We illustrate the principal ideas by applying them to a particular problem, the classic eigenvalue problem for the Fourier functions.

The simplest eigenvalue problem that is always used as the introduction eigenvalue problems will be the basis for this example. This involves the

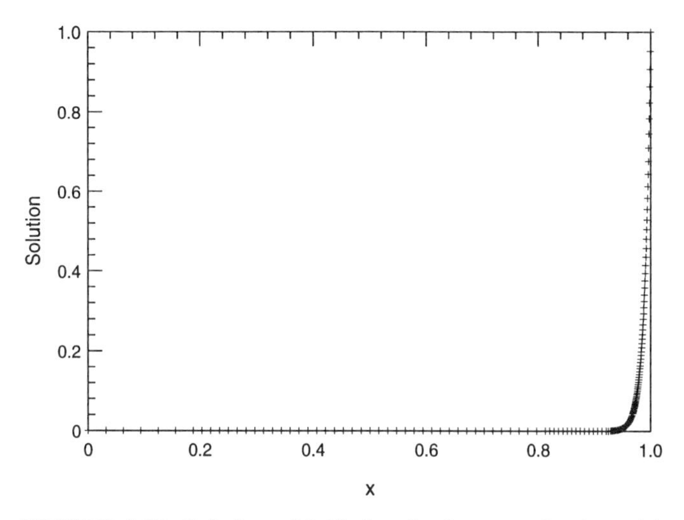

FIGURE 5.10. Solution of $0.01y'' + y' = 0$ on an adaptive grid.

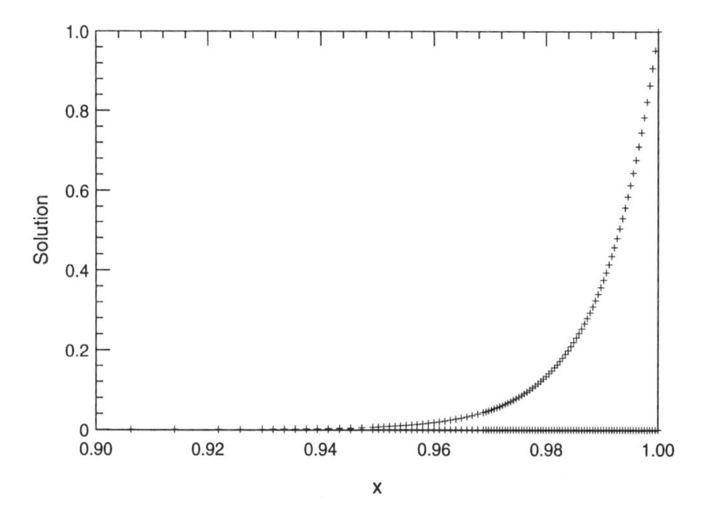

FIGURE 5.11. Detail of the solution shown in Figure 5.10.

differential equation that arises in the separation of variables solution of the wave equation for free vibrations:

$$y'' + k^2 y = 0 \tag{5.39}$$

with the boundary conditions

$$y(0) = y(1) = 0 \tag{5.40}$$

The solution to this problem is well known. There are no nontrivial solutions of this problem unless k is one of the eigenvalues:

$$k_n = n\pi \qquad n = 1,2,3,\ldots \tag{5.41}$$

and the corresponding solutions are the eigenfunctions:

$$y_n(x) = \sin n\pi x \tag{5.42}$$

5.8.1. Direct Methods

As stated above, we shall first attack the eigenvalue problem presented above using the direct approach and then discuss the application of shooting to eigenvalue problems. We shall try to assess the merits of each approach.

To obtain a finite difference approximation to the problem defined by Eqs. (5.39) and (5.40), we divide the domain $(0,1)$ into $N+1$ equal intervals and apply the second-order central difference approximation (4.13) to the second derivative at each interior point $x_i = i\Delta x$, $i = 1,2,\ldots,N$, where $\Delta x = N+1$. The resulting finite difference equations can be written

$$y_{i-1} + (k^2h^2 - 2)y_i + y_{i+1} = 0 \qquad i = 1,2,\ldots,N \tag{5.43}$$

At the points x_1 and x_N, the equations are identical to Eq. (5.43) except that the terms containing y_0 and y_{N+1}, respectively, are zero.

This set of equations can be written in matrix form as

$$A\mathbf{y} - \lambda\mathbf{y} = (A - \lambda I)\mathbf{y} = 0 \tag{5.44}$$

where A is the tridiagonal matrix

$$A = \begin{pmatrix} -2 & 1 & 0 & 0 & \cdots & \cdots & 0 \\ 1 & -2 & 1 & 0 & \cdots & \cdots & 0 \\ 0 & 1 & -2 & 1 & \cdots & \cdots & 0 \\ \vdots & \vdots & \vdots & \ddots & \vdots & \vdots & \vdots \\ \vdots & \vdots & \vdots & \vdots & \ddots & \vdots & \vdots \\ & & & & 1 & -2 & 1 \\ 0 & \cdots & \cdots & \cdots & 0 & 1 & -2 \end{pmatrix} = \mathrm{Tr}(1,-2,1) \tag{5.45}$$

and $\lambda = -k^2h^2$; here Tr is shorthand for a tridiagonal matrix with constant diagonals.

Equation (5.44) is the standard eigenvalue problem of linear algebra. Thus application of a finite difference approximation to a differential equation eigenvalue problem yields a linear algebra eigenvalue problem. Other methods of discretizing the differential equation, for example, the finite element method, also lead to algebraic eigenvalue problems, but with different matrices.

When second-order finite difference approximations are applied to eigenvalue problems arising from second-order ordinary differential equations, the result is a tridiagonal matrix eigenvalue problem. In the particular example used here, the matrix is also symmetric, so it is an ideal problem for treatment with the standard method of solving symmetric tridiagonal matrix eigenvalue problems—the shifted tridiagonal symmetric QR algorithm. This routine is quite fast, easy to use, and can produce the eigenvectors as well as the eigenvalues. Subroutines embodying this method (or ones similar to it) are available at most scientific computing centers as well as in commercially available programs such as MATHEMATICA and MATLAB and can also be found at the Netlib site on the Internet. Other ODE eigenvalue problems may not lead to symmetric matrices; for them, it is necessary to use a method designed for asymmetric matrices; the QZ method is a good choice. Subroutines for solving this type of problem are available from the sources listed above.

In this particular case it is actually possible to solve the matrix eigenvalue problem in closed form. This is an unusual situation, but this solution will prove valuable in the following chapters. The clue is provided by the solution of the differential equation (5.42). It suggests that the solution of the algebraic problem might be

$$y_j = \sin j\alpha \tag{5.46}$$

The cosine is not included because it does not satisfy the boundary condition at $x = 0$. Plugging Eq. (5.46) into Eq. (5.43) for $j = 2, 3, \ldots, N - 1$ and using the trigonometric identity

$$\sin(j \pm 1)\alpha = \sin j\alpha \cos j\alpha \pm \cos j\alpha \sin \alpha \tag{5.47}$$

we find

$$(2\cos\alpha - 2 - \lambda)\sin j\alpha = 0$$

so that (5.46) is the solution to the difference equation, provided that the term in parentheses is zero. Since $\lambda = -k^2h^2$, this is equivalent to

$$k^2 = \frac{2(1 - \cos\alpha)}{h^2}$$

and we have the solution if we can find the allowed values of α. Equations (5.43) for $j = 1$ and $j = N$ are the same as the other equations if we set $y_0 = 0$ and $y_{N+1} = 0$, respectively. The condition $y_0 = 0$ is automatically satisfied by Eq. (5.46), which is the reason why the sine function was chosen.

To satisfy the condition $y_{N+1} = 0$, we must have

$$\sin(N + 1)\alpha = 0 \tag{5.48}$$

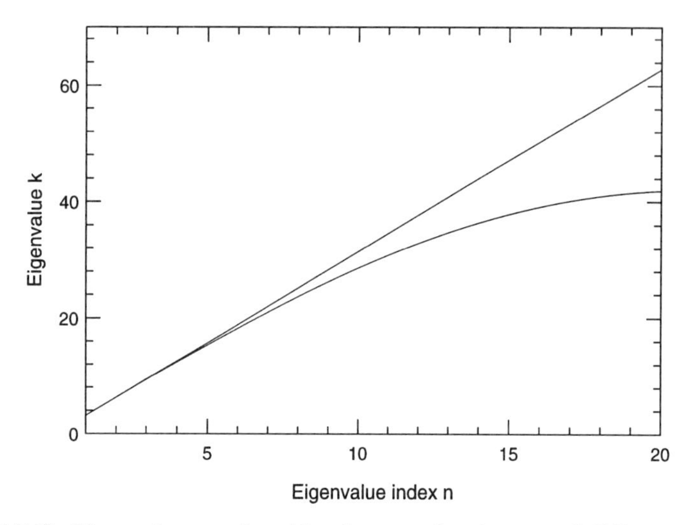

FIGURE 5.12. Eigenvalues produced by the second-order central difference approximation and the exact eigenvalues.

The solutions to this equation are $(N+1)\alpha = \pi, 2\pi, \ldots$ and thus

$$\alpha = \frac{\pi}{N+1}, \frac{2\pi}{N+1}, \ldots, \frac{N\pi}{N+1}, \ldots \tag{5.49}$$

It looks as if there are an infinite number of eigenvalues, but that is impossible for a matrix of finite size. The difficulty is resolved by noting that $\alpha_m = m\pi/(N+1)$ and, if we replace m by $2(N+1) - m$, which is another eigenvalue according to the above equations, we get

$$y_j = \frac{\sin[2(N+1) - m]\pi j}{N+1} = -\frac{\sin m\pi j}{N+1} \tag{5.50}$$

which is just the negative of the solution belonging to the eigenvalue α; this solution is not independent of the one belonging to eigenvalue α and cannot be included in the set of eigenvectors. Thus only the first N values of the set given by Eq. (5.49) produce independent solutions. The eigenvalues of the finite difference equation of Eq. (5.43) are therefore

$$k^2 = \frac{2}{h^2}\left(1 - \frac{\cos n\pi}{N+1}\right) \qquad n = 1, 2, \ldots, N-1 \tag{5.51}$$

It is interesting to compare these eigenvalues with the exact eigenvalues $k = n\pi$. This is done in Figure 5.12, in which we treat n as a continuous variable for display purposes. We see that, for small n, the exact and numerically derived eigenvalues agree quite well, but after $n \approx N/2$ the two diverge considerably. If a larger value of N had been used, the curve would have the same shape; the only difference would be a stretching of the x axis.

TABLE 5.6. Computed Eigenvalues of $y'' + k^2y = 0$

Index	Exact Eigenvalue	Computed Eigenvalue $N = 4$	$N = 6$	$N = 9$	Error $N = 4$	$N = 6$	$N = 9$
1	1	0.9836	0.9916	0.9959	0.0164	0.0084	0.0041
2	2	1.8710	1.9335	1.9673	0.1390	0.0665	0.0337
3	3	2.5752	2.7785	2.8902	0.4248	0.2215	0.1098
4	4	3.0273	3.4841	3.7420	0.9727	0.5159	0.2580
5	5		4.0150	4.5016		0.9850	0.4984
6	6		4.3446	5.1504		1.6554	0.8496
7	7			5.6723			1.3277
8	8			6.0546			1.9454
9	9			6.2878			2.7122

The ratio of the largest and smallest eigenvalues of the finite difference equations is $\lambda_N/\lambda_1 \approx 4N^2/\pi^2$ and is quite large so the system of differential equations

$$\frac{d\mathbf{x}}{dt} = A\mathbf{x} \tag{5.52}$$

is quite stiff. Unfortunately, as we shall see in the following chapter, this is essentially the system of equations that arises from applying finite difference approximations to parabolic partial differential equations. The stiffness is a major factor in the choice of methods for solving such problems and is the reason why this problem was treated in detail here.

The accuracy of the computation of the eigenvalues can be improved somewhat by using a higher-order difference formula. The methods and ideas of Section 5.2 can be applied to eigenvalue problems with very little change.

Example 5.12: Direct Solution of an Eigenvalue Problem Typical numerical results are given in Table 5.6 along with the errors in the computed eigenvalues. The lowest eigenvalues are computed relatively accurately and the accuracy improves as the number of points is increased. For the approximation used here the errors are proportional to h^2 or $1/N^2$. As the number of points is increased, the number of eigenvalues is increased, but the "new" eigenvalues are poor approximations to the eigenvalues of the ODE.

Direct methods have the advantage that they produce approximations to a large number of the eigenvalues and eigenfunctions of the differential equation. In many engineering problems, knowledge of the distribution of the eigenvalues (called the *spectrum*) is needed so direct methods provide the preferred approach. In other problems, for example, problems in the stability of fluid flows or structures, it is important to compute one eigenvalue and its eigenfunction very accurately. In those cases, it is usually the smallest or largest eigenvalue in magnitude or the one with largest or smallest real or imaginary part that is important, and it is often important to compute the eigenfunction accurately as well. For problems of that type, the shooting method is usually more effective.

5.8.2. Shooting Methods

The shooting method is capable of computing a single eigenvalue and its eigenfunction accurately, but it is not effective for computing the entire spectrum of eigenvalues of a differential equation. To use shooting, one must start with a good estimate of the desired eigenvalue. A good approach is to use the direct method to locate the desired eigenvalue approximately and then use shooting to improve the accuracy.

The shooting method for solving eigenvalue problems differs in a few important ways from shooting methods used for inhomogeneous problems. Because eigenvalue problems are homogeneous, any multiple of a solution is also a solution to the problem. As a result, the guessed initial condition, for example, the initial slope $y'(0)$, which plays a key role in inhomogeneous problems simply serves as a normalization factor in eigenvalue problems. The eigenvalue is the parameter that must be adjusted to allow the solution to satisfy the second boundary condition. Furthermore, even though eigenvalue problems are usually linear, the value of the solution at the second boundary is always a nonlinear function of the eigenvalue. Hence, an iterative method similar to the one used for the nonlinear problem in Example 5.5 must be used, even for linear eigenvalue problems.

Another difficulty is that eigenvalue problems have multiple solutions. Which solution is found depends on the initial guess. If a particular eigenvalue and eigenfunction are required, it is important to start the shooting calculation with an eigenvalue estimate that is not too far from the desired one. Such a guess might be obtained by using the direct method with a relatively small number of points.

We illustrate these points by solving the eigenvalue problem posed above with the shooting method.

Example 5.13: Shooting Solution of an Eigenvalue Problem We again consider the eigenvalue problem defined by the differential equation (5.39) and the boundary conditions (5.40). Suppose that we guess the eigenvalue k and compute the solution; the solution is normalized by requiring that $y'(0) = 1$. We shall use analytical solutions for illustration here, but any numerical method for initial value problems could be used. The solution of the differential equation that satisfies both the condition that $y'(0) = 1$ and the boundary condition $y(0) = 0$ is

$$y(x;k) = \frac{\sin kx}{k}$$

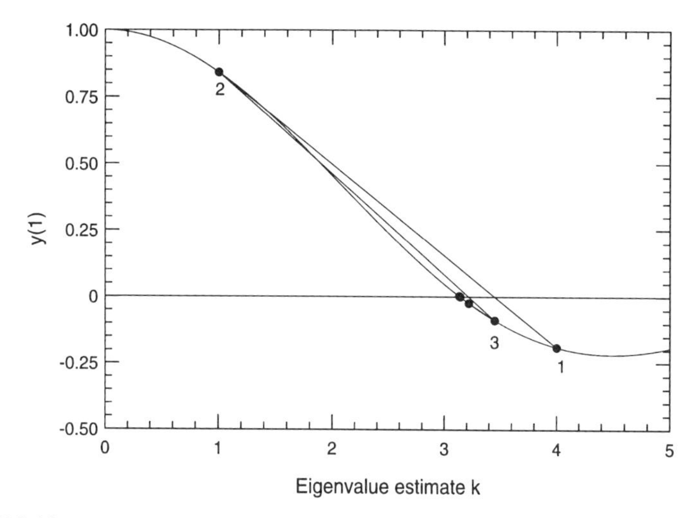

FIGURE 5.13. Relation between the final value $y(1)$ and the eigenvalue in the shooting solution of an eigenvalue problem.

For this example, we deliberately chose two rather poor starting guesses, $k = 4$ and $k = 1$; the values of $y(1)$ obtained are -0.189 and 0.841, respectively. By drawing a straight line between these points (see Figure 5.13) and determining where it crosses the line $y(1) = 0$, we find new estimate of the eigenvalue, 3.45, labeled 3 in Figure 5.13. Doing a new calculation with this estimated eigenvalue produces a new value of $y(1) = -0.088$. By locating the point of intersection of the line connecting the points 2 and 3, we obtain the next estimate of the eigenvalue, $k = 3.13$; this is quite close to the correct answer. Two more iterations produce the correct answer to within an accuracy of 10^{-5}. Still another iteration would decrease the error to less than 10^{-8}. Of course, to achieve this accuracy for the eigenvalue with a numerical method, the method needs to be capable of solving differential equations with at least this accuracy.

To show what can happen with this method, we ran the problem again with the initial guess 4 replaced by 5. The method converged, but the resulting eigenvalue was 4π, not the value π we were looking for or either of the two eigenvalues, 2π and 3π, that lie in between. Other initial guesses produce convergence to other eigenvalues. The pattern is difficult to predict, as it often is in problems with multiple solutions, making it hard to know which eigenvalue will be found. The only way to be sure to find a particular eigenvalue is to use initial guesses that are close enough to the desired eigenvalue. Unfortunately, what constitutes "close enough" is problem dependent, thus no recipe can be given.

As expected, the shooting method is capable of producing any desired degree of accuracy. For the highest accuracy, a differential equation solver with automatic error control should be used; the demanded accuracy will directly affect the accuracy of the eigenvalue.

Stiffness of the differential equation in eigenvalue problems can produce all of the difficulties that were found earlier for stiff inhomogeneous problems. It almost does not need to be said that stiff eigenvalue problems can be very difficult to solve accurately.

PROBLEMS

1. Solve the original shooting problem. Assume that the projectile leaves the gun at a speed of 600 m/s and is to hit a target 15,000 m away. At what angle does the gun barrel need to be pointed to hit the target? For purposes of this problem, neglect air friction.

   Note: You can either solve Newton's equations of motion describing the horizontal and vertical position of the projectile separately or, because the gravity force is independent of time, time can be eliminated to obtain an equation for the trajectory in terms of x and y.

2. Solve the problem

$$y'' - k^2 y = 1$$

$$y(0) = 0 \qquad y(1) = 0$$

   using just three points (two boundaries and one at $x = \xi$). The object is to compute the effect of a nonuniform grid on the error. Use the approximations

$$y'' = \frac{y_{j-1} - 2y_j + y_{j+1}}{(\Delta_j + \Delta_{j+1})^2/4}$$

   and

$$y'' = \frac{2y_{j-1}}{\Delta_j(\Delta_j + \Delta_{j+1})} - \frac{2y_j}{\Delta_j \Delta_{j+1}} + \frac{2y_{j+1}}{\Delta_{j+1}(\Delta_j + \Delta_{j+1})}$$

   and plot the error in $y(\xi)$ (the numerical result minus the exact result) as a function of ξ for $0 < \xi < 0.5$. Which formula is better and why?

3. The temperature distribution in a nuclear reactor fuel element is described by the differential equation

$$k\left(\frac{d^2T}{dr^2} + \frac{1}{r}\frac{dT}{dr}\right) = -S(r)$$

   where S is the energy source. Using $S(r)/k = 1 + 0.5(r/R)^2$ and the boundary conditions $T'(0) = 0$, $T(R) = T_0$, solve for $T(r)$. Work with a nondimensional temperature defined by $[T(r) - T_0]/T_0$.

4. The velocity profile in a laminar hydrodynamic stagnation point flow is related to the solution of the ordinary differential equation

$$\frac{d^3 f}{d\eta^3} + f\frac{d^2 f}{d\eta^2} - f'^2 + 1 = 0$$

with the boundary conditions

$$f(0) = 0 \qquad f'(0) = 0$$

$$f'(\eta) \to 1 \qquad \eta \to \infty$$

[f is the dimensionless stream function and the velocity u is proportional to $f'(\eta)$.] Solve this problem for $f(\eta)$ and plot the velocity profile.

CHAPTER 6

PARTIAL DIFFERENTIAL EQUATIONS: I. PARABOLIC EQUATIONS

Partial differential equations (PDEs) occur in almost all areas of science and engineering. In fact, many ordinary differential equations (ODEs) arise either as approximations to or as part of the solution process for partial differential equations.

As we did with ordinary differential equations, we begin by classifying the problems. Most of the classification scheme given in the Chapter 4 for ODEs applies to PDEs as well. Thus, PDEs can be of first or higher order; there may be a single equation or a system of equations; they may be linear or nonlinear; and they may be homogeneous or inhomogeneous. These classifications are as important for PDEs as they were for ODEs. Much of what was said in Chapters 4 and 5 applies to PDEs as well as ODEs and need not be repeated here.

There is, however, another classification scheme that is unique to PDEs. This classification is very important and influences the choice of method for solution of the equation enormously. We shall review it only briefly here. For a more thorough discussion of these issues, the reader is referred to textbooks on partial differential equations such as those by Zauderer (1983) and Courant and Hilbert (1953).

After reviewing the classification of PDEs and its importance in the selection of solution method, we shall take up the solution of the type of PDE most closely related to initial value problems of ODEs (parabolic PDEs) in the remainder of this chapter. Chapters 7 and 8 consider the solution of equations of elliptic and hyperbolic type, respectively.

6.1. CLASSIFICATION OF PARTIAL DIFFERENTIAL EQUATIONS

The distinction between initial and boundary value problems that is so important in the numerical solution of ODEs is more complicated for PDEs. The reason is that there are now two or more independent variables, so a problem may be an initial value problem with respect to one variable and a boundary value problem in another. It is possible to have pure initial value problems, pure boundary value problems, and mixed initial boundary value problems. The question of what are appropriate initial and/or boundary conditions for a given equation, and thus the question of the numerical method used to solve the PDE is intimately tied to the type of equation. Thus a short discussion of the classification of PDEs must be given.

First-order partial differential equations occur only occasionally in physics and engineering. Systems of first-order equations are more common but can be discussed along with higher-order equations. We note in passing that first-order partial differential equations with real coefficients have real characteristics (see below) and thus behave much like hyperbolic equations of higher order. In fact, the standard method for solving such equations (both analytically and numerically) is the method of characteristics (see Chapter 8).

Many PDEs occurring in applications are of second-order in at least one of the independent variables. Fourth-order equations are also fairly common; they can frequently be treated by methods similar to those applied to elliptic second-order equations in Chapter 7. Thus we concentrate on second-order equations and say only a bit about equations of other orders.

The classification scheme for second-order PDEs depends on the nature of their characteristics, so we need to review some of the properties of the characteristics of PDEs.

6.1.1. Characteristics

For equations with two independent variables, characteristics are lines in the plane of the independent variables along which "signals" can propagate. For equations with more independent variables, the characteristics become surfaces or hypersurfaces. For example, the characteristics might represent the possible positions of wavefronts in space as a function of time. They are also the locations of possible discontinuities in the solution of the equation. These properties are closely connected with the fact that along a characteristic, a PDE takes a particularly simple form; in two dimensions, a PDE behaves very much like an ordinary differential equation on the characteristics. The presence of real characteristics means that there are particular directions in which signals or information can propagate. Knowing the characteristics is often a valuable aid to finding the solution of the PDE.

The remainder of this book is devoted mainly to methods of solving partial differential equations of second-order in two independent variables. This choice has been made because second-order equations occur far more often in applications than any other kind and because these equations have been the focus of a great deal of analytical work and are therefore the best understood type of PDEs. Many of the methods presented can be applied to equations of first, third, and fourth order and to equations with a larger number of independent variables. We shall remark on these applications where it is appropriate to do so.

The most general second-order PDE in two independent variables that is linear in all of the highest derivatives is

$$a\phi_{xx} + b\phi_{xy} + c\phi_{yy} = f \tag{6.1}$$

where a, b, c, and f may all be functions of x, y, ϕ, ϕ_x, and ϕ_y; an equation of this type is called quasi linear. The classification of this equation depends on the sign of $b^2 - 4ac$. If $b^2 - 4ac > 0$, the equation is called *hyperbolic*; hyperbolic equations have two sets of real characteristics and arise in systems with finite propagation speeds. If $b^2 - 4ac = 0$, the equation is *parabolic*; parabolic equations have only a single set of real characteristics (or, more correctly, two identical sets). Finally, if $b^2 - 4ac < 0$, the equation is *elliptic*; elliptic equations have no real characteristics. All three cases occur commonly in physics and engineering problems. The most important effect that the type of equation has on the solution is connected with the way information propagates through space, and this has an important influence on the initial and boundary conditions that can be applied to the PDE.

Equations with more than two independent variables may not fit so neatly into this classification scheme, but understanding how information propagates often leads to a good choice of solution method.

As just noted, hyperbolic equations possess two families of real characteristics. Physical systems that are governed by hyperbolic equations are ones in which signals propagate at finite speed or over a finite region; examples include problems in which the propagation of electromagnetic or sound waves is important. The characteristics are illustrated schematically in Figure 6.1. The lines α = constant and β = constant represent the two families of characteristics along which signals are allowed to propagate. An observer at point P can "feel" the effects of what has happened in the vertically cross-hatched region, but disturbances outside this region cannot be felt. This region is known, therefore, as the *domain of dependence* of the point P. Similarly, the effect of a disturbance created at point P can be felt only in the horizontally cross-hatched region known as the *domain of influence*. It is important that numerical methods for solving hyperbolic equations recognize this behavior.

The most common hyperbolic equation (sometimes called the canonical hyperbolic equation) is the well-known wave equation

$$\phi_{tt} - c^2\phi_{xx} = 0 \tag{6.2}$$

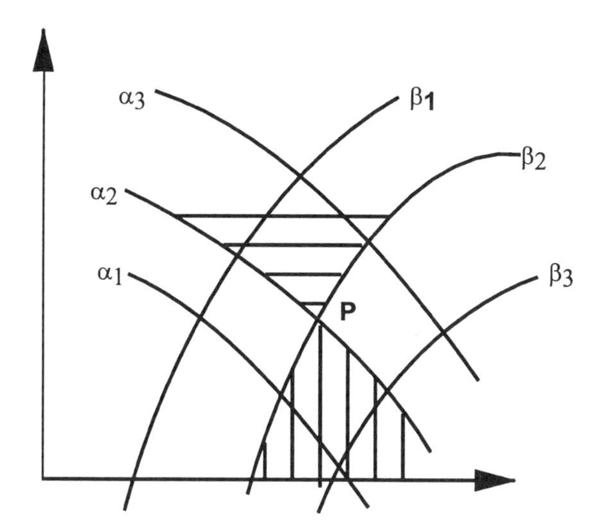

FIGURE 6.1. Characteristics of a hyperbolic partial differential equation. The horizontally cross-hatched region is the domain of dependence of the point P and the vertically cross-hatched region is its domain of influence. Signals propagate at speeds that are the slopes (dx/dt) of the characteristic lines.

Many of its close relatives are also quite common. We shall call the variables "space" and "time" even though there are hyperbolic problems in which all the coordinates are spatial.

Hyperbolic equations are always posed in domains that extend to infinity in the timelike coordinate. The spatial domain may or may not be bounded. In either case, one typically specifies two initial conditions at $t = 0$. If the spatial region is bounded, there may also be boundary conditions (one at each boundary); otherwise, we have a pure initial value problem. For a more general treatment of the initial and boundary conditions that one is allowed to apply, the reader is referred to such standard books on the subject as Zauderer (1990), Garabedian (1964), and Courant and Hilbert (1953). More will be said about this issue in Chapter 8.

Parabolic equations can be regarded as the limiting case of hyperbolic equations in which the propagation speed of the signals becomes infinite. Such equations are often obtained when it is assumed that the speed of light (or sound or some other type of wave) is so large compared to any other parameter in the problem that it may be set to infinity (e.g., the incompressible limit in gas dynamics or heat conduction in materials). With respect to Figure 6.1, this means that all the characteristic lines become horizontal, so only one independent family of characteristics remains. Consequently, the behavior of the solution at the point P is influenced by everything that happened earlier, and what happens at P influences everything that happens in the future. The numerical method should reflect this property.

The most common parabolic equation is the heat equation

$$\phi_t = \alpha\phi_{xx} \tag{6.3}$$

which governs the distribution of heat or other properties that diffuse. There are many other parabolic equations of importance, including some nonlinear ones. The initial and boundary conditions typically applied to parabolic equations are similar to those for hyperbolic problems with one important exception. Since Eq. (6.3) is only first order in time, only one initial condition at $t = 0$ is necessary. This is also related to the fact that there is only one set of characteristics. The domain of solution is again open in the time dimension. These properties mean that parabolic equations behave very much like initial value problems for ordinary differential equations, which is the primary reason why they are treated before the other types of PDEs. The spatial domain may be open or closed. In the case of a closed spatial domain, boundary conditions are essentially the same as those applied in the hyperbolic case.

Finally, we come to the elliptic case. Elliptic PDEs have no real characteristics. They often arise as the time-independent limit of an unsteady equation. For example, the steady version of the heat equation (6.3) in two spatial coordinates is the canonical elliptic equation known as Laplace's equation:

$$\phi_{xx} + \phi_{yy} = 0 \tag{6.4}$$

which is the best known elliptic equation. It usually represents steady-state diffusion, but there are a number of other problems that lead to this equation. In elliptic problems, complex characteristics can be introduced, leading to the well-known and important connection between the solutions of Laplace's equation (6.4) and functions of complex variables. There are no domains of dependence or influence for elliptic equations; every point in the solution domain is affected by disturbances at every other point. This property makes elliptic problems particularly difficult to solve.

The solution domain for elliptic equations is normally closed, that is, bounded on all sides but it is possible to have open domains. Furthermore, it can be shown that necessary and sufficient boundary conditions are provided by giving one datum (the value of the dependent variable, its normal derivative, or a linear combination of the two) at each point on the surface of the domain.

To give this a more physical interpretation, we note that Laplace's equation describes the diffusion of heat in a conducting medium. The temperature distribution on a flat plate of any shape should depend only on the temperature distribution around the periphery or on the heat flux distribution there, but not both. In fact, in the most common type of elliptic problem, one of these is given and the other is to be calculated.

Parabolic equations are treated first because the methods used for solving them are closely related to the methods used for ODEs; they are the subject of the remainder of this chapter. Elliptic equations are usually solved by iterative procedures, some of which are derived from methods used for parabolic equations, so they are treated in the next chapter. Finally, treatment of hyperbolic equations, for which some of the methods (especially the method of characteristics) rely heavily on the properties of the PDEs, are presented in Chapter 8.

6.2. EXPLICIT METHODS

Parabolic partial differential equations behave more like initial value problems for ordinary differential equations than either elliptic or hyperbolic equations. With respect to the time variable, the diffusion or heat equation:

$$\frac{\partial \phi}{\partial t} = \alpha \frac{\partial^2 \phi}{\partial x^2} \tag{6.5}$$

is first order, requires only a single initial condition, usually given at $t = 0$, and is usually to be solved for all $t > 0$. Equation (6.5) will be used as the example of a parabolic PDE, but many others arise in applications, including some important nonlinear ones. Fortunately, much of what we learn about the heat equation applies to other equations, so it is sensible to begin with it. Equations with more than one spatial independent variable are also very important and are discussed later.

Let us suppose that the problem is to solve Eq. (6.5) with the initial condition

$$\phi(x,0) = f(x) \tag{6.6}$$

and the boundary conditions

$$\phi(0,t) = \phi(1,t) = 0 \tag{6.7}$$

Of course, the boundary conditions need not be homogeneous.

Before solving Eq. (6.5) numerically, we note that the problem is very much like an initial value problem in time but is a boundary value problem with respect to the spatial coordinate. For this reason, we shall begin by approximating the spatial derivative in much the same way as it is treated in direct methods of solving boundary value problems for ODEs. We first create a grid of points in space on which the solution shall be computed, that is, we shall discretize the equation with respect to the spatial variable. To keep things as simple as possible, in the first problem that we deal with, we shall use a uniform grid. With N intervals, the grid size is $\Delta x = 1/N$ so that points $x_j = jh$, $j = 0,1,2,\ldots,N$. We approximate the second spatial derivative with the second-order central difference formula (4.13). Since the boundary values of the solution are given, this approximation is applied only at the interior grid points $x_j = jh$, $j = 1,2,\ldots,N-1$. Use of this approximation replaces the

PDE by the system of ordinary differential equations:

$$\frac{d\phi_j(t)}{dt} = \frac{\alpha}{h^2}(\phi_{j+1} - 2\phi_j + \phi_{j-1}) \qquad j = 1,2,\ldots,N-1 \tag{6.8}$$

where we have used the abbreviated notation $\phi_j(t)$ for $\phi(x_j,t)$. Where ϕ_0 occurs in the first equation, it is replaced by the boundary value. Similarly, where ϕ_N occurs in the last equation, it is replaced by the given value. Both of the latter are zero in this example.

An approach, such as this, that uses a discretization in only one direction is called a *semidiscrete* method. Discretization of the partial differential equation with respect to the spatial variable replaces it by a set of ordinary differential equations. The set of equations (6.8) is very similar to some of the systems of equations that were studied in the preceding chapter. Thus, finite differencing the heat equation reduces the problem to one that we have already considered; this is why parabolic equations have been taken up first.

The system of equations (6.8) can be written in matrix form as

$$\frac{d\phi}{dt} = \frac{\alpha}{h^2}A\phi \tag{6.9}$$

where the A is the tridiagonal matrix $\mathrm{Tr}(1,-2,1)$ that was encountered in Chapter 5. Here, $\phi(t)$ is a vector whose components are the values of the solution at the grid points.

Equation (6.8) or (6.9) can be solved without using a computer. Although this is an unusual approach, quite a bit can be learned from it, so we shall pursue it. We begin by noting that the eigenvalues and eigenvectors of the matrix A were given in Section 5.8. Having them allows us to diagonalize Eq. (6.9). As stated, we do this mainly for pedagogical reasons. For parabolic equations more complex than this one, diagonalization might be possible, but it would be so expensive that no one should consider using it. Furthermore, it would need to be redone if the boundary conditions are changed.

Rather than using the standard linear algebra approach to diagonalizing a matrix, we shall use a formal method more commonly used in solving differential equations. Let the eigenvectors and eigenvalues of A be defined by

$$A\psi^{(k)} = \lambda_k\psi^{(k)} \qquad k = 1,2,\ldots,N-1 \tag{6.10}$$

Explicit expressions for the eigenvalues and eigenvectors were given in Section 5.8. Since A is symmetric, its eigenvectors form a complete set so that any vector of size $N-1$ can be represented as a linear combination of them. In particular,

$$\phi(t) = \sum_{k=1}^{N-1} c_k(t)\psi^{(k)} \tag{6.11}$$

It is also important to note that, because the matrix is symmetric, the eigenvectors are orthogonal; that is,

$$\sum_{j=1}^{N-1} \psi_j^{(k)} \psi_j^{(m)} = \delta_{km} \tag{6.12}$$

Now we substitute the expansion (6.11) (which is a generalized discrete Fourier series) into the finite difference form of the partial differential equation (6.8) and take the scalar product of the result with $\psi^{(m)}$, another of the eigenvectors of A. Taking advantage of the orthogonality property, we find an ordinary differential equation for the Fourier coefficient c_m:

$$\frac{dc_m}{dt} = \frac{\alpha\lambda_m}{h^2} c_m \tag{6.13}$$

which has the solution $c_m = c_m(0)\exp(\lambda_m \alpha t/h^2)$.

The solution of Eq. (6.8) can thus be written

$$\phi(t) = \sum_k c_k(0) e^{\alpha\lambda_i t/h^2} \psi^{(k)} \tag{6.14}$$

If we can determine the constants $c_k(0)$, the solution is complete. The only remaining unused data is the initial condition. We therefore evaluate the solution (6.14) at $t = 0$ and substitute it into the initial condition (6.6) to get

$$\phi(0) = \sum_k c_k(0) \psi_j^{(k)} = \mathbf{f} \tag{6.15}$$

where $\mathbf{f}$ is a vector whose elements are the values of $f(x)$ at the grid points. The constants $c_k(0)$ can be found by making use of the orthogonality property [Eq. (6.12)] of the eigenvectors. We take the scalar product of Eq. (6.15) with another eigenvector $\psi^{(m)}$ to obtain an explicit expression for $c_m(0)$:

$$c_m(0) = \mathbf{f}(x_j) \cdot \psi_j^m = \sum_j f(x_j) \psi_j^{(m)} \tag{6.16}$$

This completes the formal solution of the problem. We emphasize that this is a formal solution. The cost of computing the solution in this way is much greater than the cost of using some of the methods given below even when (as here) the eigenvalues and eigenvectors are known. For a more complex problem, the cost of the computation required to carry out the indicated steps would be prohibitive. This solution is useful for studying the behavior of numerical methods, however, which is why we derived it.

The reader is invited to show that the procedure that we have followed here is entirely equivalent to the formal linear algebra method of diagonalizing the matrix and solving a system of linear ordinary differential equations.

The eigenvalues λ_k are all negative and increase in magnitude as the index k increases. Although the initial conditions may be such that some of the constants $c_k(0)$ are fairly large, the terms in Eq. (6.14) with large indices k decay rapidly and, after a relatively short time, only the low-k components of the series make significant contributions to the solution. In many cases, only the long-term component of the solution is required. In other, rarer, cases it is important to compute the precise short-term behavior of the solution.

The smallest eigenvalue of the matrix A is λ_1, which is approximately $-\pi^2 h^2 \approx -\pi^2/N^2$, governs the very long-term behavior of the solution. The largest eigenvalue, λ_{N-1}, is approximately -4. The ratio of these two eigenvalues is approximately $(2N/\pi)^2$, so for large N, the system of ordinary differential equations (6.8) or (6.13) is quite stiff. This situation is not at all unusual; *almost all parabolic PDEs are equivalent to stiff systems of ODEs*. This is a key observation and plays a determining role in the selection of numerical methods for solving parabolic PDEs.

Due to the stiffness of the semidiscretized system of equations, the stability requirement will force methods that are explicit with respect to the time variable to take very small time steps. To make the point more strongly, suppose we try to solve the discretized diffusion equation (6.8) with the Euler method. We would have

$$\phi_j^{n+1} = \phi_j^n + \frac{\alpha\,\Delta t}{h^2}(\phi_{j+1}^n - 2\phi_j^n + \phi_j^{n-1}) \tag{6.17}$$

This method is also known as the forward time, centered space, or FTCS, method.

Recall that the Euler method is stable for Eq. (6.13) only if

$$\frac{|\lambda_l|\alpha\,\Delta t}{h^2} < 2 \tag{6.18}$$

This must be true for each eigenvalue and, in particular, for the eigenvalue with the largest magnitude. Since $\lambda_{\max} \approx 4$, this method will be unstable unless

$$\frac{\alpha\,\Delta t}{h^2} < \frac{1}{2} \tag{6.19}$$

The fact that the stability criterion contains the parameter $\alpha\,\Delta t/h^2$ could have been anticipated through the application of dimensional analysis; this parameter will govern the stability of any method. An immediate consequence of this observation is that the allowable time step for any conditionally stable method is proportional to the square of the spatial mesh size. Improving accuracy by increasing the spatial resolution will cost dearly. For example, if the spatial step size is cut in half, the time step will need to be cut by a factor of 4, the cost of computing to a given time will increase by a factor of 8, and the error will be reduced by a factor of 4. This is quite expensive and is a consequence of the stiffness of the semidiscrete equations. The situation will be even worse if there is more than one spatial variable.

It is clear that the Euler method, or, indeed, any explicit method, is not well suited to the solution of parabolic PDEs. A much better choice is the use of implicit methods, except, possibly, when a need for extreme time accuracy dictates the use of small time steps.

We should also note that the fact that $\lambda_{\max} = 4$ is a consequence of the spatial differencing method used. If we were to approximate $\partial^2\phi/\partial x^2$ accurately (using Fourier methods, e.g.), $\lambda_{\max}$ could be as large as π^2. The factor 2 in Eq. (6.19) would be replaced by $\pi^2/2$, and the stability limit on the time step would be about two and a half times tighter. Thus the stability criterion depends on both the temporal and spatial differencing methods, and the result (6.19) applies only for the second-order central difference approximation.

As we have seen, the stability condition is derived from the largest eigenvalue of the matrix A. Each element of the corresponding eigenvector is of opposite sign to the one preceding it, that is, it corresponds to a solution that changes sign at each grid point. This component of the solution is most amplified when a calculation based on the Euler method becomes unstable. Consequently, in an unstable calculation, the solution oscillates spatially; the nature of the instability of the Euler method means that the solution also changes sign at each point at every time step. The presence of such a rapid oscillation in both space and time is the best indicator that a calculation has become unstable. We shall illustrate this with an example below.

An alternative approach to studying the stability of numerical methods for PDEs was developed by von Neumann for whom the method is named. This method is similar to the one used above but is simpler to apply in many cases. In the von Neumann method one ignores the boundary conditions, assumes a solution of the form $\phi(x,t) = e^{ikx} f(t)$, and substitutes this into the discretized partial differential equation to obtain an ODE for $f(t)$ that contains k as a parameter. The stability criterion is found by choosing the value of k that gives the strictest stability limit.

If we substitute $\phi(x,t) = e^{ikx} f(t)$ into the semi-discretized heat equation, (6.8), we find after a little calculation that the von Neumann approach to studying the stability of methods for the heat equation leads to

$$\frac{df}{dt} = \frac{2\alpha\Delta t}{h^2}(\cos kh - 1)f \tag{6.20}$$

The maximum absolute value of $\cos kh - 1$ is 2 and, from the stability criterion for Euler's method for ODEs, we again find that Euler's method applied to the heat equation is stable only if Eq. (6.19) holds.

Another important issue is the question of accuracy, which is easily studied for the Euler method. From the results of Chapter 4 we know that

$$\frac{\phi_{j+1}^n - 2\phi_j^n + \phi_{j-1}^n}{h^2} \approx \frac{\partial^2 \phi_j^n}{\partial x^2} + \frac{h^2}{12}\frac{\partial^4 \phi_j^n}{\partial x^4} \tag{6.21}$$

and

$$\frac{\phi_j^{n+1} - \phi_j^n}{\Delta t} \approx \frac{\partial \phi_j^n}{\partial t} + \frac{\Delta t}{2}\frac{\partial^2 \phi_j^n}{\partial t^2} \tag{6.22}$$

Note the sub- and superscripts carefully. The derivations of these equations are based on the use of Taylor series; when there are two or more independent variables, terms involving derivatives with respect to both coordinates may occur. Which error terms occur, and especially the leading error terms, depend on the point about which the Taylor series expansions are made [(x_j, t_n) in the case given above]. For consistency, every term in the finite difference approximation must be expanded about the same point. We see from these equations that the principal errors in Euler's method applied to the heat equation are proportional to Δt and Δx^2.

Another interpretation of these results is that approximating the heat equation with Euler's method in time and second-order central differences in space is equivalent to replacing the original PDE by

$$\frac{\partial \phi}{\partial t} - \alpha \frac{\partial^2 \phi}{\partial x^2} = \frac{\alpha h^2}{12}\frac{\partial^4 \phi}{\partial x^4} - \frac{\Delta t}{2}\frac{\partial^2 \phi}{\partial t^2} \tag{6.23}$$

which again shows that the error is proportional to h^2 and Δt. Thus the Euler method in time with central differences in space is first-order accurate in time and second-order accurate in space. It is typical that finite difference approximations to PDEs cannot be characterized by a single-order parameter in all of the coordinates.

Equation (6.23) is called the *modified equation*; it is particular to the heat equation and the numerical method used to solve it. What the modified equations tells us is that the numerical method produces a solution to an equation that is closer to Eq. (6.23) than the original one, that is, the numerical approximations have the effect of replacing the actual heat equation (6.3) by the modified one. Modified equations are interesting in their own right and can be used to investigate the stability of numerical methods.

The right-hand side of Eq. (6.23) represents the truncation error of the method. For the heat equation, the term $\partial^2\phi/\partial t^2$ can be evaluated by differentiating Eq. (6.5) with respect to time. The right-hand side of Eq. (6.23) can then be rewritten

$$\left(\frac{\alpha h^2}{12} - \frac{\alpha^2 \Delta t}{2}\right)\frac{\partial^4 \phi}{\partial x^4} \tag{6.24}$$

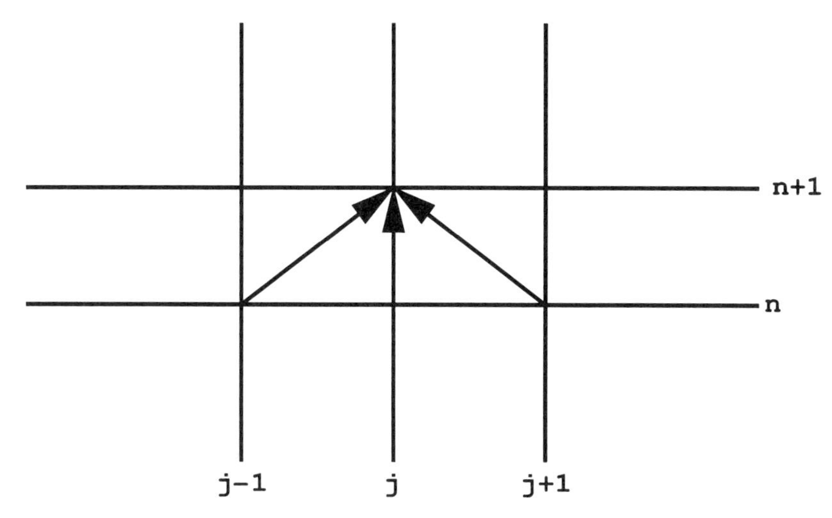

FIGURE 6.2. Computational molecule for the Euler method applied to the heat equation.

which is an alternate form of the truncation error for this problem. It is remarkable that the expression for the error (6.24) can be made to be zero by choosing $\alpha \Delta t = h^2/6$. When this is done, the leading errors exactly cancel out everywhere in space and time, and the method becomes second-order accurate in time and fourth-order accurate in space; considerable accuracy is gained at relatively little cost. However, this is a fortuitous circumstance that is possible only for an equation as simple as this one and only with uniform steps in both time and space.

In general, the time and space differencing errors cannot be made to cancel, and it is not usually known in advance whether they reinforce or cancel each other. In the absence of knowledge of how the errors interact, the best approach is to try to make the error arising from time discretization as nearly equal to the error arising from the space discretization as possible. If one of the errors is much larger than the other, we could use a larger step size in the coordinate with the smaller error without increasing the total error much; the computational cost would be reduced with little loss of accuracy. Exact balance at every point in space and time is not possible (except in the unusual case presented above), but it is best to keep the errors in approximate balance.

Another way to describe numerical methods is by drawing a diagram called a *computational molecule* that displays which points in space and time contribute to the difference formula at a particular point. The computational molecule for the Euler method applied to the heat equations is shown in Figure 6.2. The value of the solution at time t_{n+1} at point x_j depends on the values of the function at three spatial points at the earlier time t_n, but not on any of the values at the current time. This is behavior expected of hyperbolic equations, not parabolic ones. The hyperbolic nature of the approximation can also be

TABLE 6.1. Solution of the Heat Equation with the Explicit Euler Method

Δt	β	$\phi(0.5,0.1)$
0.00050	0.0500	0.4732
0.00100	0.1000	0.4721
0.00200	0.2000	0.4698
0.00500	0.5000	0.4745
0.00526	0.5263	0.3766
0.00556	0.5556	0.9779
0.00588	0.5882	−2.1087
0.00667	0.6667	−36.0113
0.01000	1.0000	1100.9990

seen from Eq. (6.23). The presence of a second time derivative (albeit with a small coefficient) makes the modified equation hyperbolic.

Example 6.1: Solution of the Heat Equation with the Euler Method As our example, we take the heat equation (6.5) with $\alpha = 1$. For the initial condition we take

$$\phi(x,0) = 1 \qquad 0 \leq x \leq 1$$

with the boundary conditions

$$\phi(0,t) = \phi(1,t) = 0$$

Physically, this problem might represent the cooling of a heated slab of material, an idealization of a metallurgical heat treatment process. The interesting result from an engineering point of view is the rate of cooling of the center of the slab. The problem of heating a cold slab is also of considerable interest (e.g., in predicting the time to cook a roast). We apply a finite difference approximation to this problem with $\Delta x = 0.1$ and 0.05; Δt will be allowed to vary. The solutions we shall obtain for even the smallest Δt contains some error due to the spatial differencing error.

The program used to solve this problem can be found on the Internet site described in Appendix A and is called HEATEUL. As are most of the other programs, this one is partially interactive and requests some of the data. A major part of the calculation has been put into a subroutine so that the same program can be used in conjunction with subroutines for other methods given in the following sections, provided that the subroutine name is changed in the CALL statement.

Table 6.1 gives the centerline temperature at time $t = 0.1$ for various values of $\beta = \alpha \Delta t/h^2$, which is $100\Delta t$ for the parameters of this problem. For $\beta <$ 0.5000, the stability limit for this method, there is little change in the result

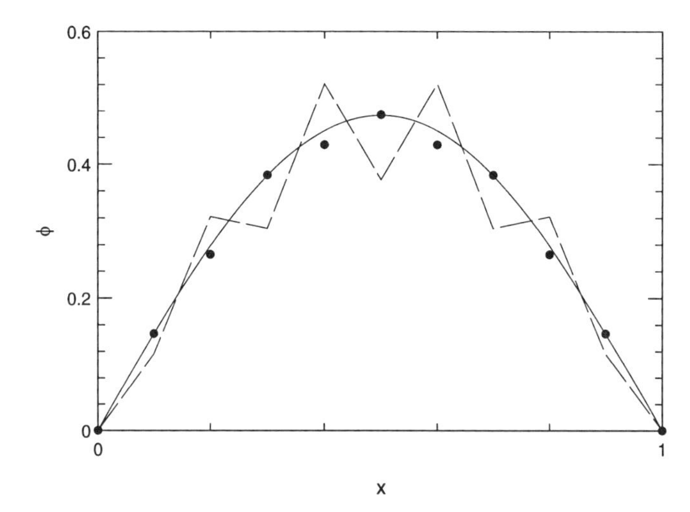

FIGURE 6.3. Solution of the heat equation with the Euler method; all results are at $t = 0.1$. The solid line is the solution obtained with $\Delta t = 0.01$ ($\beta = 0.1$); the solution for $\beta = 0.5$ is represented by the solid dots; and the solution with $\beta = 0.5263$ is the dashed line.

as β changes. The explanation is that since the exact solution at $t = 0.1$ is smooth only the first few eigenvectors contribute significantly to it. For these, reasonable accuracy (better than 1%) is obtained for any time step that obeys the stability limit. If the stability limit is exceeded by even a small amount, the resulting instability produces a solution that is seriously in error. Again, note that the instability is associated with the terms connected with the largest eigenvalues; these terms make very little contribution to the solution at $t = 0.1$.

A further illustration of this is provided in Figure 6.3. With $\beta = 0.5000$, which is exactly the stability limit, a reasonably good result is obtained. At the stability limit ($\beta = 0.5$), the component of the solution that changes sign at every mesh point is neutrally stable. As a result, the solution oscillates about the smooth curve (but only slightly). Increasing β to 0.5263, in other words, using 19 time steps instead of 20 to get to $t = 0.1$, produces the jagged curve shown in Figure 6.3 from which the presence of the instability is obvious. The amplification of the solution component that oscillates at every mesh point is obvious. Increasing β further produces results that are completely awful, as can be seen in Table 6.1.

In this example, the limitation on the size of the time step is derived from the stability criterion. If the method had remained stable, a larger time step could have been used and reasonable accuracy could have been expected. We conclude that the Euler explicit method or, indeed, any other explicit method is not well-suited to the solution of the heat equation, a conclusion that holds for other parabolic equations.

Example 6.2: Effect of Grid Nonuniformity Meshes with variable grid size are frequently used when the solution is known to vary more rapidly in one region than another (see Section 5.5). In order to see the effect of a variable mesh on the behavior of the method, we take a simple case: the problem of Example 6.1 with only one interior point that is not centered. With the central difference formula (4.12) and the boundary conditions $\phi_0 = \phi_2 = 0$, the problem reduces to a single ODE

$$\frac{d\phi_1}{dt} = -\frac{2\alpha}{h_1 h_1}\phi_1$$

where $h_1 = x_1 - x_0$ and $h_2 = x_2 - x_1$. From the results of Chapter 5, we know that the stability condition for Euler's method is

$$\frac{\alpha \Delta t}{h_1 h_2} < 1$$

Since $h_1 + h_2 = 1$, $h_1 h_2 = h_1(1 - h_1)$, the stability parameter is smallest when $h_1 = h_2$ and grows when h_1 is either smaller or larger than h_2. Thus the method becomes less stable as the mesh variation increases. It is also worth noting that, with $h_1 = h_2$, this stability criterion is less stringent than Eq. (6.19); the difference arises because Eq. (6.19) is a result applicable when the number of points is large.

Analysis of the stability of problems with large numbers of nonuniformly spaced grid points is difficult. We can be sure, however, that the greater the variation in grid size, the stiffer the problem, and the more stringent the stability condition. A good rule of thumb is that one should always choose parameters so that $\alpha \Delta t / h_{\min}^2 < c$ where c is a constant dependent on the method (2 for the Euler method). This criterion may be stricter than necessary, but it is not far from the actual limit and is on the safe side. This stability condition makes use of explicit methods in problems with nonuniform grids an even poorer choice than they are for uniform grids.

6.3. CRANK–NICOLSON METHOD

In Chapter 4, we found that the best way to treat stiff problems in ordinary differential equations is through the use of an implicit method. It follows that it should be possible to deal with the stiffness associated with parabolic problems by using an implicit method. In doing so, we must remember that, in the PDE case, we are dealing with a set of N equations, where N may be large. This could make the cost of an implicit method prohibitive. For parabolic PDEs with only one spatial variable, this problem is not very severe. In the two- or three-dimensional cases that we will treat later, large arrays are needed to represent the solution. This may result in a need for large amounts of both storage and computation. From a practical point of view, multistep methods may be ruled out.

We thus come to the conclusion that a reasonable method of advancing parabolic PDEs in time is through the use one of the simpler implicit methods. A particularly good choice, because it is second-order accurate and unconditionally stable, is the trapezoid rule. Applying it to the time advancement of the semidiscretized diffusion equation (6.8), in which second-order central differences have been used to approximate the spatial derivative, we arrive at the discretized equations:

$$\phi_j^{n+1} - \phi_j^n = \left(\frac{\alpha\,\Delta t}{2h^2}\right)[(\phi_{j+1}^n - 2\phi_j^n + \phi_{j-1}^n) + (\phi_{j+1}^{n+1} - 2\phi_j^{n+1} + \phi_{j-1}^{n+1})] \tag{6.25}$$

which holds for $j = 1,2,\ldots,N-1$. With the introduction of the definition $\beta = \alpha\,\Delta t/h^2$, these equations can be rewritten:

$$-\beta\phi_{j-1}^{n+1} + 2(1+\beta)\phi_j^{n+1} - \beta\phi_{j+1}^{n+1} = \beta\phi_{j-1}^n + 2(1-\beta)\phi_j^n + \beta\phi_{j+1}^n \tag{6.26}$$

Computing the solution at the new time step $n+1$ with this method requires solving a set of linear algebraic equations for the variables ϕ_j^{n+1}, $j = 1,2,\ldots,N-1$. The system of equations is tridiagonal, so the cost is directly proportional to the number of grid points. In other words, the cost per grid point is independent of the number of points, just as it is for the explicit Euler method. In fact, the cost per time step for this method is approximately double that of the explicit Euler method. As might be expected, this method, which is known as the *Crank–Nicolson* method when applied to PDEs, is unconditionally stable; this will be demonstrated below. The combination of increased accuracy and stability allows one to use a larger time step with the Crank–Nicolson method than with the Euler method (or higher-order explicit methods, for that matter), so, for a given accuracy, the implicit Crank–Nicolson method produces the solution at a lower computational cost than the Euler method.

This statement remains true for nonlinear equations, although the advantage may be reduced. In the nonlinear case, one has to solve a set of coupled nonlinear equations at each time step, but the set of equations is tridiagonal. A reasonable first guess of the solution at the new time step is the converged solution at the preceding time step, so iterative procedures such as the Newton–Raphson and secant methods converge rapidly; usually a few iterations suffice, and Crank–Nicolson remains the method of choice. Methods for nonlinear problems will be given in Section 6.11.

The computational molecule for this method is shown in Figure 6.4. We see that not only does the Crank–Nicolson method have the advantages described

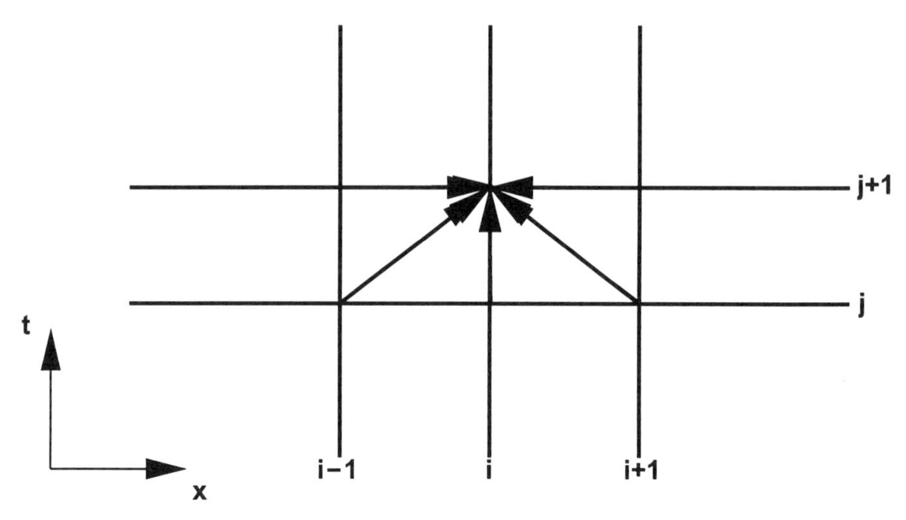

FIGURE 6.4. Computational molecule for the Crank–Nicolson method applied to the heat equation.

above but it remains faithful to the parabolic character of the partial differential equation—the solution at point x_j at time t_{n+1} is influenced not only by all the earlier values but by the solution at all other points at time t_{n+1} as well.

From the character of the methods on which it is based, one might guess that the Crank–Nicolson method is second-order accurate with respect to both independent variables, but it is instructive to derive the modified equation to demonstrate this. Note that the trapezoid rule is second-order with respect to time $t_n + \Delta t/2 = t_{n+1/2}$, so it is important to make the Taylor series expansions about the point $(x_j, t_{n+1/2})$. The easiest way to do this is to use results already obtained. For a uniform spatial grid, we have

$$\frac{\phi_{j+1}^n - 2\phi_j^n + \phi_{j-1}^n}{h^2} \approx \frac{\partial^2 \phi_j^n}{\partial x^2} + \frac{h^2}{12}\frac{\partial^4 \phi_j^n}{\partial x^4} \tag{6.27}$$

$$\frac{\phi_{j+1}^{n+1} - 2\phi_j^{n+1} + \phi_{j-1}^{n+1}}{h^2} \approx \frac{\partial^2 \phi_j^{n+1}}{\partial x^2} + \frac{h^2}{12}\frac{\partial^4 \phi_j^{n+1}}{\partial x^4} \tag{6.28}$$

$$\frac{\phi_j^{n+1} - \phi_j^n}{\Delta t} \approx \frac{\partial \phi_j^{n+1/2}}{\partial t} + \frac{\Delta t^2}{24}\frac{\partial^3 \phi_j^{n+1/2}}{\partial t^3} \tag{6.29}$$

For any function ψ, Taylor series can be used to show that

$$\frac{\psi^{n+1} + \psi^n}{2} \approx \psi^{n+1/2} + \frac{\Delta t^2}{8}\frac{\partial^2 \psi^{n+1/2}}{\partial t^2} \tag{6.30}$$

and applying this to the spatial derivative terms in Eq. (6.25), we find that the modified equation for the method is

$$\frac{\partial \phi}{\partial t} - \alpha \frac{\partial^2 \phi}{\partial x^2} = -\frac{\Delta t^2}{24}\frac{\partial^3 \phi}{\partial t^3} + \alpha \left(\frac{\Delta t^2}{8}\frac{\partial^2}{\partial t^2}\frac{\partial^2}{\partial x^2} + \frac{h^2}{12}\frac{\partial^4}{\partial x^4} \right) \phi$$
$$= \frac{\alpha}{12}\left(\alpha^2 \Delta t^2 \frac{\partial^6 \phi}{\partial x^6} + h^2 \frac{\partial^4 \phi}{\partial x^4} \right) \tag{6.31}$$

In deriving the last version of the right-hand side of this equation, we used the heat equation to eliminate the time derivatives. From Eq. (6.31), the second-order accuracy of the method is obvious. Note also that as h and Δt approach zero, the modified equation reduces to the original PDE. This property, which is shared by both methods discussed so far in this chapter, is called *convergence* and is clearly desirable. We note that the Crank–Nicolson method is convergent because we shall encounter a method for which this property does not obtain.

As suggested in the preceding section, one should attempt to balance the errors arising from the time and space differencing approximations. Doing this requires that we estimate the magnitudes of $\partial^4\phi/\partial x^4$ and $\partial^6\phi/\partial x^6$ and make them equal. This cannot be done at every point in space and time, but it is possible to estimate each component of the error using the Richardson method. To do this, one repeats a calculation with the spatial grid size halved but the time step unchanged; one third the difference between the two solutions is an estimate of the error due to spatial differencing approximation in the more accurate solution. The error on the original grid is four times as large. A calculation with the time step halved and the spatial grid size unchanged will yield an estimate of the contribution of the time advancement method to the error.

A simpler approach is to construct a rule of thumb by noting that the ratio of these two derivatives is the square of a length scale. Unless the solution is sharply peaked (which is rarely the case for the heat equation), this length scale must be some fraction of the size of the computational region, which we shall take to be $L/2 = Nh/2$. We then find that the errors are approximately balanced if $\beta = \alpha \Delta t/h^2 \approx N/2$. This is somewhat generous as we shall see later, but it is clear that the Crank–Nicolson method can use time steps that are quite a bit larger than explicit methods without sacrificing much accuracy; this is what makes it an effective tool for parabolic PDEs.

The Crank–Nicolson method has been the overwhelming favorite for the solution of parabolic PDEs in one spatial dimension for some time. It does have difficulties in more than one spatial dimension but, as we shall see later, these problems can be surmounted, and the Crank–Nicolson method provides the basis for some of the better methods of solving diffusion equations in two and three dimensions.

It is instructive to consider how the Crank–Nicolson method deals with oscillating solutions because Fourier analysis often proves useful in analyzing physical problems. We essentially perform a von Neumann stability analysis of this method. We assume a solution of the form $\phi_j^n = \varphi^n e^{ikx_j}$ and plug it into Eq. (6.25) to obtain

$$\frac{\varphi^{n+1}}{\varphi^n} = \frac{(1-2\beta) + 2\beta\cos k\,\Delta x}{(1+2\beta) - 2\beta\cos k\,\Delta x} \tag{6.32}$$

Because the cosine is periodic, the range $0 < k\,\Delta x < \pi$ produces all the possible values of φ^{n+1}/φ^n. This ratio is never greater than unity. Its minimum value is $(1-4\beta)/(1+4\beta)$, which is attained when $k\,\Delta x = \pi$. Since β is positive, this quantity is never less than -1, so the method is unconditionally stable according to the von Neumann criterion.

It is useful to investigate this a bit further. If $\beta < \frac{1}{4}$, φ^{n+1}/φ^n lies between 0 and 1 for all values of k. In this case, all waves, including those with large values of k, which oscillate rapidly in space, are damped. When β is greater than $\frac{1}{4}$, some of the waves have negative amplification factors (6.32); that is, they change sign at each successive time step but they remain damped. Finally, when β is very large, some waves are multiplied by factors close to -1 and thus change sign at each time step with little damping. This observation plays an important role in the following example and in the development of methods for solving elliptic partial differential equations.

Example 6.3: Crank–Nicolson Solution of the Heat Equation We now compute the solution to the problem of Example 6.1 using the Crank–Nicolson method. The program (CRANIC) used for this calculation is available on the Internet site described in Appendix A. Table 6.2 gives the results for the case of 10 spatial intervals ($\Delta x = 0.1$) and 20 spatial intervals ($\Delta x = 0.05$) at time $t = 0.1$ for various time steps; the time steps were chosen so that the same values of β occur in the two sets of results. We see that, as anticipated, the second-order Crank–Nicolson method produces accurate solutions for time steps such that $\beta = \alpha\,\Delta t/\Delta x^2 \le N/2$, or about four times the stability limit of the first-order explicit Euler method. For larger values of β, the method remains stable, but the results contain oscillations due to inadequate damping of the higher modes. Larger values of β can be used as the number of spatial points increases as can be seen from the table; this is in agreement with the rule of thumb suggested above.

Note that the solution at the midpoint behaves reasonably well even for β as large as 5, but the solution near the boundary begins to deteriorate when $\beta > 2$. As already noted, this behavior is due to the rapidly oscillating components of the solution. Near the boundary, these modes make a larger relative contribution to the solution; this is the reason the results behave more erratically there.

The error due to the temporal approximation may be assumed negligible for the smallest time steps shown in the table. By comparing the results obtained with the two spatial grid sizes at the smallest time steps, we see that the error due to the spatial finite difference approximation is about 10^{-4}

TABLE 6.2. Crank–Nicolson Solution of the Heat Equation

Δt	β	$\phi(0.1,0.1)$	$\phi(0.5,0.1)$
	Ten Intervals ($\Delta x = 0.1$)		
0.0010	0.10	0.14670	0.47435
0.0020	0.20	0.14669	0.47434
0.0050	0.50	0.14666	0.47428
0.0100	1.00	0.14655	0.47403
0.0200	2.00	0.14146	0.47375
0.0250	2.50	0.16501	0.47188
0.0333	3.33	0.08283	0.46045
0.0500	5.00	0.32688	0.46625
0.1000	10.00	−0.30495	0.56972
	Twenty Intervals ($\Delta x = 0.05$)		
0.00025	0.10	0.14669	0.47447
0.00050	0.20	0.14669	0.47445
0.00125	0.50	0.14669	0.47446
0.00250	1.00	0.14668	0.47444
0.00500	2.00	0.14666	0.47438
0.00625	2.50	0.14659	0.47433
0.00833	3.33	0.14516	0.47422
0.01250	5.00	0.13015	0.47402
0.02500	10.00	0.12186	0.47143

in the 10-interval calculation; it is this small because the solution is very smooth, allowing the second-order finite difference approximation to be accurate.

Solutions of the heat equation (and other parabolic equations) tend to become smoother with the passage of time. It is therefore possible to use small time steps for the earliest time steps and larger time steps later on. The small time steps allow the decay of the rapidly varying components of the solution to be computed accurately. Later on, the time step can be increased and the method still predicts the smooth component of the solution accurately.

Although solutions of the heat equation are generally smooth, especially for large values of t, there are problems in which the use of nonuniform meshes is advantageous. For problems in which the solution varies rapidly in some parts of the spatial domain, this approach is economical in terms of the amount of computation needed to obtain a given accuracy. It is difficult to analyze problems of this kind, but, as we saw in the preceding section, an approximate and conservative criterion for the stability of the Euler method (and other explicit methods) is to use the smallest Δx in the stability criterion (see Example 6.2). This can be a very severe constraint if the grid is highly nonuniform, so the relative advantage of implicit methods is increased when nonuniform meshes are used.

6.4. DUFORT–FRANKEL METHOD

The Crank–Nicolson method has two advantages over the Euler method: stability and improved accuracy. This suggests that we might look for still other methods. In Chapter 4, we saw that there are higher-order methods for ordinary differential equation initial value problems that have good stability properties. These could provide the basis for methods of solving parabolic problems. Among the better second-order explicit methods at our disposal are the Adams–Bashforth and leapfrog methods. The former has the same stability limitation as the Euler method. The latter, when applied to the semidiscretized heat equation (6.8), yields

$$\phi_j^{n+1} - \phi_j^{n-1} = \frac{2\alpha\,\Delta t}{h^2}(\phi_{j+1}^n - 2\phi_j^n + \phi_{j-1}^n) \tag{6.33}$$

This method is unconditionally unstable because von Neumann analysis shows that it is equivalent to applying the leapfrog method to $y' = \alpha y$ with real α. It does not appear to be a very good candidate for solving parabolic equations.

Dufort and Frankel pointed out that the leapfrog method is second-order accurate (for what little that is worth when it is unstable) and that the approximation:

$$\phi_j^n \approx \tfrac{1}{2}(\phi_j^{n+1} + \phi_j^{n-1}) \tag{6.34}$$

is also second-order accurate [compare with Eq. (6.30)]. Substituting Eq. (6.34) into Eq. (6.33) for ϕ_j^n, but leaving the terms involving ϕ_{j-1}^n and ϕ_{j+1}^n as they are, we have

$$(1 + 2\beta)\phi_j^{n+1} = (1 - 2\beta)\phi_j^{n-1} + 2\beta(\phi_{j-1}^n + \phi_{j+1}^n) \tag{6.35}$$

where, as before, $\beta = \alpha\,\Delta t/h^2$. This method is rather surprising. To begin with, it is both explicit and unconditionally stable. (The proof of this statement is not difficult to construct but, as it involves a fair amount of tedious algebra, we shall not present it. The interested reader is encouraged to try his/her hand at it.)

Another surprising result is discovered when one derives the modified equation for this method (again, a lot of tedious algebra is required):

$$\frac{\partial \phi}{\partial t} - \alpha\frac{\partial^2 \phi}{\partial x^2} = -\frac{\Delta t^2}{6}\frac{\partial^3 \phi}{\partial t^3} + \frac{\alpha h^2}{12}\frac{\partial^4 \phi}{\partial x^4} - \frac{\alpha\,\Delta t^2}{h^2}\frac{\partial^2 \phi}{\partial t^2} - \frac{\alpha\,\Delta t^4}{12h^2}\frac{\partial^4 \phi}{\partial t^4} \tag{6.36}$$

What is unusual here is the presence of a term with a coefficient proportional to $\Delta t^2/h^2$. This means that when one takes the limit in which Δt and h both go to zero, the original PDE is not recovered unless $\Delta t/h$ also goes

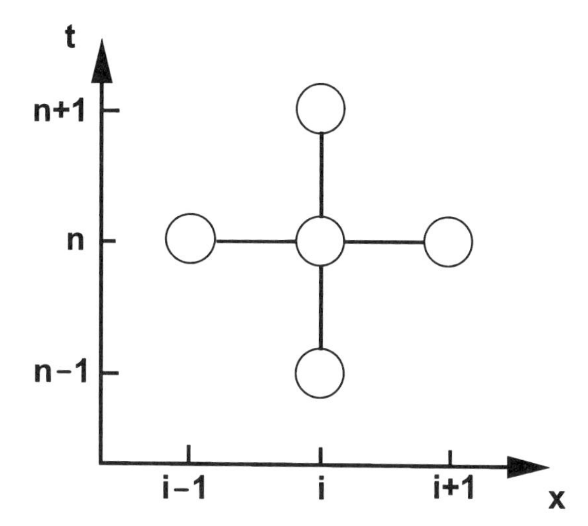

FIGURE 6.5. Computational molecule for the Dufort–Frankel method applied to the heat equation.

to zero. In terms of the definition given in the preceding section, the Dufort–Frankel method may or may not be convergent. In other words, unless Δt and h are properly chosen, the method is not even first-order accurate! If $\alpha \Delta t^2/h^2$ is large, Eq. (6.36) becomes a version of the wave equation. The hyperbolic character of the approximation might be anticipated by looking at the computational molecule shown in Figure 6.5.

These results demonstrate the need for caution in choosing a numerical method. Surprises are possible and are usually undesirable. We note in passing that L. F. Richardson (of extrapolation and dimensionless number fame) did computations of atmospheric flows using the leapfrog method on mechanical desk calculators in the 1920s; after two years of many people running these machines, he discovered that the method was unstable.

Despite these difficulties, the Dufort–Frankel method was fairly popular in the past, one reason being that it can be applied to problems in two or three space dimensions with little more effort than is required in one dimension. Since the Crank–Nicolson method has difficulty with these problems, there was a niche for the Dufort–Frankel method to fill. In recent years, it has been largely supplanted by the methods described later in this chapter. The Dufort–Frankel method was presented here mainly to point out the possibility that a very reasonable looking method can, in fact, produce completely erroneous results.

Example 6.4: Dufort–Frankel Solution of the Heat Equation Since the Dufort–Frankel method is a variation of the leapfrog method, it requires another method to start it. For this example, the Euler method is used for the

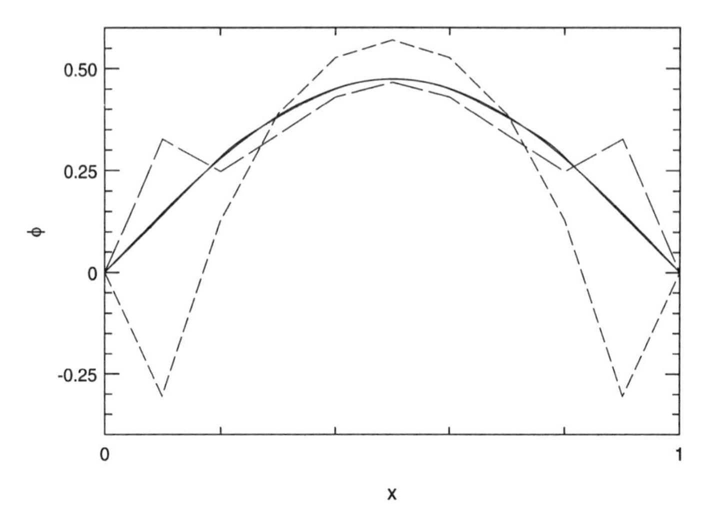

FIGURE 6.6. Solution of the heat equation with the Dufort–Frankel method. The solid curve is the solution obtained with $\beta = 0.5$, the long dashed curve is for $\beta = 1.0$, and the shorter dashed curve is for $\beta = 1.667$.

first time step. The program used (DUFORT) can be found on the Internet site described in Appendix A; the only change from the preceding subroutines is that, since the Dufort–Frankel method requires a different starting method, a parameter is needed to indicate the first call.

The results obtained are shown in Figure 6.6. When β is small, the results obtained are accurate and smooth. As β is increased, however, the results deteriorate. The oscillatory behavior seen for $\beta = 1$ and $\beta = 1.667$ in Figure 6.6 can be explained in at least two different ways.

First, since the method is based on the leapfrog approximation, there is a tendency for the solution to oscillate about the exact result at alternate time steps. Thus the solutions shown in Figure 6.6 would look quite different at the following time step; the points at which the solution is low would become the ones at which it is high and vice versa.

Another way to look at the results is that the Euler starting method for $\beta > 1$ produces a negative "spike" near each boundary at the first time step. Then since the Dufort–Frankel method has hyperbolic character, the spikes propagate inward as time increases. A "motion picture" of the solution would show this clearly.

Even though the Dufort–Frankel method is not unstable, it can produce very poor solutions. The results depend on the starting method and, in general, the method is not as well behaved as the implicit Crank–Nicolson method. As a result, it has lost considerable favor in the last few years. As stated above, it was presented here mainly to demonstrate that it is possible to construct methods that are not convergent, that is, as a caution.

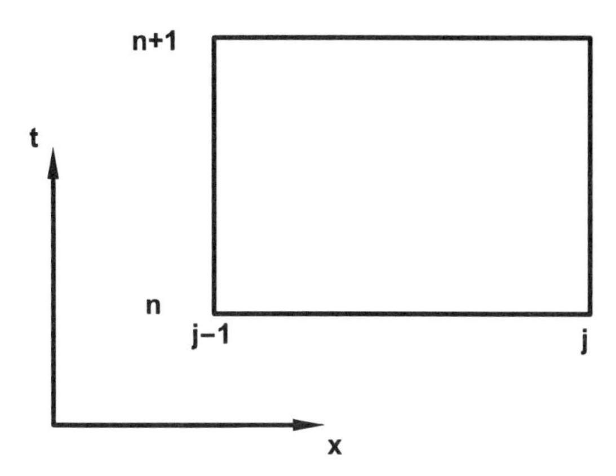

FIGURE 6.7. Domain of integration in the Keller box method.

6.5. KELLER BOX METHOD

Another method that has achieved some popularity in recent years is the box method developed by Keller. It is an alternative to the Crank–Nicolson method to which it is closely related. The basic difference between the methods is that Keller chooses to write the PDE as a system of first-order equations. For the heat equation (6.5), this is easily done. Defining

$$\psi = \frac{\partial \phi}{\partial x} \tag{6.37}$$

Eq. (6.5) can be written

$$\frac{\partial \phi}{\partial t} = \alpha \frac{\partial \psi}{\partial x} \tag{6.38}$$

Equations (6.37) and (6.38) are a coupled set of first-order equations equivalent to the heat equation.

To discretize these equations, Keller integrates Eq. (6.38) over the box shown in Figure 6.7; this gives the method its name. Using the trapezoid rule to approximate integrals whenever necessary, we have

$$\int_{t_n}^{t_{n+1}} \int_{x_{j-1}}^{x_j} \frac{\partial \phi}{\partial t} dx\, dt = \int_{x_{j-1}}^{x_j} [\phi(x,t_{n+1}) - \phi(x,t_n)]\, dx$$

$$\approx \frac{h}{2}[(\phi_j^{n+1} + \phi_{j-1}^{n+1}) - (\phi_j^n + \phi_{j-1}^n)] \tag{6.39}$$

$$\int_{t_n}^{t_{n+1}} \int_{x_{j-1}}^{x_j} \frac{\partial \psi}{\partial x} dx\, dt = \int_{t_n}^{t_{n+1}} [\psi(x_j,t) - \psi(x_{j-1},t)]\, dt$$

$$\approx \frac{\Delta t}{2}[(\psi_j^{n+1} + \psi_j^n) - (\psi_{j-1}^{n+1} + \psi_{j-1}^n)] \tag{6.40}$$

so that Eq. (6.38) is approximated by

$$(\phi_j^{n+1} + \phi_{j-1}^{n+1}) - \gamma(\psi_j^{n+1} + \psi_{j-1}^{n+1}) = (\phi_j^n + \phi_{j-1}^n) + \gamma(\psi_j^n + \psi_{j-1}^n) \tag{6.41}$$

where $\gamma = \alpha\,\Delta t/h$. Equation (6.37), which is essentially an ODE, can be approximated by

$$\tfrac{1}{2}h(\psi_j^{n+1} + \psi_{j-1}^{n+1}) = \phi_j^{n+1} + \phi_{j-1}^{n+1} \tag{6.42}$$

Equations (6.41) and (6.42) are a coupled set of linear algebraic equations for the dependent variables ϕ_j^{n+1} and ψ_j^{n+1} for $j = 1, 2, \ldots, N$, where N is the number of interior grid points. We now have two dependent variables at each grid point, but each equation contains the variables at only two grid points. The matrix corresponding to this set of equations thus has a different structure than the ones we encountered earlier, that is, it is not tridiagonal but, by taking advantage of the structure of this matrix, it is possible to construct a subroutine for solving this system of equations that is almost as efficient as the tridiagonal solvers used in the Crank–Nicolson method.

For an equation as simple as the heat equation, we can manipulate the discretized equations of the Keller box method into a form that facilitates comparison with the Crank–Nicolson method. Subtracting Eq. (6.42) with index j from the same equation with $j+1$, we obtain

$$\tfrac{1}{2}h(\psi_{j+1}^{n+1} + \psi_{j-1}^{n+1}) = \phi_{j+1}^{n+1} + 2\phi_j^{n+1} + \phi_{j-1}^{n+1} \tag{6.43}$$

This equation also holds with $n+1$ replaced by n. On the other hand, if Eq. (6.41) with indices j and $j+1$ are added, we find

$$\begin{aligned}(\phi_{j+1}^{n+1} + 2\phi_j^{n+1} + \phi_{j-1}^{n+1}) &- \gamma(\psi_{j+1}^{n+1} + \psi_{j-1}^{n+1}) \\ &= (\phi_{j+1}^n + 2\phi_j^n + \phi_{j-1}^n) + \gamma(\psi_{j+1}^n + \psi_{j-1}^n)\end{aligned} \tag{6.44}$$

Finally, if Eq. (6.43) with indices n and $n+1$ are substituted into Eq. (6.44), we have

$$\begin{aligned}(\phi_{j+1}^{n+1} + 2\phi_j^{n+1} + \phi_{j-1}^{n+1}) &- 2\beta(\phi_{j+1}^{n+1} - 2\phi_j^{n+1} + \phi_{j-1}^{n+1}) \\ &= (\phi_{j+1}^n + 2\phi_j^n + \phi_{j-1}^n) + 2\beta(\phi_{j+1}^n - 2\phi_j^n + \phi_{j-1}^n)\end{aligned} \tag{6.45}$$

where, as before, $\beta = \alpha\,\Delta t/\Delta x^2$.

These equations would be the Crank–Nicolson equations if, in the first term on each side of Eq. (6.45), $\phi_{j+1} + 2\phi_j + \phi_{j-1}$ is replaced by $4\phi_j$. Thus the results of the Keller box method are very similar to those of the Crank–Nicolson method; the error terms are slightly different but both have second-order accuracy.

As we noted earlier, the Keller box method is a competitor to Crank–Nicolson. It requires slightly more computation per point but is also slightly more accurate, so it may require fewer mesh points. More significantly, the box method lends itself very well to the use of extrapolation methods (e.g., Richardson extrapolation) and can thus be used as the basis for high-order methods that do not require an inordinate amount of computer storage. It is also more easily applied on nonuniform grids. It is these properties that have been most responsible for its popularity.

6.6. SECOND-ORDER BACKWARD METHOD

Although the Crank–Nicolson method and variations on it have been among the most popular methods for parabolic equations, they are not without problems. Let us consider some of the difficulties associated with the Crank–Nicolson and Keller box methods.

As we will see in the next chapter, elliptic problems are very difficult to solve. They usually arise from steady-state problems, which suggests that one way to solve an elliptic partial differential equation is to introduce an unphysical time derivative term into the equation and advance the resulting parabolic equation in time until a steady state is reached. Since the objective is to reach steady state as quickly as possible, it is a good idea to use the largest possible time step, and this suggests use of an implicit method such as Crank–Nicolson. However, when used with a large time step, the Crank–Nicolson method tends to produce a solution that changes sign on each time step with very little reduction in magnitude. In fact, the method is just barely stable for very large time steps. This means that, when it is used to solve steady-state problems, the Crank–Nicolson method may converge very slowly. Because it is very close to being unstable, when it is applied to nonlinear problems, the Crank–Nicolson method may actually become unstable. There is a clear role for a method that does not suffer from these problems.

To be able to use large time steps, we need an implicit method that is unconditionally stable but produces a solution that decays relatively rapidly when the time step is very large. One way to achieve this is to use a second-order backward method based on the approximation (4.11). If this is applied to the semidiscretized heat equation (6.8), we obtain

$$\frac{\phi_j^{n-1} - 4\phi_j^n + 3\phi_j^{n+1}}{2\,\Delta t} = \frac{\alpha}{h^2}(\phi_{j-1}^{n+1} - 2\phi_j^{n+1} + \phi_{j+1}^{n+1}) \tag{6.46}$$

which can be rearranged to

$$2\beta\phi_{j-1}^{n+1} - (3 + 4\beta)\phi_j^{n+1} + 2\beta\phi_{j+1}^{n+1} = -4\phi_j^n + \phi_j^{n-1} \tag{6.47}$$

where, as usual, $\beta = \alpha\,\Delta t/h^2$.

This is a two-step method and therefore requires that a different method be used at the first time step. However, both characteristic roots of this method (found by applying it to $y' = \alpha y$) become zero for large time steps. It is therefore an excellent method for parabolic and elliptic partial differential equations.

Although this method is second-order accurate in time, its error is four times larger than the Crank–Nicolson method error. It is therefore not as good a choice as the latter method when time accuracy is the primary concern.

Example 6.5: Second-Order Backward Solution of the Heat Equation Because the problem has been dealt with in previous examples, we shall give only a very brief overview of results obtained by applying the second-order backward method to the heat equation. As noted above, a starter method is needed; the Crank–Nicolson method is used for the first step. The code used, BACK2, is found on the Internet site described in Appendix A. The only unusual point about the code that should be noted is that, because two old fields are required by Eq. (6.46), it is necessary to retain the solutions at the last two time steps in memory; this complicates the programming just a little. The even and odd time steps are treated differently in the code, but other means of dealing with the problem could be used.

The results are given in Table 6.3. By comparing this table with Table 6.2, we see that, as expected, the error introduced by the discretization in time is much larger for the fourth-order backward method. For a problem in which time accuracy is essential, small time steps are required, and one is better off using the Crank–Nicolson method. For larger time steps, the solution degrades much more slowly with the second-order backward method, making it the better choice for problems with slowly changing solutions. As noted above, it is also a better choice for elliptic problems.

6.7. HIGHER-ORDER METHODS

It is possible to construct methods of higher-order spatial accuracy for parabolic PDEs from the higher-order methods that were applied to boundary value problems for ODEs in Section 5.5. Everything said about those methods applies here. Specifically, it is difficult to maintain the high-order accuracy near the boundaries, where it is important to do so. It is also difficult to construct high-order methods for nonuniform grids. Finally, when these methods are used with methods that are implicit in time for parabolic equations (which

TABLE 6.3. Solution of the Heat Equation with the Second-Order Backward Method

Δt	β	$\phi(0.1,0.1)$	$\phi(0.5,0.1)$
	Ten Intervals ($\Delta x = 0.1$)		
0.0010	0.10	0.14526	0.47973
0.0020	0.20	0.14382	0.46512
0.0050	0.50	0.13952	0.45140
0.0100	1.00	0.13247	0.42882
0.0200	2.00	0.11909	0.38514
0.0333	3.33	0.10022	0.33401
0.0500	5.00	0.06808	0.27997
0.1000	10.00	−0.52850	0.14904
	Twenty Intervals ($\Delta x = 0.05$)		
0.00025	0.10	0.14633	0.47325
0.00050	0.20	0.14597	0.47214
0.00125	0.50	0.14488	0.46864
0.00250	1.00	0.14308	0.46285
0.00500	2.00	0.13949	0.45134
0.00833	3.33	0.13476	0.43614
0.01250	5.00	0.12898	0.41725

we have seen to be necessary), special routines are required for solving the systems of equations for the variables at the new time step.

The most effective high-order method was found to be the compact fourth-order method (see Section 5.4), so we apply it to the heat equation here. Applying the compact fourth order in space method to the heat equation (6.3) and using the trapezoid or Crank–Nicolson method in time, we obtain

$$\begin{aligned} &\tfrac{1}{6}(\phi_{j+1}^{n+1} + 10\phi_j^{n+1} + \phi_{j-1}^{n+1}) - \beta(\phi_{j+1}^{n+1} - 2\phi_j^{n+1} + \phi_{j-1}^{n+1}) \\ &\quad = \tfrac{1}{6}(\phi_{j+1}^{n} + 10\phi_j^{n} + \phi_{j-1}^{n}) + \beta(\phi_{j+1}^{n} - 2\phi_j^{n} + \phi_{j-1}^{n}) \end{aligned} \tag{6.48}$$

where, as earlier, $\beta = \alpha\Delta t/\Delta x^2$. This method is no more difficult to use than the Crank–Nicolson method. Use of the fourth-order compact method clearly reduces the error arising from the spatial discretization. We have noted that the time step should be chosen so that the temporal and spatial differencing errors are approximately equal, so to obtain the maximum benefit from this method, it is also necessary to reduce the time step, thereby increasing the cost somewhat.

Finding methods that have higher-order accuracy in time that can be applied to parabolic problems is a more difficult issue. The major problem is that there are no unconditionally stable methods of order higher than 2. We could use

Adams–Moulton methods, which have reasonable stability properties, but that would require keeping a number of time steps and special starter methods are needed. A good alternative is to use the fourth-order Runge–Kutta method, which has good stability properties and is very easy to use.

We now proceed to the more difficult question of finding numerical methods for parabolic equations in two or three spatial dimensions.

6.8. TWO AND THREE SPATIAL DIMENSIONS: ALTERNATING DIRECTION IMPLICIT METHODS

6.8.1. Heat Equation in Two Dimensions

We now consider parabolic PDEs in two and three spatial dimensions. Again, special attention is given to the heat equation. From the numerical point of view, the most important changes occur in going from one dimension to two. Most of the ideas developed for two-dimensional problems generalize to three-dimensional ones without too much difficulty. For this reason, the two-dimensional case is considered in some detail.

In two dimensions the heat equation becomes

$$\frac{\partial \phi}{\partial t} = \alpha \left(\frac{\partial^2 \phi}{\partial x^2} + \frac{\partial^2 \phi}{\partial y^2} \right) = \alpha \nabla^2 \phi \tag{6.49}$$

where ∇^2 is the Laplacian operator.

As in the one-dimensional case, spatial derivatives are approximated by the second-order central difference formula. (Other approximations can be used, of course, but this one is the workhorse and others are used only occasionally.) To prevent the notation from becoming too cumbersome, we introduce the abbreviations

$$\left.\frac{\delta^2 \phi}{\delta x_1^2}\right|_{i,j} = \frac{\phi_{i+1,j} - 2\phi_{i,j} + \phi_{i-1,j}}{h_1^2} \tag{6.50}$$

$$\left.\frac{\delta^2 \phi}{\delta x_2^2}\right|_{i,j} = \frac{\phi_{i,j+1} - 2\phi_{i,j} + \phi_{i,j-1}}{h_2^2} \tag{6.51}$$

We call the independent spatial variables x_1 and x_2 in order to facilitate later extension to three dimensions. The semidiscretized heat equation in two dimensions,

$$\frac{\partial \phi}{\partial t} = \alpha \left(\frac{\delta^2 \phi}{\delta x_1^2} + \frac{\delta^2 \phi}{\delta x_2^2} \right) \tag{6.52}$$

can be represented in linear algebra fashion by creating the vector

$$\phi = \begin{pmatrix} \phi_{1,1} \\ \phi_{1,2} \\ \vdots \\ \phi_{1,M} \\ \phi_{2,1} \\ \vdots \\ \phi_{2,M} \\ \phi_{3,1} \\ \vdots \\ \phi_{N,M} \end{pmatrix} \tag{6.53}$$

We have made the elements of the solution that belong to the leftmost vertical line the topmost elements in the vector; they are followed by the elements on the next vertical line to the right, and so on. We have also assumed that there are M interior mesh points in the x_2 direction and N interior points in the x_1 direction and that the region is rectangular. The elements could just as well have been entered by horizontal lines rather than vertical lines—in this instance, the choice is not critical—but, as we see later, it may be important in elliptic problems. The treatment of nonrectangular geometries, an important problem, is taken up in Section 6.9.

With the vector written as in Eq. (6.53), the matrix form of the operator appearing in Eq. (6.52) becomes

$$A = \begin{pmatrix} -2(h_1^{-2} + h_2^{-2}) & h_2^{-2} & 0 & \cdots & 0 & h_1^{-2} & 0 & 0 & \cdots \\ h_2^{-2} & -2(h_1^{-2} + h_2^{-2}) & h_2^{-2} & \cdots & 0 & 0 & h_1^{-2} & 0 & \cdots \\ 0 & h_2^{-2} & \cdots & \cdots & & & & & \\ 0 & 0 & \cdots & & & & & & \\ \vdots & \vdots & \ddots & & & & & & \\ h_1^{-2} & 0 & \cdots & & & & & & \\ 0 & h_1^{-2} & \cdots & & & & & & \\ 0 & 0 & \cdots & & & & & & \\ \cdots & \cdots & \cdots & & & & & & \end{pmatrix} \tag{6.54}$$

Only five diagonals of this $MN \times MN$ matrix contain elements that are not zero: the main diagonal [all the elements of which are $-2(h_1^{-2} + h_2^{-2})$], the diagonals immediately above and below it (with elements h_2^{-2}), and the diagonals M rows above and below the main diagonal (with elements h_1^{-2}). Although this matrix has a very simple band structure, two of the diagonals of nonzero elements are far from the main diagonal. The cost of solving a system of equations containing a matrix of this form by Gauss elimination is proportional NM^3. If this method were used, the cost per grid point (the only fair measure by which to compare methods) would be more than M^2 times the cost per grid point for a one-dimensional calculation, making two-dimensional problems much more expensive to solve than one-dimensional ones. In three dimensions, the per variable cost would be something like M^4 times that of the one-dimensional case, which is completely prohibitive.

Before we look for methods of overcoming this difficulty, let us consider another issue. We already found that the one-dimensional problem is very stiff; it is important to determine whether the two-dimensional problem is even stiffer. In order to assess the stiffness of the two-dimensional problem, let us consider the eigenvalue problem

$$A\psi = \lambda\psi \tag{6.55}$$

which can also be written as

$$\frac{\delta^2\psi}{\delta x_1^2} + \frac{\delta^2\psi}{\delta x_2^2} = \lambda\psi \tag{6.56}$$

The eigenvalues and eigenvectors can again be obtained by making an educated guess. The partial differential equation from which this problem was derived has as its eigenfunctions products of eigenfunctions of the corresponding one-dimensional problems. Thus we expect the solution to Eq. (6.56) to have the form

$$\psi_{i,j}^{m,n} = \sin\frac{m\pi}{M}i \sin\frac{n\pi}{N}j \tag{6.57}$$

This turns out to be correct and, with some calculation, we can show that the eigenvalues are the sum of the one-dimensional eigenvalues

$$\lambda_{m,n} = \frac{2}{h_1^2}\left(1 - \cos\frac{m\pi}{M+1}\right) + \frac{2}{h_2^2}\left(1 - \cos\frac{n\pi}{N+1}\right) \tag{6.58}$$

where m takes the values $1, 2, \ldots M$ and $n = 1, 2, \ldots N$. From the expression (6.58) for the eigenvalues we find that the problem in two dimensions has about twice the stiffness ratio of the one-dimensional problem. The extensions of these results to the three-dimensional case is, to use a time-worn phrase, tedious but straightforward, and the stiffness is about three times that of the one-dimensional problem. Since the stiffness ratio can vary by orders of magnitude, we can say that the semidiscretized multidimensional heat equation is only a little stiffer than the one-dimensional one.

We cannot expect to use explicit methods, except possibly the Dufort–Frankel method, which has the difficulties that were pointed out earlier. The Crank–Nicolson method is a natural one to try. Applying it leads to

$$-\left(\frac{\alpha\,\Delta t}{2}\right)\left(\frac{\delta^2\phi_{i,j}^{n+1}}{\delta x_1^2}+\frac{\delta^2\phi_{i,j}^{n+1}}{\delta x_2^2}\right)+\phi_{i,j}^{n+1}=\left(\frac{\alpha\,\Delta t}{2}\right)\left(\frac{\delta^2\phi_{i,j}^{n}}{\delta x_1^2}+\frac{\delta^2\phi_{i,j}^{n}}{\delta x_2^2}\right)+\phi_{i,j}^{n} \tag{6.59}$$

The left-hand side of this equation is a discrete approximation to the quantity $-(\alpha\,\Delta t/2)\nabla^2\phi+\phi$, so Eq. (6.59) is the discretized form of a two-dimensional elliptic problem. The matrix corresponding to Eq. (6.59) has the same structure (but not the same elements) as the matrix (6.54). As already noted, the cost of solving a system of linear equations with a matrix of this type is much higher than the cost of solving a pentadiagonal system in which the nonzero elements are found on the diagonals closest to the main diagonal. This is the primary difficulty in solving elliptic equations and, if no way out of it were available, the cost of solving parabolic equations in two and three dimensions would be prohibitive. Fortunately, a means of avoiding this difficulty was discovered by Peaceman and Rachford in the 1950s; their method is the subject of the next subsection.

6.8.2. Peaceman–Rachford Method

We appear to be in a bind as far as solving the heat equation in two or three dimensions is concerned. The choice appears to be between explicit methods, which require many time steps due to stability limitations, and implicit methods, which require the use of an expensive solver for the linear equations at each time step.

What is needed is a method that provides the stability benefits of implicit methods without the large penalty in computation time associated with solving a two-dimensional elliptic problem. One might try a compromise that treats one direction implicitly and the other explicitly. Methods of this kind turn out to be conditionally stable (if one is lucky); the time step restrictions are similar to those for explicit methods, so almost nothing is gained by using them.

A way to avoid this dilemma was provided by a method that has proven to be one of the great surprises and advances in numerical analysis. It has, in fact, led to a whole class of powerful new methods. The idea is to first treat one direction implicitly and the other explicitly and then reverse the roles. This method, which has very nice properties, is the famous alternating direction implicit (ADI) method originally developed by Peaceman and Rachford (1956).

Applied to the two-dimensional heat equation, the Peaceman–Rachford method consists of two steps. In the first, the x_1 direction is treated by backward Euler method and the x_2 direction is treated by the explicit Euler method;

the equation is advanced a half time step

$$\phi_{i,j}^{n+1/2} - \phi_{i,j}^{n} = \frac{\alpha\,\Delta t}{2}\left(\frac{\delta^2\phi_{i,j}^{n+1/2}}{\delta x_1^2} + \frac{\delta^2\phi_{i,j}^{n}}{\delta x_2^2}\right) \tag{6.60}$$

This is followed by a step in which the roles of the two methods are reversed:

$$\phi_{i,j}^{n+1} - \phi_{i,j}^{n+1/2} = \frac{\alpha\,\Delta t}{2}\left(\frac{\delta^2\phi_{i,j}^{n+1/2}}{\delta x_1^2} + \frac{\delta^2\phi_{i,j}^{n+1}}{\delta x_2^2}\right) \tag{6.61}$$

As stated above, this method has surprising properties. Considered as a method in its own right, each of the two steps is second-order accurate in space and first-order accurate in time (as might be expected from the properties of the ODE methods from which they are derived) and conditionally stable. The combined method, however, has properties that are much better than those of either of its components. We shall show later it is second-order accurate in both space and time and unconditionally stable. To get some idea of why this might be, we eliminate $\phi^{n+1/2}$ from these equations by adding them to obtain

$$\phi_{i,j}^{n+1} - \phi_{i,j}^{n} = \alpha\,\Delta t\left[\frac{\delta^2\phi_{i,j}^{n+1/2}}{\delta x_1^2} + \frac{1}{2}\left(\frac{\delta^2\phi_{i,j}^{n}}{\delta x_2^2} + \frac{\delta^2\phi_{i,j}^{n+1}}{\delta x_2^2}\right)\right] \tag{6.62}$$

which shows that the overall method is equivalent to treating the x_1 direction by the midpoint rule and the x_2 direction by the trapezoid rule. (The order of the two steps could be reversed, which would also interchange the methods of treatment of the two directions.) Since both the midpoint rule and the trapezoid rule are second-order accurate, the total method is second-order accurate, even though each of its components has only first-order accuracy.

We next investigate the stability of the method in some detail because parts of the analysis will prove important later. This is most easily done with the von Neumann approach, which allows us to avoid the question of boundary conditions. We assume that the difference equations have solutions of the form

$$\begin{aligned}\phi^{n} &= \rho^{n} e^{ik_1x_1} e^{ik_2x_2}\\ \phi^{n+1/2} &= \xi\rho^{n} e^{ik_1x_1} e^{ik_2x_2}\end{aligned} \tag{6.63}$$

where ξ is a constant and the spatial indices are suppressed for convenience. We note that

$$\frac{\delta^2}{\delta x_\ell^2} e^{ik_\ell x_\ell} = \frac{2}{h_\ell^2}(\cos k_\ell h_\ell - 1)e^{ik_\ell x_\ell} \tag{6.64}$$

where $\ell = 1$ or 2. Substituting Eqs. (6.63) into Eqs. (6.60) and (6.61), using the definition

$$\beta_\ell = \frac{\alpha \Delta t}{h_\ell^2}(\cos k_\ell h_\ell - 1) \tag{6.65}$$

and Eq. (6.64), and after canceling out common factors, we have

$$\xi - 1 = \xi\beta_1 + \beta_2 \tag{6.66}$$

$$\rho - \xi = \xi\beta_1 + \beta_2 \tag{6.67}$$

Finally, solving for ρ, we have

$$\rho = \frac{1 + \beta_1}{1 - \beta_1}\frac{1 + \beta_2}{1 - \beta_2} \tag{6.68}$$

This is essentially a product of two functions each similar to the one that determines the stability of the trapezoid rule for ODEs. Since β_1 and β_2 are both negative, the ADI method is unconditionally stable.

The first step of the Peaceman–Rachford ADI method, represented by Eq. (6.60), requires the solution of a tridiagonal system of equations on each line of constant y (or constant j); that is, we have to solve N systems of M tridiagonal equations; this can be done in parallel if necessary. The second step [Eq. (6.61)] requires solving M sets of N tridiagonal equations. Since the number of operations required to solve a tridiagonal system is linearly proportional to the number of equations in the system, the total cost of this method is proportional to MN, that is, to the total number of grid points. Thus the cost per grid point is independent of the number of grid points. This is as good as one could hope for and the ADI method is, therefore, hard to beat. The two-dimensional ADI method requires about twice as much computation per grid point as the one-dimensional Crank–Nicolson method. It has all of the desirable properties of a numerical method—speed, accuracy, and stability—so it is hardly surprising that the ADI method, but not necessarily the version presented here, is the preferred method for parabolic problems in two and three dimensions. Indeed, these properties have made it a favorite method for elliptic equations as well.

We next look at some of the variants of the ADI method, which are valuable in constructing other methods.

6.8.3. Approximate Factorization

The ADI method was the first of a large class of methods that are now called *splitting* or *approximate factorization* methods. The success of ADI prompted an search for other methods with similar properties; this search has been enormously successful. It is, therefore, worth looking into the Peaceman–Rachford ADI method and some of its relatives further.

Essentially, the concept of splitting is as follows. Equation (6.59), the basis for the Crank–Nicolson method in two dimensions, can be written in the symbolic operator form:

$$\left[1 - \frac{\alpha\,\Delta t}{2}(\delta_{xx} + \delta_{yy})\right]\phi^{n+1} = \left[1 + \frac{\alpha\,\Delta t}{2}(\delta_{xx} + \delta_{yy})\right]\phi^{n} \tag{6.69}$$

where δ_{xx} and δ_{yy} represent the second-order central finite difference approximations to $\partial^2/\partial x^2$ and $\partial^2/\partial x^2$, respectively. On the other hand, Eqs. (6.60) and (6.61) can be written in this notation as

$$\left(1 - \frac{\alpha\,\Delta t}{2}\delta_{xx}\right)\phi^{n+1/2} = \left(1 + \frac{\alpha\,\Delta t}{2}\delta_{yy}\right)\phi^{n} \tag{6.70}$$

and

$$\left(1 - \frac{\alpha\,\Delta t}{2}\delta_{yy}\right)\phi^{n+1} = \left(1 + \frac{\alpha\,\Delta t}{2}\delta_{xx}\right)\phi^{n+1/2} \tag{6.71}$$

By multiplying Eq. (6.70) by the operator $(1 + \alpha\,\Delta t\,\delta_{xx}/2)$ from the left and Eq. (6.71) by the operator $(1 - \alpha\,\Delta t\,\delta_{xx}/2)$ (again form the left) and adding the results, the last two equations can be combined to give a single equation:

$$\left(1 - \frac{\alpha\,\Delta t}{2}\delta_{xx}\right)\left(1 - \frac{\alpha\,\Delta t}{2}\delta_{yy}\right)\phi^{n+1} = \left(1 + \frac{\alpha\,\Delta t}{2}\delta_{yy}\right)\left(1 + \frac{\alpha\,\Delta t}{2}\delta_{xx}\right)\phi^{n} \tag{6.72}$$

We used the fact that all of the operators commute in deriving this equation. If the multiplication of the operators in this last equation are carried out, we find that Eq. (6.72) is the same as Eq. (6.69) except for an extra term on the left-hand side, which is $(\alpha^2\,\Delta t^2/4)\,\delta_{xx}\,\delta_{yy}(\phi^{n+1} - \phi^{n}) \approx (\alpha^2\,\Delta t^3/4)\,\delta_{xx}\,\delta_{yy}\,\partial\phi/\partial t$. If Δt is small, this term is smaller than the truncation error of the Crank–Nicolson method [Eq. (6.59)], which is of order Δt^2 and can be neglected. If we do so, we find that the split method is equivalent to the Crank–Nicolson method. Thus, we have a method that is as accurate and as stable as the Crank–Nicolson method but is much less expensive to run. The approximate factorization concept actually provides an important benefit at very little cost. The only other technique that might be put into this category is Richardson extrapolation.

We can regard Eq. (6.72) as an approximate factorization of Eq. (6.69). This is an alternative name for this class of methods and provides insight into ways of finding other methods of this type. The concept at the heart of all approximate factorization methods is the replacement of an operator whose inversion is expensive by the product of two (or more) operators whose inversion is computationally simpler and cheaper; the factorization must be one that does not decrease the order of accuracy of the method. The savings in computational effort can be spectacular.

6.8.4. Other Splitting Methods

The Peaceman–Rachford method given above has been very popular, but other approximate factorizations of Eq. (6.69) can be and have been devised. On a closer look, we see that the Peaceman–Rachford factorization treats the discretized differential operators in the horizontal and vertical directions differently. The discretized Laplacian operator $(\delta^2/\delta x^2 + \delta^2/\delta y^2)$ can be written as $H/\Delta x^2 + V/\Delta y^2$ where H is the matrix arising from the horizontal derivative (the one with respect to x_1) and V is the matrix arising from vertical derivative (the one with respect to x_2). The Peaceman–Rachford splitting is based on this decomposition. In this notation, Eqs. (6.70) and (6.71) can be written

$$[1 - (\alpha\Delta t/2\Delta x_1^2)H]\phi^* = [1 + (\alpha\Delta t/2\Delta x_2^2)V]\phi^n \tag{6.73}$$

$$[1 - (\alpha\Delta t/2\Delta x_2^2)V]\phi^{n+1} = [1 + (\alpha\Delta t/2\Delta x_1^2)H]\phi^* \tag{6.74}$$

This is why we called the program based on this algorithm ADIHV; it can be found at the Internet site described in Appendix A.

Once the idea of splitting is available, it is not difficult to devise other splitting methods. One alternative to the Peaceman–Rachford, or HV, splitting for the heat equation can be derived by writing:

$$\Delta x_1^2\delta^2/\delta x_1^2 + \Delta x_2^2\delta^2/\delta x_2^2 = L + U \tag{6.75}$$

where L is a lower triangular matrix and U is upper triangular. (This additive decomposition should not to be confused with the multiplicative LU decomposition discussed in Chapter 1.) In this decomposition, the treatment of the terms on the main diagonal of the original unsplit matrix is somewhat arbitrary, but we will assign half of each diagonal term to each matrix. We then have an alternative ADI method:

$$[1 - (\alpha\Delta t/2\Delta x_1^2)L]\phi^* = [1 + (\alpha\Delta t/2\Delta x_2^2)U]\phi^n \tag{6.76}$$

$$[1 - (\alpha\Delta t/2\Delta x_1^2)U]\phi^{n+1} = [1 + (\alpha\Delta t)/2\Delta x_2^2)L]\phi^* \tag{6.77}$$

which may be called the ADILU method; a code embodying this method can be found at the Internet site described in Appendix A.

This method has the advantage that the matrices L and U are both triangular and therefore easily inverted. In fact, each sweep of this method can be carried out in a marching manner, starting at one corner of the domain, proceeding along a line (either horizontal or vertical) and then moving to the next line, until the entire grid has been visited. There is no need for a tridiagonal solver, making this version of the method less costly in terms of computation time than the ADIHV method. The solution procedure in each step of this method is similar to a Gauss–Seidel iteration, which is discussed in the next chapter.

Another advantage of this method is that it is much easier to apply on an irregular solution domain than the ADIHV method.

It is difficult to extend the Peaceman–Rachford method to three dimensions. (The ADILU method, which was discovered much later than the ADIHV method, is readily extended to three dimensions.) A scheme similar to the Peaceman–Rachford method needs to contain three steps or factors; such a method was developed by Douglass and Gunn. One version of their method is

$$\left(1 - \frac{\alpha\,\Delta t}{2}\delta_{xx}\right)\phi^{n+1^*}\left[1 + \frac{\alpha\,\Delta t}{2}(\delta_{xx} + 2\delta_{yy} + 2\delta_{zz})\right]\phi^n \tag{6.78}$$

$$\left(1 - \frac{\alpha\,\Delta t}{2}\delta_{yy}\right)\phi^{n+1^{**}} = \phi^{n+1^*} - \frac{\alpha\,\Delta t}{2}\delta_{yy}\phi^n \tag{6.79}$$

$$\left(1 - \frac{\alpha\,\Delta t}{2}\delta_{zz}\right)\phi^{n+1} = \phi^{n+1^{**}} - \frac{\alpha\,\Delta t}{2}\delta_{zz}\phi^n \tag{6.80}$$

Showing that this is a factorization of the three-dimensional version of Eq. (6.59) is tedious but not difficult. The two-dimensional version obtained by dropping Eq. (6.80) completely and δ_{zz} from Eq. (6.78) is essentially equivalent to the Peaceman–Rachford method. Other approximate factorizations are possible including, of course, the *LU* factorization mentioned above.

Example 6.6: ADI Solution of the Two-Dimensional Heat Equation As a test of the ADI method, we use the two-dimensional analog of the problem treated in Example 6.1. Specifically, we solve

$$\frac{\partial\phi}{\partial t} = \frac{\partial^2\phi}{\partial x^2} + \frac{\partial^2\phi}{\partial y^2} \tag{6.81}$$

with the initial condition

$$\phi(x,y,0) = 1 \tag{6.82}$$

and the boundary conditions

$$\phi(0,y,t) = \phi(1,y,t) = \phi(x,0,t) = \phi(x,1,t) = 0 \tag{6.83}$$

A few notes on the programs used in this example (ADIHV and ADILU, available from the Internet site described in Appendix A) are in order. Since the solution of a two-dimensional problem requires more data than does a one-dimensional problem, a number of parameters are set internally rather than being requested as input. To avoid using new data where they should not be used, it is necessary to store the results of odd-numbered sweeps in a different array (FF) than the results of even-numbered sweeps (F).

TABLE 6.4. ADI Solution of Two-Dimensional Heat Equation

Δt	β	$\phi(0.25,0.25)$	$\phi(0.5,0.25)$	$\phi(0.5,0.5)$
		Four Intervals in Each Direction		
0.0005	0.008	11.2195	15.8079	22.2728
0.0010	0.016	11.2193	15.8077	22.2726
0.0020	0.032	11.2190	15.8075	22.2727
0.0050	0.080	11.2144	15.8028	22.2684
0.0100	0.160	11.1988	15.7865	22.2536
0.0200	0.320	11.1379	15.7213	22.1908
0.0500	0.800	11.0196	15.2555	21.1196
		Eight Intervals in Each Direction		
0.0005	0.032	11.2587	15.9129	22.4911
0.0010	0.064	11.2582	15.9123	22.4904

In the first calculation, five grid points (including the boundaries) are used in each direction. Some of the results obtained at time $t = 0.1$ are shown in Table 6.4. The symmetry in this problem makes it necessary to look at the results at only three grid points. Although an exact solution is not available for this problem, we can accept the solution obtained using the smallest time step as an exact solution of the semidiscretized problem. We find that the error increases quadratically with the time step, as expected for a method that is second-order in time.

In order to assess the error due to spatial differencing, the calculation is repeated using half the interval size, that is, nine grid points in each direction. The results at the points used in the first calculation are given at the bottom of Table 6.4. The difference between $\phi(0.25,0.25)$ obtained in this calculation and the value obtained using the same time step in the first calculation is approximately three-fourths of the spatial error in the first calculation. The spatial error is thus approximately 5×10^{-4}, and the time step that produces the same error due to time differencing is a bit smaller than 0.02; the corresponding value of β is about 0.3. Using an argument similar to the one used in Section 6.3, we would predict that the errors would balance with β about twice as large as this. In view of the crudity of the argument, this is not bad, but a little conservatism (in using a smaller time step) is not out of place.

A problem associated with ADI or splitting methods is the construction of boundary conditions for the intermediate stages of the method. Since the intermediate results are created strictly for numerical reasons, they have no direct connection with the solution of the differential equation. When the boundary conditions are time independent, as in the above example, there is usually no problem. The exact boundary conditions can be used and excellent results are obtained.

In a time-dependent problem, however, the problem arises from the fact that the intermediate results (e.g., $\phi^{n+1/2}$ in the Peaceman–Rachford ADI method)

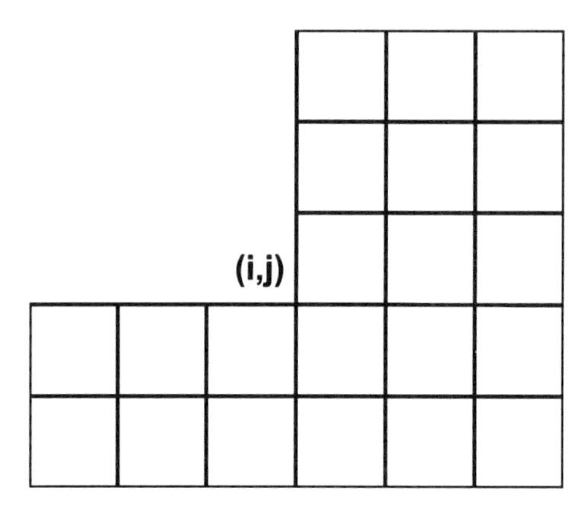

FIGURE 6.8. Domain with a concave corner.

are less accurate than the values at the end of a complete step (e.g., ϕ^n or ϕ^{n+1}). If accurate values of the boundary data are given, an extra error can be introduced by the fact that the boundary conditions and the approximations are "out of synch." (This situation is reminiscent of the problem with the leapfrog method; better results were obtained when the initial conditions were picked to match the method rather than the exact solution.)

To see how the boundary conditions ought to be chosen, we subtract Eq. (6.70) from Eq. (6.71) to get

$$\phi^{n+1/2} = \frac{1}{2}\left(1 + \frac{\alpha\,\Delta t}{2}\delta_{yy}\right)\phi^n + \frac{1}{2}\left(1 - \frac{\alpha\,\Delta t}{2}\delta_{yy}\right)\phi^{n+1} \tag{6.84}$$

Then, given the values of ϕ^n and ϕ^{n+1} on the boundary, the value of $\phi^{n+1/2}$, which are properly "synchronized" with the method, can be found.

A more serious problem occurs at a corner such as the one shown in Figure 6.8. At the corner point (i, j), computing the right-hand side of Eq. (6.84) requires the value of $\phi_{i-1,j}^{n+1}$, which would not normally be computed until after all of the values of $\phi^{n+1/2}$ have been computed. To get out of this difficulty, we can compute $\phi^{n+1/2}$ on the horizontal lines $1, 2, \ldots, j-1$. We then have the information needed to compute ϕ^{n+1} on vertical lines $1, 2, \ldots, i-1$. Specifically, this provides the value of $\phi_{i-1,j}^{n+1}$ and removes the difficulty, allowing the calculation to proceed. Obviously, more complex geometries require still more sophisticated logic to handle the boundary conditions accurately.

6.9. OTHER COORDINATE SYSTEMS

Frequently, we need solutions to problems in geometries that are not well suited to the use of Cartesian coordinates. Using another coordinate system might make the problem simpler. This is especially true if the boundaries are coordinate surfaces in the new system. When the geometry is irregular, the problem is much more difficult. The treatment of irregular geometries is taken up in the following chapter.

By far the most used coordinate system other than the Cartesian system is cylindrical coordinates. (Problems involving spherical geometry are less common but can be treated in a similar way.) In cylindrical geometry, if the solution is independent of the angular (or azimuthal) coordinate, the heat equation takes the form

$$\frac{\partial \phi}{\partial t} = \alpha \frac{1}{r} \frac{\partial}{\partial r}\left(r \frac{\partial \phi}{\partial r}\right) \tag{6.85}$$

where r is the radial coordinate. Typically, one is given an initial condition

$$\phi(r,0) = f(r) \tag{6.86}$$

The partial differential equation (6.85) is singular at $r = 0$ and has solutions that are not bounded as $r \to 0$. In such a case, the condition that that solution remain bounded as $r \to 0$ plays the role of a boundary condition. The most convenient way to enforce this condition is to require

$$\frac{\partial \phi}{\partial r}(0,t) = 0 \tag{6.87}$$

which is not a true boundary condition because it can be derived from the differential equation by taking the limit $r \to 0$.

Only one further boundary condition can be provided. It is usually given at some finite radius R that represents the physical boundary. The boundary condition might specify ϕ or its derivative $\partial\phi/\partial r$. The most general linear boundary condition is

$$\frac{\partial \phi(R,t)}{\partial r} + k\phi(R,t) = q \tag{6.88}$$

which contains the others as special cases.

To discretize the operator on the right-hand side of Eq. (6.85), we may proceed in two different ways. In the first, we note that

$$\frac{1}{r} \frac{\partial}{\partial r}\left(r \frac{\partial \phi}{\partial r}\right) = \frac{\partial^2 \phi}{\partial r^2} + \frac{1}{r} \frac{\partial \phi}{\partial r} \tag{6.89}$$

and then finite difference each term using formulas developed earlier. For example, we might use central difference approximations:

$$\left.\frac{\partial^2 \phi}{\partial r^2}\right|_{r_j, t_n} = \frac{\phi_{j+1}^n - 2\phi_j^n + \phi_{j-1}^n}{\Delta r^2} \tag{6.90}$$

$$\frac{1}{r}\frac{\partial \phi}{\partial r} = \frac{\phi_{j+1}^n - \phi_{j-1}^n}{2r_j \Delta r} \tag{6.91}$$

where a uniform grid has been assumed in order to keep the notation from becoming too cumbersome. These formulas are second-order accurate and are satisfactory, but this method is usually not as good as the one given below.

In the second approach, we first approximate the first derivatives at the midpoints of the intervals by using finite differences; central difference approximations are preferred and give

$$\left. r\frac{\partial \phi}{\partial r}\right|_{r_{j+1/2}, t_n} = r_{j+1/2}\frac{\phi_{j+1}^n - \phi_j^n}{\Delta r} \tag{6.92}$$

$$\left. r\frac{\partial \phi}{\partial r}\right|_{r_{j-1/2}, t_n} = r_{j-1/2}\frac{\phi_j^n - \phi_{j-1}^n}{\Delta r} \tag{6.93}$$

These formulas are second-order accurate because they use centered differences. Then

$$\begin{aligned}\left.\frac{1}{r}\frac{\partial}{\partial r}\left(r\frac{\partial \phi}{\partial r}\right)\right|_{r_j, t_n} &= \frac{1}{r_j \Delta r}\left[\left(r\frac{\partial \phi}{\partial r}\right)_{r_{j+1/2}, t_n} - \left(r\frac{\partial \phi}{\partial r}\right)_{r_{j-1/2}, t_n}\right] \\ &= \frac{1}{r_j \Delta r^2}(r_{j+1/2}\phi_{j+1}^n - 2r_j\phi_j^n + r_{j-1/2}\phi_{j-1}^n)\end{aligned} \tag{6.94}$$

The boundary condition (6.87) at $r = 0$ can be satisfied by introducing an artificial point $r_{-1} = -\Delta r$ and letting $\phi_{-1} = \phi_1$. At $r = 0$ the PDE then reduces to

$$\frac{\partial \phi_0}{\partial t} = \frac{2}{\Delta r^2}(\phi_1 - \phi_0) \tag{6.95}$$

The boundary condition (6.88) at the outer boundary is treated by introducing an artificial or phantom point at $r_{N+1} = R + \Delta r = (N+1)\Delta r$. Then

$$\frac{\phi_{N+1}^n - \phi_{N-1}^n}{2\Delta r} + k\phi_N^n = q \tag{6.96}$$

represents the boundary condition. This equation is then solved together with the $N + 1$ equations that represent the approximations to the PDE at the points

$r_0, r_1, \ldots r_N$. We thus have $N+2$ equations containing the same number of unknowns.

The preference for the difference approximation (6.94) over Eqs. (6.90) and (6.91) is a consequence of a conservation property of the differential equation. If we multiply Eq. (6.85) by r and integrate from 0 to R we find

$$\frac{d}{dt}\int_0^R r\phi(r,t)\,dr = \alpha r\frac{\partial \phi}{\partial r}\bigg|_R \tag{6.97}$$

Physically, this is a global conservation statement for the entire cylinder. In the heat transfer context, it states that the rate of change of thermal energy contained in the cylinder is equal to the energy flux into the cylinder. It is desirable (and in some cases crucial) that the numerical approximation retain this property. The numerical analog of this property is obtained by multiplying the finite difference equation for $\phi(r_j,t)$ by r_j and summing over j. When this is done, we find that the first method does not produce anything simple, but the second one gives

$$\frac{d}{dt}\sum_{j=1}^{N} r_j\phi_j = \frac{\alpha}{\Delta r^2}\sum_{j=1}^{N}(r_{j+1/2}\phi_{j+1} - 2r_j\phi_j + r_{j-1/2}\phi_{j-1}) \tag{6.98}$$

Most of the terms in the summation drop out, and we have

$$\begin{aligned}\frac{d}{dt}\sum_{j=1}^{N} r_j\phi_j\,\Delta r &= \alpha\frac{r_{N+1/2}\phi_{N+1} - (r_{N-1/2} - 2r_N)\phi_N}{\Delta r}\\ &= \alpha r_{N+1/2}\frac{\phi_{N+1} - \phi_N}{\Delta r}\end{aligned} \tag{6.99}$$

which is a numerical approximation to the conservation property (6.97). Note, however, that if $k = 0$ in the boundary condition (6.88) or (6.96), which means that the heat flux is specified, then the right-hand side of Eq. (6.99) is different from the derivative approximation used in Eq. (6.96). In this case, the rate of accumulation of energy in the domain will equal the heat flux only to within an error of order Δr.

There are, of course, still other coordinate systems. We cannot cover them all here for reasons of space, but many of these methods are similar to one used in the cylindrical case.

If we have a problem in cylindrical geometry that is not azimuthally symmetric, it can be handled by the methods already presented. The radial derivative should be treated in the manner just suggested, and the azimuthal (θ) derivative can be treated by central differences in a straightforward way. The major difference is that, due to the geometry, the θ coordinate is periodic. As a result, there are two elements in the matrix for the equations in cylindrical geometry that are not present in Cartesian geometry. The ADI method can be applied to the solution of these equations; the major difference is that the equations on the azimuthal lines have what is called periodic tridiagonal structure rather than the standard tridiagonal structure found in Cartesian coordinates. This requires the use of a modified solver.

The important question of handling truly complex geometries is a very important issue that will be discussed in the next chapter. The methods described there can be applied to parabolic equations as well.

6.10. NONLINEAR PROBLEMS

So far, we have dealt only with linear problems, but many problems arising in applications are in fact nonlinear. Nonlinearity is a difficult subject to treat with any generality because there is not just one kind of nonlinearity. We will therefore treat a particular example and present some ideas that can be applied to other nonlinear problems. The particular case we will deal with is a simplification of a problem that arises in fluid mechanics; in the boundary layer equations two of the most important terms are contained in the equation:

$$\frac{\partial u^2}{\partial x} = \nu \frac{\partial^2 u}{\partial y^2} \tag{6.100}$$

where ν is a viscosity.

The fact that the nonlinearity of this equation is quadratic can be used to advantage in devising methods for its solution. The boundary conditions might require that the velocity u be given at two locations in the y coordinate; for example,

$$u(x,0) = 0 \qquad u(x,h) = U \tag{6.101}$$

The initial condition is given at some location upstream, for example:

$$u(0,y) = f(y) \qquad 0 < y < h \tag{6.102}$$

Equation (6.100) can be discretized using the Crank–Nicolson method to give

$$u_{i+1,j}^2 - u_{i,j}^2 = \frac{\nu \Delta x}{\Delta y^2}[(u_{i,j+1} - 2u_{i,j} + u_{i,j-1}) + (u_{i+1,j+1} - 2u_{i+1,j} + u_{i+1,j-1})]$$

$$j = 1,2,\ldots,N-1 \tag{6.103}$$

where the boundary conditions may be used to provide $u_{i+1,0}$ and $u_{i+1,N}$. The solution can be obtained by marching in x in the usual way, so we require only a method for advancing the solution a single step in x.

If the values of ν at $x_i = i\,\Delta x$ are assumed known, Eqs. (6.103) for the solution on the line $x = x_{i+1}$ are a tridiagonal set of nonlinear algebraic equations. The solution of the system of equations (6.103) can be carried out quite efficiently by doing a few iterations of the Newton–Raphson method for systems of equations; at each iteration, a tridiagonal linear system must be solved, which this can be done with the standard tridiagonal technique for linear equations. If the step size Δx is not too large, the solution at the preceding step, $x_i = i\,\Delta x$, provides a good initial guess at the solution at the new step x_{i+1}; this helps the convergence considerably. Note that there is no point in iterating the solution to Eq. (6.103) to complete convergence; the result obtained if this were done will still contain an error due to the truncation error introduced when Eq. (6.100) was discretized to Eq. (6.103).

An alternative method is applicable to this particular case (due to the quadratic nature of the nonlinearity). We can linearize the equations directly. That is, we write

$$u_{i+1,j} = u_{i,j} + (u_{i+1,j} - u_{i,j}) = u_{i,j} + \Delta u_{i+1,j} \tag{6.104}$$

which defines $\Delta u_{i+1,j}$. Equation (6.103) then becomes

$$2u_{i,j}\,\Delta u_{i+1,j} = \frac{\nu\,\Delta x}{\Delta y^2}[2(u_{i,j+1} - 2u_{i,j} + u_{i,j-1}) + (\Delta u_{i+1,j+1} - 2\,\Delta u_{i+1,j} + \Delta u_{i+1,j-1})] \tag{6.105}$$

The nonlinear term $(\Delta u_{i+1,j})^2 \approx \Delta x^2 (\partial u/\partial x)^2_{i+1,j}$, which has been neglected in writing this equation, is of the same order as the finite difference truncation error in Eq. (6.103). Neglecting this term increases the truncation error but, as the extra error is of the same order (Δx^2) as the original truncation error, the difference is not large. The sacrifice in accuracy (or the increase in the number of steps needed to compensate for it) is well worth the advantage obtained by replacing a nonlinear system of equations by a linear one. Equations (6.105) can be solved with the standard tridiagonal solution method.

This is known as the Δ-form of the method and can be applied to linear as well as nonlinear problems. It is essentially equivalent to the first iteration of an iterative solution method. What we have shown is that, for equations with quadratic nonlinearity, second-order accuracy can be achieved without iteration.

In the linear case the choice of whether to use the standard or Δ-form of the method is mostly a matter of taste.

6.11. FINAL REMARKS—OTHER METHODS

There are many other methods that can be applied to parabolic partial differential equations. Among the most valuable of these are adaptive grid and multigrid methods. These methods can be very useful but, as we noted earlier, parabolic equations are the easiest type of partial differential equations to solve numerically, and rarely does the solution of one of them require much computer time. These methods can be useful in programs that require solution of many sets of parabolic equations; one example of such an application is the so-called viscous interaction method that is often used in aerodynamic design.

These methods will not be covered here. They are presented in forms applicable to elliptic partial differential equations in the next chapter. We simply note that these methods can indeed be applied to parabolic equations as well.

PROBLEMS

1. Consider the application of the leapfrog method to the solution of the heat equation.
   a. Show that the leapfrog method in time applied to the heat equation is unstable.
   b. Compute the solution to the heat equation

$$\frac{\partial \phi}{\partial t} = \alpha \frac{\partial^2 \phi}{\partial x^2}$$

   up to $t = 10$ with the initial condition $\phi(x,0) = 1$, and boundary conditions $\phi(0,t) = 0$ and $\phi(1,t) = 0$ using the leapfrog method with just one interior point. Use the Euler method as a starter and let $\beta = \alpha \Delta t / \Delta x^2 = 1$. Does reducing β to 0.5 improve the result?

2. In many systems of interest, the material properties are functions of spatial position. The heat of diffusion equation used as an example in this chapter is then replaced by

$$\frac{\partial \phi}{\partial t} = \frac{\partial}{\partial x}\left(\alpha(x)\frac{\partial \phi}{\partial x}\right)$$

   This equation has the property that

$$\frac{d}{dt}\int_0^1 \phi(x,t)\,dx = \alpha(1)\frac{\partial \phi}{\partial x}(1,t) - \alpha(0)\frac{\partial \phi}{\partial x}(0,t)$$

   as can be shown by integration. It states that the amount of property ϕ in the region of interest changes only due to flow through the boundaries.

a. Derive a finite difference version of this PDE.

b. Determine whether your method has the finite difference analog of the conservation property just given. If it does not, find one that does.

3. Solve the heat equation with $\alpha = 0.01\ \text{cm}^2/\text{s}$ subject to the initial condition

$$\phi(x,0) = 100x(1-x)$$

and the boundary conditions $\phi(0,t) = \phi(1,t) = 0$.

a. Find the temperature distribution at $t = 1$ s using four spatial intervals ($\Delta x = 0.25$ cm). Use any numerical method you like, but read part (b) before proceeding.

b. Estimate (in any way you choose) the accuracy of the solution obtained in part (a). Present your results as a table.

4. The following is a method of solving the heat equation:

$$\phi^{n+1*} - \phi^n = \alpha\,\Delta t \frac{\delta^2 \phi^{n+1*}}{\delta x^2} + \alpha^2\,\Delta t^2 \frac{\delta^2}{\delta x^2}\frac{\delta^2}{\delta y^2}\phi^n$$

$$\phi^{n+1} - \phi^{n+1*} = \alpha\,\Delta t \frac{\delta^2 \phi^{n+1}}{\delta y^2}$$

where $\delta^2/\delta x^2$ and $\delta^2/\delta y^2$ represent the second-order central finite difference approximations to $\partial^2/\partial x^2$ and $\partial^2/\partial y^2$.

a. Eliminate ϕ^{n+1*} from these equations to get an equation for ϕ^{n+1} directly in terms of ϕ^n.

b. Of what method is this an approximate factorization?

c. What can you say its accuracy and stability?

5. The equation

$$\frac{\partial \phi}{\partial t} + c\frac{\partial \phi}{\partial x} = \alpha \frac{\partial^2 \phi}{\partial x^2}$$

is called the convection–diffusion equation; with c and α both constant, this represents a very simple model of the dispersion of pollution in the atmosphere or the ocean; it is also used as a model for fluid mechanics.

a. Classify this equation (elliptic, parabolic, hyperbolic) and give a set of simple boundary conditions that could be applied to it.

b. Suggest a numerical method for solving this problem. Be specific as to difference method, treatment of boundary conditions, spatial mesh size, and time step. Give your rationale for the choices made.

We will meet this equation again in Chapter 8.

CHAPTER 7

PARTIAL DIFFERENTIAL EQUATIONS: II. ELLIPTIC EQUATIONS

In Chapter 6 we saw that elliptic partial differential equations (PDEs) possess no real characteristics. This means that there are no preferred directions of information propagation—a source of disturbance anywhere in the domain influences the solution everywhere. Hence there are no timelike coordinates or directions. These properties make elliptic equations much more difficult to solve than parabolic ones. There is no obvious place to start the solution procedure; one has to solve for the dependent variable at every point in the domain simultaneously. Furthermore, the matrices that result from finite difference approximations to elliptic partial differential equations do not have a structure that allows easy solution. As a result, with a few exceptions, elliptic equations are solved by iterative methods.

Elliptic equations often arise from parabolic or hyperbolic PDEs in which the time dependence has an assumed form. Steady-state situations provide another source of elliptic problems; these are often the stationary limit of problems whose unsteady behavior is governed by a hyperbolic or parabolic equation. The simplest and most common elliptic PDE is Laplace's equation, which, in three dimensions, is

$$\nabla^2 \phi = \frac{\partial^2 \phi}{\partial x_1^2} + \frac{\partial^2 \phi}{\partial x_2^2} + \frac{\partial^2 \phi}{\partial x_3^2} = 0 \tag{7.1}$$

This equation arises in electrostatics, heat conduction, and solid and fluid mechanics, among other fields. In two dimensions, the last term is absent.

Other well-known elliptic equations are Poisson's and Helmholtz's equations. The former is the inhomogeneous form of Laplace's equation; it rep-

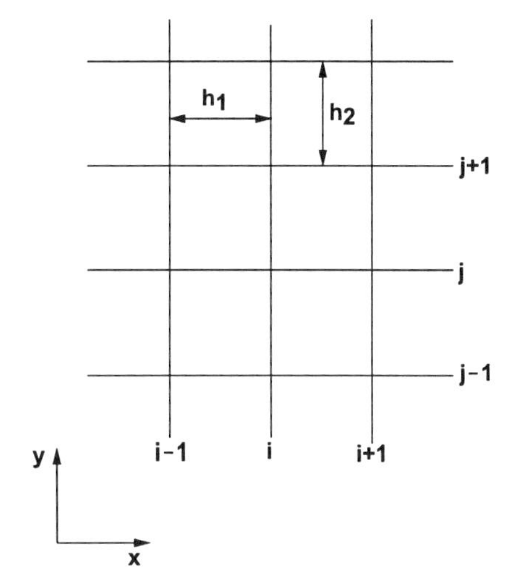

FIGURE 7.1. Rectangular grid on which Laplace's equation is discretized.

resents systems in which sources (electrical charges, heat sources, or fluid sources) are found in the domain. Helmholtz's equation is Laplace's equation with a term proportional to ϕ included. It arises most commonly when one seeks sinusoidal solutions to the wave equation but may also occur in problems in which there is a mechanism for destroying or creating ϕ within the domain. There are other important elliptic equations but, for our purposes, the differences are less important than the similarities, and numerical methods that work for Laplace's equation can be applied to the other equations, although there may be some difficulties. Nonlinear elliptic equations are also common, arising from problems in which material properties depend on the solution (temperature-dependent thermal conductivity, e.g.) and in fluid mechanics.

In this chapter, we shall concentrate on the solution of Laplace's equation and will discuss how the methods can be applied to the other equations.

7.1. DISCRETIZATION

7.1.1. Finite Differences

The simplest numerical approximation to Laplace's equation is what one might expect from the preceding chapters. For the present, we assume that the region in which the solution is desired is rectangular, that the boundaries are grid lines, and that a uniform grid is used. Under these circumstances, one can lay a uniform rectangular mesh over the region on which the solution is desired (see Figure 7.1). (The more general case is discussed later.) The most obvious

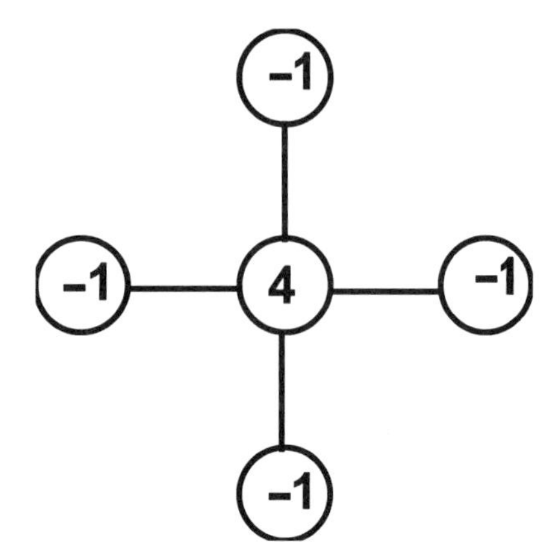

FIGURE 7.2. Computational molecule for the five-point approximation to Laplace's equation.

way to discretize the equation is to approximate the derivatives in Laplace's equation by the second-order central finite difference approximation that has been used frequently in the preceding two chapters:

$$\left.\frac{\partial^2 \phi}{\partial x_1^2}\right|_{i,j} \approx \frac{\phi_{i+1,j} - 2\phi_{i,j} + \phi_{i-1,j}}{h_1^2} \tag{7.2}$$

$$\left.\frac{\partial^2 \phi}{\partial x_2^2}\right|_{i,j} \approx \frac{\phi_{i,j+1} - 2\phi_{i,j} + \phi_{i,j-1}}{h_2^2} \tag{7.3}$$

These are substituted into Eq. (7.1). For convenience, and because the general case differs mainly by being more cumbersome, we make the further simplifying assumption that $h_1 = h_2$. The resulting finite difference form of Laplace's equation resulting at any interior point of the mesh is

$$\phi_{i,j} - \tfrac{1}{4}(\phi_{i+1,j} + \phi_{i-1,j} + \phi_{i,j+1} + \phi_{i,j-1}) = 0 \tag{7.4}$$

This equation states that the solution at any point is the average of the values at the four nearest neighboring points. The computational molecule for this method is shown in Figure 7.2. Values of the coefficients are also shown in the figure as this is a convenient way to display finite difference schemes for elliptic equations. This approximation is so commonly used that it is known as the *five-point difference operator.*

An alternate way of representing the discretized equations is via a kind of geographic notation. In place of the grid indices used above, we call the central node P and the neighbors W, N, E, and S, respectively. In this notation, Eq.

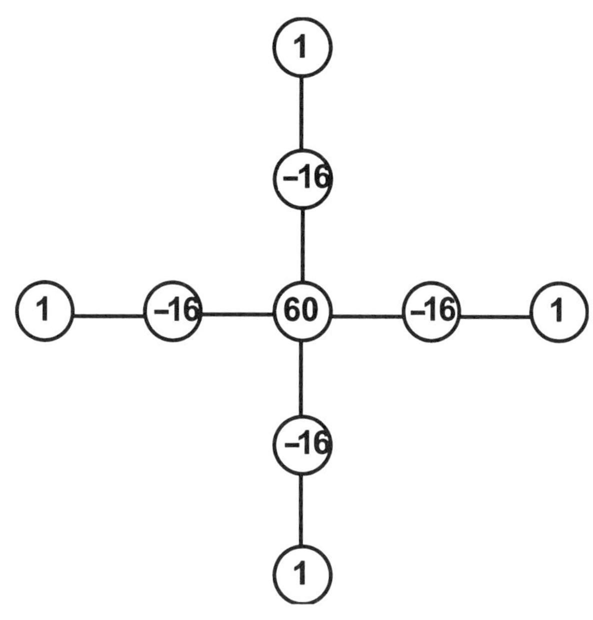

FIGURE 7.3. Computational molecule for the standard fourth-order approximation to Laplace's equation.

(7.4) becomes

$$\phi_P - \tfrac{1}{4}(\phi_E + \phi_W + \phi_N + \phi_S) = 0 \tag{7.5}$$

or, for a different equation for which the coefficients are not all equal,

$$A_P\phi_P - (A_E\phi_E + A_W\phi_W + A_N\phi_N + A_S\phi_S) = 0 \tag{7.6}$$

In three dimensions, two additional points must be included in the formula. They are usually labeled T and B for "top" and "bottom." Other nearby points can be labeled using similar notation such as EE for the point $(i+2, j)$ and SW for the point at $(i-1, j-1)$, and so on.

Of course, this notation can only be used on regular (also called structured) grids, but it will prove handy in some of the work to follow.

Equation (7.4) is not the only way in which Laplace's equation can be approximated using finite differences, but it is the most common choice. Higher accuracy can be obtained by using either a smaller grid size or a higher-order difference approximation. The latter is a better choice in terms of cost for a given accuracy, but it is difficult to apply near boundaries.

Fourth-order accuracy can be obtained via use of the fourth-order estimate of the second derivative given in Table 4.1. For Laplace's equation this leads to the difference scheme represented by Figure 7.3. The problems this method

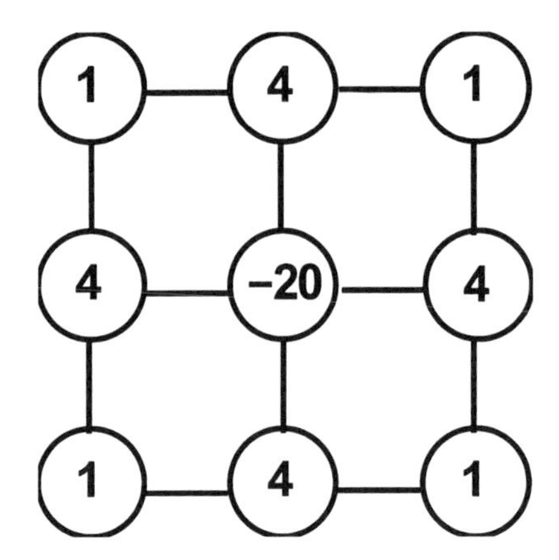

FIGURE 7.4. Computational molecule for the compact fourth-order approximation to Laplace's equation.

has near boundaries has led to its fall from favor. An alternative is the method represented by Figure 7.4, which is obtained by applying the compact fourth-order method described in Section 5.4. As the molecule shows, the result is a nine-point operator but it has a compact structure. Its application to the Poisson or Helmholtz equations (or other elliptic equations) requires extra care.

In recent years use of Fourier transforms for the estimation of derivatives and for the solution of elliptic PDEs has gained considerable popularity (see Sections 7.14 and 7.15). These methods are extremely powerful for problems in which the boundaries are rectangular or nearly so.

7.1.2. Finite Volume Approximations

It is possible to derive approximations to Laplace's equation by integrating the equation over a small but finite areas. These become volumes in three dimensions, giving the method its name, the *finite volume* method. Finite volume methods have become a very popular method of deriving discretizations of partial differential equations because they preserve the conservation properties of the differential equation better than finite difference methods.

To illustrate this approach, we assume that the values of the dependent variable are to be computed at the centers of small areas such as the one shown in Figure 7.5. The derivation is simpler if Laplace's equation (7.1) is written in the form:

$$\nabla \cdot \nabla \phi = 0 \tag{7.7}$$

This version of the equation has the interpretation that the divergence of the flux of ϕ is everywhere zero. The flux is assumed proportional to the gradient of ϕ through Fourier's law of heat conduction or Fick's law of diffusion.

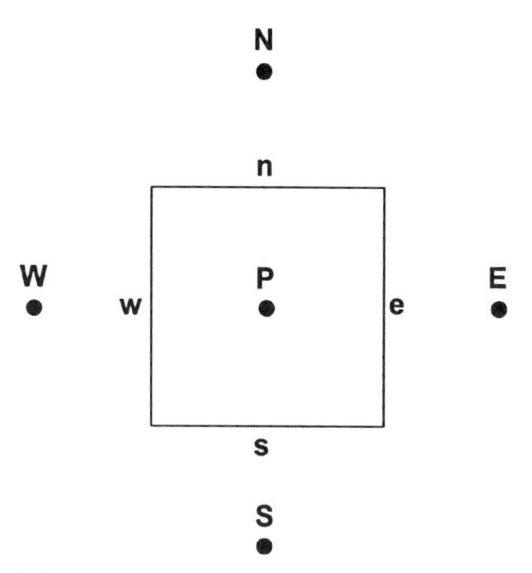

FIGURE 7.5. Typical finite volume. The values of the dependent variable are defined at the centers of the volumes, which are marked in capital letters using geographic notation. The fluxes are defined at the centers of the faces, which are marked with lower case letters.

When Equation (7.7) is integrated over the control volume, Gauss's theorem may be used to write the result as:

$$\int_S \nabla\phi \cdot \mathbf{n}\, dS \tag{7.8}$$

where S is the surface of the control volume. This equation states that the net flux of ϕ into the control volume is zero. For the rectangular control volume shown in Figure 7.5, the integral can be broken into the sum of four integrals, one over each face. We consider the integral over just one face in detail. The integral over the right (or east) face is

$$\int_{x_{2,j-1/2}}^{x_{2,j+1/2}} \frac{\partial\phi}{\partial x_1}(x_{1,i+1/2}, x_2)\, dx_2 \tag{7.9}$$

and must be approximated in terms of the values of ϕ at the control volume centers in order to close the equations. If the integral with respect to x_2 is approximated by the midpoint rule and the derivative with respect to x_1 is approximated by the usual central difference formula, we find

$$\begin{aligned} \int_{x_{2,j-1/2}}^{x_{2,j+1/2}} \frac{\partial\phi}{\partial x_1}(x_{1,i+1/2}, x_2)\, dx_2 &\approx (x_{2,j+1/2} - x_{2,j-1/2}) \frac{\partial\phi}{\partial x_1}(x_{1,i+1/2}, x_{2,j}) \\ &= (x_{2,j+1/2} - x_{2,j-1/2}) \frac{\phi(x_{1,i+1}, x_{2,j}) - \phi(x_{1,i}, x_{2,j})}{x_{1,i+1} - x_{1,i}} \end{aligned} \tag{7.10}$$

Applying similar approximations to the other three faces of the control volume and assuming that the grid size is uniform and the same in both directions, we recover the approximation (7.4).

The approximations used in this derivation are the simplest ones; others could be used. The finite volume method has the following advantage. Laplace's equation is a statement of conservation of the quantity ϕ. Integration over a control volume produces a conservation statement for the finite volume; the net flux through its surface must be zero. So long as the fluxes through a cell face are evaluated in the same way for each of the two cells that share the face, then, when equations for the control volumes are summed over all of the control volumes in the domain, the internal fluxes will cancel out and we will have a global conservation statement. Thus, global conservation is built into the method. Note that conservation and accuracy are separate issues.

An alternative is to write Laplace's equation as a system of three first-order equations before integrating, which gives an elliptic version of the Keller box method. Another important alternative is the *finite element* method, which has gained considerable popularity in recent years because it has the ability to handle oddly shaped regions with relatively little difficulty. This method is described briefly in Section 7.13.

7.1.3. Boundary Conditions

Next, we consider the question of boundary conditions. Appropriate boundary conditions for second-order elliptic problems provide one datum at every point on the boundary. This may be the value of the function, the derivative of the function in the direction normal to the boundary, or a linear combination of the two. For the second-order approximation [Eq. (7.4)], the boundary conditions may be treated in the same way they were in Chapter 5. When the value of the function is given at a boundary point, no difference approximation is required at that point, and the given value is used wherever it occurs in the equation(s) at the neighboring point(s). When the derivative or a combination of derivative and function is given, one introduces a fictitious (phantom) point outside the boundary and uses an equation analogous to Eq. (5.10) to represent the boundary condition; the finite difference equation at the boundary point becomes part of the system of equations to be solved. Not all of the boundary values can be zero (if they were, the solution would be zero everywhere) so at least some of the discretized equations contain inhomogeneous terms.

The method represented by Figure 7.3 has difficulty with boundary conditions. The problem was encountered in Section 5.3 and can be dealt with by the methods introduced there. The best method, as was found there, is to use a third-order formula at the point next to the boundary. This reduces the overall accuracy somewhat but produces reliable results. This method is rarely used.

7.1.4. System of Equations

The result of the discretization process applied to Laplace's equation is a large set of linear algebraic equations. It is natural to put the problem into a standard linear algebra context. We first note that, although the dependent variables are doubly subscripted, they should be regarded as components of a vector. As we saw in the preceding chapter, there are two natural ways of entering the unknowns $\phi_{i,j}$ as the elements of a vector—we can list the elements in x_1 = constant lines sequentially or we can do the same for x_2 = constant lines. The choice is arbitrary, but if the number of points in the 1 direction, M, is smaller than the number of points in the 2 direction, N, it is preferable to use the ordering

$$\phi = \begin{pmatrix} \phi_{1,1} \\ \phi_{2,1} \\ \vdots \\ \phi_{M,1} \\ \phi_{1,2} \\ \vdots \\ \phi_{M,2} \\ \vdots \\ \phi_{M,N} \end{pmatrix} \tag{7.11}$$

This is known as lexicographical (dictionary) ordering. The matrix corresponding to the system of equations of Eq. (7.4) is then

$$A = \begin{pmatrix} -4 & 1 & 0 & 0 & \cdots & \cdots & 1 & 0 & \cdots & 0 \\ 1 & -4 & 1 & 0 & \cdots & \cdots & 0 & 1 & \cdots & 0 \\ 0 & 1 & -4 & 1 & & & & & & 0 \\ \vdots & \vdots & \vdots & \ddots & & & & & & \vdots \\ 1 & 0 & 0 & 0 & & & & & & 0 \\ 0 & 1 & 0 & 0 & & & & & & 0 \\ \vdots & \vdots & \ddots & \vdots & & & & & & \vdots \\ \vdots & \vdots & \vdots & \ddots & & & & & & \vdots \\ & & & & 1 & & 1 & -4 & & 1 \\ & & & & & 1 & 0 & 1 & & -4 \end{pmatrix} \tag{7.12}$$

The lowest 1 in the first column occurs in the $(M+1)$st row and the last 1 in the first row is in the $(M+1)$st column. The set of equations to be solved can be written

$$A\phi = \mathbf{b} \tag{7.13}$$

where $\mathbf{b}$ is a vector, most of whose elements are zero; the nonzero elements arise from the boundary conditions.

In terms of the geographical notation introduced earlier, the matrix becomes

$$A = \begin{pmatrix} A_P & A_E & 0 & 0 & \ldots & \ldots & A_W & 0 & \ldots & 0 \\ A_W & A_P & A_E & 0 & \ldots & \ldots & 0 & A_W & \ldots & 0 \\ 0 & A_W & A_P & A_E & & & & & & 0 \\ \vdots & \vdots & \vdots & \ddots & & & & & & \vdots \\ A_S & 0 & 0 & 0 & & & & & & 0 \\ 0 & A_S & 0 & 0 & & & & & & 0 \\ \vdots & \vdots & \ddots & \vdots & & & & & & \vdots \\ \vdots & \vdots & \vdots & \ddots & & & & & & \vdots \\ & & & & A_S & & & A_W & A_P & A_E \\ & & & & & A_S & & 0 & A_W & A_P \end{pmatrix} \tag{7.14}$$

The method of solving the system of equations (7.13) depends to a large extent on the properties of the matrix A. The size of this matrix is $MN \times MN$, which means that it is quite large even for fairly small values of M and N. However, it is very sparse (most of its elements are zero) and an efficient solution method will take advantage of this fact. In fact it is more than sparse; it is banded, meaning that all of its nonzero elements lie on a few particular diagonals that are not too distant from the principal diagonal.

This allows use of simplified storage of the matrix in computer memory; we can store the five nonzero diagonals as arrays of size $1 \times MN$. As noted in Chapter 1, Gauss elimination methods designed for the solution of systems involving banded matrices require a number of arithmetic operations proportional to the product of the size of the matrix (MN in the present case) and the bandwidth (M) squared. Thus to solve this system would require NM^3 operations or M^2 operations per point. This is quite a bit and it seems natural, therefore, to seek other more efficient methods. In particular, except for problems in very regular geometries and simple boundary conditions, the best way to the solve this type of problem is to use iterative methods, to which we shall devote most of this chapter.

7.1.5. Complex Geometry

We end this section by noting that many problems in engineering practice require the solution of equations in domains that have irregular boundaries, and it is necessary to develop methods for solving them. The treatment of complex geometry is one of the most difficult problems in computational engineering. There are several methods available for dealing with this issue; all of them have disadvantages severe enough that there is room for further improvement.

One approach uses a coordinate transformation to convert the solution domain into a rectangle. One can then easily create a regular or structured grid in the transformed coordinate system. This is a good approach for geometries that are not too complex, but it is almost impossible to apply in multiply-connected domains. The other difficulty is that the transformation introduces geometric coefficients (Jacobians) that may be difficult to compute accurately.

Another alternative is to simply use a Cartesian grid with special treatment of the regions near the boundaries. We will say more about both of these methods later.

The finite element method, which was described briefly for one-dimensional problems in Chapter 4, allows considerable flexibility in the choice of grids and can be used. This has made it the method of choice for many problems, especially in the field of solid mechanics, which it dominates. Finite elements are discussed in Section 7.13.

Before considering those methods, we look at methods of solving the systems of equations generated by the discretization process in simple geometries.

7.2. INTRODUCTION TO ITERATIVE METHODS AND THEIR PROPERTIES

7.2.1. Construction of Iterative Methods

In the preceding section we saw that direct solution of the finite difference equations for elliptical PDEs is expensive. An exception is the case in which the boundaries are rectangular and the boundary conditions are of simple type; one can then use fast methods (see Section 7.14) to obtain the solution quickly. When direct solution is too expensive, the only reasonable alternative is to use an iterative method, several of which are described in this chapter. We begin with a short general discussion of iterative methods and their properties.

Iterative methods can be applied to the solution of any set of linear or nonlinear algebraic equations; indeed, for nonlinear equations, they are the only choice (with only a few exceptions). In discussing these methods, it is sometimes easier to use the matrix form of the equations. At other times, the component form of the equations is more convenient. We will use both ways of writing the equations. An iterative method applied to Eq. (7.13) is a procedure of the type

$$M\phi^{(p+1)} = N\phi^{(p)} + \mathbf{d} \tag{7.15}$$

in which an improved estimate of the solution $\phi^{(p+1)}$ is computed from an existing one, $\phi^{(p)}$. Here M and N are matrices, $\mathbf{d}$ is a vector and p is an index denoting the iteration number. To use this method, some initial guess at the solution, $\phi^{(0)}$, is constructed, and Eq. (7.15) is used to compute $\phi^{(1)}$. Then we use $\phi^{(1)}$ to compute $\phi^{(2)}$ and so on. If this is to be a satisfactory means of solving the original equation, this scheme should have the following properties:

- It should converge to the exact solution of the original equation. That is, we should have

$$\lim_{p\to\infty} \phi^{(p)} = \phi \tag{7.16}$$

  where ϕ is the solution of Eq. (7.13). This requires

$$A = P(M - N) \qquad \mathbf{d} = P\mathbf{b} \tag{7.17}$$

  where P is called a preconditioning matrix whose only purpose is to accelerate the convergence of the method. (It may be the identity matrix.)
- The convergence should be rapid, that is, the limit [Eq. (7.16)] should be reached quickly.
- So that each step does not require much computation, the matrix M should be easy to invert. In an ideal case, M is diagonal; other good ones have tridiagonal or triangular M. The matrix N should have simple structure in order to facilitate computation of $N\phi^{(p)}$. Generally, this means that N should be sparse, which is possible only if A is also sparse. Thus iterative methods are best for sparse matrices. Fortunately, matrices that represent discretizations of partial differential equations are always very sparse.

An alternative form of the generic iterative method just presented is sometimes useful. If we subtract $M\phi^{(p)}$ from both sides of Eq. (7.15), we obtain

$$M(\phi^{(p+1)} - \phi^{(p)}) = (N - M)\phi^{(p)} + \mathbf{d} = \mathbf{d} - A\phi^{(p)} \tag{7.18}$$

Now we define

$$\rho^{(p)} = \mathbf{d} - A\phi^{(p)} \tag{7.19}$$

as the residual at the pth iteration and

$$\delta^{(p+1)} = \phi^{(p+1)} - \phi^{(p)} \tag{7.20}$$

as the update for the $(p+1)$st iteration. The alternative method then consists of the following steps. First, given the solution at the pth iteration, $\phi^{(p)}$, we compute the residual from Eq. (7.19). Then we solve Eq. (7.18) for the update. Finally, we update the solution using

$$\phi^{(p+1)} = \phi^{(p)} + \delta^{(p+1)} \tag{7.21}$$

This is a more complicated version of the method, but it will prove useful.

It is not difficult to construct iterative methods, but constructing one with all of the desired properties is another story. In this chapter, we shall begin with the simplest iterative scheme and systematically improve upon it.

7.2.2. Errors in Iterative Methods

Another important issue is the evaluation and control of the errors in the solution. In the numerical solution of any partial differential equation, three sources of error need to be distinguished:

- **Geometry Error.** The method of discretizing the domain may introduce errors. These errors may be due to not fitting the boundary exactly or to inexact evaluation of coefficients introduced by a transformation. This type of error is not present in the rectangular domains used in most of this chapter. Some people prefer to group it with the discretization error discussed next.
- **Discretization Error.** As we have seen, the discretization process, which reduces the partial differential equation to a system of algebraic equations, is based on the use of approximations that introduce error. This error is usually proportional to some power of the grid size. Even if the algebraic equations are solved exactly, the solution will contains an error from this source. The discretization error may be defined as the difference between the *exact* solution of the partial differential equation and the *exact* solution of the discretized equations.
- **Convergence Error.** When an iterative method is used to solve the discretized equations, we must use some criterion for deciding to stop the iteration process; the choice of such a criterion is discussed below. The difference between the solution achieved by the iterative method and the *exact solution of the discretized equations* is the convergence error. It should be considerably smaller than the discretization error; otherwise, it is impossible to evaluate the latter.

If the convergence error has been made smaller than the discretization error, it is possible to evaluate the latter by solving the problem on two grids of different sizes and computing the difference, that is, using the Richardson method for evaluating the error. This requires that the order of accuracy of the method be known and that the grid sizes be small enough to allow the leading order error term to dominate. If the Richardson method is applied on grids that are too large to allow the leading order error term to dominate, the error estimate obtained will not be very accurate, but both the estimate and the actual error will be quite large and this is sufficient to indicate that a finer grid is required.

7.2.3. Convergence Error

The standard method for investigating the convergence of iterative methods begins by rewriting Eq. (7.15) in the form:

$$\phi^{(p+1)} = M^{-1}(N\phi^{(p)} + \mathbf{d}) \tag{7.22}$$

Next, we introduce an error vector (which represents the convergence error) defined as the difference between the pth iterate produced by the iterative method and the exact solution of the discretized system of equations:

$$\epsilon^{(p)} = \phi^{(p)} - \phi \tag{7.23}$$

If the method converges, then for large p, $\phi^{(p)} = \phi^{(p+1)} = \phi$ so that

$$\phi = M^{-1}(N\phi + \mathbf{d}) = B\phi + \mathbf{g} \tag{7.24}$$

where $B = M^{-1}N$ is called the iteration matrix and $\mathbf{g} = M^{-1}\mathbf{d}$.

Subtracting this equation from Eq. (7.22), we get a homogeneous equation for the error vector

$$\epsilon^{(p+1)} = B\epsilon^{(p)} \tag{7.25}$$

In terms of the error, convergence requires

$$\lim_{p\to\infty} \epsilon^{(p)} = 0 \tag{7.26}$$

We must now try to prove that this is so.

In the investigation of the convergence of an iterative method, it is convenient to introduce the eigenvectors and eigenvalues of the matrix B, which are formally defined by

$$B\psi^{(k)} = \lambda_k \psi^{(k)} \qquad k = 1,2,\ldots,K \tag{7.27}$$

where K is the size of the matrix.

We are now in a position to give a formal analysis of the convergence properties of an iterative method. The method is formal because calculating the eigenvalues and eigenvectors of the matrix would require more work than simply using the method. The investigation begins by writing the initial error vector $\epsilon^{(0)}$ as a linear combination of the eigenvectors; this was justified in Chapter 1. We have

$$\epsilon^{(0)} = \sum_{k=1}^{K} \epsilon_k \psi^{(k)} \tag{7.28}$$

Then, from Eq. (7.27), we find by direct computation

$$\epsilon^{(1)} = B\epsilon^{(0)} = B\sum_{k=1}^{K} \epsilon_k \psi^{(k)} = \sum_{k=1}^{K} \epsilon_k B\psi^{(k)} = \sum_{k=1}^{K} \lambda_k \epsilon_k \psi^{(k)} \tag{7.29}$$

and, continuing the calculation,

$$\epsilon^{(p)} = \sum_{k=1}^{K} \lambda_k^p \epsilon_k \psi^{(k)} \tag{7.30}$$

We may obtain the information we seek from this expression.

It is clear that a necessary and sufficient condition for the error to become zero for large p is that all of the eigenvalues of the matrix B be less than unity in magnitude. (The eigenvalues may be complex.) This condition must be obeyed by any iterative method. Since the magnitude of the largest eigenvalue of a matrix plays such a critical role in computational linear algebra, it is given a special name—the *spectral radius*. The spectral radius is a property of the iterative method and the equation to which it is applied and has to be evaluated for each case.

7.2.4. Stopping Criterion

We would like the iterative procedure to converge in as few steps as possible. However, we must assure that the convergence error is considerably smaller than the discretization error. This requires a method for estimating the convergence error. We will give the formal development of such a method and, later, a practical version of it. In this analysis, we shall assume that all of the eigenvalues of the iteration matrix B are real. They are complex for some of the methods to be described later. The generalization of the argument to arbitrary eigenvalues can be made (but is more complicated) and can be found in a recent study by Ferziger and Perić (1996).

If the largest eigenvalue (which we assume is λ_1) is real, the dominant error after many iterations is the first term of the sum in Eq. (7.30). Since the solution of the discretized equations is ϕ, the solution at the end of the pth iteration is approximately

$$\phi^{(p)} = \phi + \epsilon^{(p)} \approx \phi + a_1 \lambda_1^p \psi_1 \tag{7.31}$$

To get an expression for the convergence error, we need to eliminate the unknown exact solution, ϕ. This can be done by taking the difference of two successive iterates:

$$\chi_p = \phi^{(p+1)} - \phi^{(p)} \approx (\lambda_1 - 1)\lambda_1^p a_1 \psi_1 \tag{7.32}$$

If we can estimate the eigenvalue λ_1, this expression can be evaluated. This eigenvalue can be estimated by computing the norms of χ_p and χ_{p-1}, which we define as the root mean squares of their components (their values at each grid point), and computing the ratio:

$$\lambda_1 \approx \frac{|\chi_p|}{|\chi_{p-1}|} \tag{7.33}$$

where $|Q|$ represents the norm of a vector Q.

With this estimate of the eigenvalue, it is not difficult to estimate the convergence error. In fact, by rearranging Eq. (7.32), we find

$$\epsilon^{(p)} = \phi^{(p)} - \phi \approx a_1 \lambda_1^p \psi_1 \approx \frac{\chi_p}{\lambda_1 - 1} \tag{7.34}$$

This estimate can be computed from the iterates of the solution.

The most accurate criterion for stopping an iterative procedure requires that the error estimated in the manner presented above be smaller than some prescribed limit. Many authors stop iterating when the magnitude of difference of iterates, χ_p, is less than some limit. When the eigenvalue λ_1 is close to unity, Eq. (7.34) shows that this can be dangerous—the error may be very large even though the difference of iterates is small. An alternative to using the convergence error estimate given above can be obtained by noting that the difference of iterates decreases about as rapidly as the error. If the initial guess at the solution is zero everywhere in the domain, the initial error is approximately the same size as the solution. We can then compute the factor by which the error must be reduced and require that the difference of iterates be reduced by that amount. A reasonable practical criterion is that the difference of iterates be reduced by about four or five orders of magnitude. A better criterion is to require that the convergence error be significantly smaller than the discretization error.

We end this subsection by emphasizing the importance of reducing the convergence error to the point at which it is approximately an order of magnitude smaller than the discretization error before the iterative procedure is halted.

7.2.5. Estimation of Discretization Error

The discretization error was defined earlier as the difference between the exact solutions of the differential and discretized equations. We can estimate it using the Richardson method introduced in Chapter 3. For this method to be used, it is essential that the convergence error be much smaller that the discretization error on the finest grid used; an order of magnitude ratio is a good choice. In this subsection, we assume that this has been achieved.

Usually we are interested in the total discretization error, the errors resulting from all of the approximations used in discretizing the differential equation. That being the case, if the approximations used in the spatial discretizations are of second-order accuracy, the discretization error can be computed by solving the problem on two grids with grid sizes h and $h/2$; one third of the difference is then an approximation to the error:

$$\epsilon_{h/2} = \tfrac{1}{3}(\phi_{h/2} - \phi_h) \tag{7.35}$$

There are problems in which the errors in the two coordinate directions are unequal due to an asymmetry in the problem and there may be a need to assess the errors associated with the discretization in one or both of the coordinate directions separately. This can be done by halving the grid size in one direction while keeping the grid size in the other direction fixed. The formula (7.35) then gives an estimate of the error due to the discretization in the direction in whose grid was halved.

This error estimation technique is also a useful aid in debugging a program. If a method is known to be second-order accurate, the error should be reduced by a factor of 4 when the grid size is halved. If the error is estimated on several grids and the error is reduced in proportion to the square of the grid size, it is a good indication that both the analysis and the programming are correct. If any surprises are found, it is generally a good idea to search for the cause; most often it is a bug in the program. The technique can be refined to compute the errors arising from various terms in the equation, thus further aiding the debugging process.

The Richardson method is ideally suited to the regular grids that were applied to Laplace's equation above. In real engineering problems it is rare that the geometry or the grid are so regular. Grid irregularity makes application of the Richardson method more difficult but not impossible. If the grid is refined by halving all of the grid sizes while retaining its layout, the method can be used without difficulty. If this cannot be done, one can interpolate one of the solutions onto another regular grid; it is important that the interpolation be more accurate than the solution itself to avoid contaminating the latter.

We now go on to particular iterative methods, beginning with some well-known older methods. As in the earlier chapters, we shall analyze one of the simpler methods in some detail to gain an understanding of how methods of this class behave and then present some of the better methods with less analysis.

7.3. JACOBI ITERATION

7.3.1. The Method

The first iterative method for solving the discretized Laplace's equation that we shall consider is obtained by writing $A = -D + C$ where D is the diagonal matrix $D = 4I$ and C is the matrix obtained by zeroing out the main diagonal of the matrix (7.12). Letting the matrices of the iterative method be $M = D$ and $N = C$, we have the iterative scheme:

$$D\phi^{(p+1)} = C\phi^{(p)} - \mathbf{b} \tag{7.36}$$

This equation may be rewritten as

$$\phi^{(p+1)} = D^{-1}C\phi^{(p)} - D^{-1}\mathbf{b} = B_J\phi^{(p)} - \mathbf{d} \tag{7.37}$$

where $B_J = D^{-1}C$. This is the *Jacobi* iterative method. In component form this equation is

$$\phi_{i,j}^{(p+1)} = \tfrac{1}{4}(\phi_{i+1,j}^{(p)} + \phi_{i-1,j}^{(p)} + \phi_{i,j+1}^{(p)} + \phi_{i,j-1}^{(p)} - b_{i,j}) \tag{7.38}$$

Thus the Jacobi method computes the new value of $\phi_{i,j}$ by averaging the values of its neighbors at the preceding iteration and adding the contribution of the inhomogeneous (or boundary condition) term. This method is easily generalized to give a method for solving any set of linear algebraic equations.

In terms of the geographic notation introduced earlier, Eq. (7.38) becomes

$$\phi_P^{(p+1)} = \frac{1}{A_P}(A_E\phi_E^{(p)} + A_W\phi_W^{(p)} + A_N\phi_N^{(p)} + A_S\phi_S^{(p)} - b_P) \tag{7.39}$$

We shall not write all of the methods to be introduced later in this form.

7.3.2. Convergence

Because the matrix associated with this method is particularly simple, it is easy to investigate the convergence of this method. In fact, in component form, the eigenvectors of the matrix B_J of Eq. (7.37) are given by Eq. (6.56). The eigenvalues are easily computed and are

$$\lambda_{m,n} = \frac{1}{2}\left(\cos\frac{m\pi}{M+1} + \cos\frac{n\pi}{N+1}\right) \qquad m = 1,2,\ldots,M \quad n = 1,2,\ldots,N \tag{7.40}$$

The Jacobi method is certainly simple to use and the cost of an iteration about as low as it could possibly be; it requires only five arithmetic operations per mesh point so it satisfies some of the criteria laid out in the preceding section. It is essential to determine whether this method also satisfies the most important criterion—convergence. This is easily done. Given the explicit expression (7.40) for the eigenvalues, we see by inspection that every $\lambda_{m,n}$ is smaller than unity so the Jacobi method is convergent.

The rate at which the method converges is equally important and is easy to study. The largest eigenvalues of the matrix B_J are equal in magnitude and have opposite signs; they are the ones with $n = m = 1$ and $n = N$, $m = M$. Through use of a Taylor series expansion, it is easy to see that, for large M and N, these eigenvalues are

$$\lambda_{\max} = \frac{1}{2}\left(\cos\frac{\pi}{M+1} + \cos\frac{\pi}{N+1}\right) \approx 1 - \frac{\pi^2}{4}\left(\frac{1}{(M+1)^2} + \frac{1}{(N+1)^2}\right) \tag{7.41}$$

Thus the largest eigenvalues are both less than unity, but only very slightly so if M and N are large. As a result, the convergence of the Jacobi method is very slow. To see just how slow it is, recall that Eq. (7.30) shows that, after a large number of iterations, p, the convergence error is proportional to $\lambda_{\max}^p$ where $\lambda_{\max}$ is the largest eigenvalue of B_J. To simplify matters, let us take $M = N$. For large N, we can make the approximation

$$\lambda_{\max}^p \approx \left(1 - \frac{\pi^2}{2N^2}\right)^p \approx e^{-\pi^2 p/2N^2} \tag{7.42}$$

This quantity has to be smaller than the desired convergence error before the iteration process can be terminated. If δ is the allowable error, setting it equal to the right-hand side of Eq. (7.42), and solving for p, we find that the number of iterations required to produce the desired result is approximately

$$p \approx -\frac{2N^2}{\pi^2}\ln\delta \tag{7.43}$$

For example, if the required accuracy is 10^{-3} (a typical value), the number of iterations needed to reduce the error to that value is about $1.6N^2$. Since each iteration requires five arithmetic operations per mesh point, or a total of $5N^2$ operations, the number of arithmetic operations needed to obtain the desired results is approximately $8N^4$, which is more than the direct application of Gauss elimination would require. Thus the Jacobi method is not an attractive choice; we will need to search for better methods.

7.3.3. Connection to Heat Equation

We now look at the Jacobi method in another way that will help us find improved methods. We begin with the observation that elliptic equations often arise as the steady-state limit of problems governed by parabolic or hyperbolic equations. Thus if we solve the heat equation in two space dimensions with the given boundary conditions and an arbitrary initial condition; the solution should eventually settle down or *relax* to the desired solution of Laplace's equation. Equations other than the heat equation could be used for this purpose; all that is required is that the solution relax to a solution of Laplace's equation. There need not be any connection to a physical problem, and purely artificial time dependent problems are frequently used.

Suppose we solve the heat equation using the Euler method. Choosing $h_1 = h_2 = h$ for convenience and letting $\beta = \alpha \Delta t / h^2$, we have

$$\phi_{i,j}^{(p+1)} = (1 - \beta)\phi_{i,j}^{(p)} + \beta(\phi_{i+1,j}^{(p)} + \phi_{i-1,j}^{(p)} + \phi_{i,j+1}^{(p)} + \phi_{i,j-1}^{(p)} + \tfrac{1}{4}b_{ij}) \quad (7.44)$$

The distinction between what we are doing now and what we did in the preceding chapter is that here we have no interest in an accurate time history of the solution. We simply want to get to the steady state as quickly as possible. It seems logical, therefore, to take the biggest time step that stability allows. For Eq. (7.44) the limit is $\beta = 1$; using this value in Eq. (7.44) immediately reduces it to Eq. (7.38), the basic equation of the Jacobi method. Thus the Jacobi method for Laplace's equation is equivalent to the Euler method for the heat equation, provided the largest allowable time step is used.

A large family of iterative methods for elliptic equations are called relaxation methods because they are based on allowing the solution of the heat equation to relax to a steady state. The Jacobi method is also known as *simultaneous relaxation*, since the solution at each mesh point could be computed simultaneously from the old values. This makes the Jacobi method an ideal candidate for parallelization; we will say more about this later. In the next several sections, we will search for methods of speeding up the convergence of iterative methods.

7.3.4. Other Equations

The Jacobi method can be applied to other elliptic equations. After Laplace's equation, the next simplest such equation is Poisson's equation. It is simply Laplace's equation with a source term:

$$\nabla^2 \phi = \frac{\partial^2 \phi}{\partial x_1^2} + \frac{\partial^2 \phi}{\partial x_2^2} = \rho(x_1, x_2) \quad (7.45)$$

Any technique used to solve Laplace's equation can be applied to this equation; this includes the discretization and the solution methods. If we repeat the analysis applied to Laplace's equation for this equation, we will find that the presence of the source term has no effect on the behavior of the Jacobi method, that is, it converges at exactly the same rate for both equations.

On the other hand, a modification that introduces terms that contain the dependent variable can have an enormous effect on the behavior of a solution method. Let us consider Helmholtz's equation:

$$\nabla^2\phi + k^2\phi = \frac{\partial^2\phi}{\partial x_1^2} + \frac{\partial^2\phi}{\partial x_2^2} + k^2\phi = 0 \tag{7.46}$$

This equation arises in the solution of the wave equation when it is assumed that the solution is separable and that the behavior in time is sinusoidal. In that case, $k^2 > 0$, but there are other interesting problems (heat transfer and neutron transport theory are examples) in which $k^2 < 0$ so we need to consider both possibilities.

When Eq. (7.46) is discretized using the five-point difference formula, we have

$$(1 + k^2h^2)\phi_{i,j}^{(p+1)} = \tfrac{1}{4}(\phi_{i+1,j}^{(p)} + \phi_{i-1,j}^{(p)} + \phi_{i,j+1}^{(p)} + \phi_{i,j-1}^{(p)} - b_{i,j}) \tag{7.47}$$

where, for convenience, it is assumed that the same grid size h is used in both coordinate directions. The Jacobi method can be applied to this equation. The only difference is that the diagonal elements of the matrix are now $a_{ii} = 1 + k^2h^2$ in place of $a_{ii} = 1$, which we had for Laplace's equation.

It is no harder to find the eigenvalues of the Jacobi matrix for this case than it was for Laplace's equation. The eigenvectors are unchanged and it not difficult to show that the eigenvalues are

$$\lambda_{m,n} = \frac{1}{2(1 + k^2h^2)}\left(\cos\frac{m\pi}{M+1} + \cos\frac{n\pi}{N+1}\right)$$

$$m = 1, 2, \ldots, M \quad n = 1, 2, \ldots, N \tag{7.48}$$

If $k^2 > 0$, these eigenvalues are smaller than the corresponding eigenvalues of the Jacobi method applied to Laplace's equation. For this case, Jacobi's method converges more rapidly for the Helmholtz equation than it does for the Laplace equation.

On the other hand, if $k^2 < 0$, these eigenvalues are larger than the corresponding eigenvalues of the Jacobi method applied to Laplace's equation. Since the eigenvalues for Laplace's equation are just a little smaller than 1, it is likely that the eigenvalues of the Jacobi method for the Helmholtz equation with $k^2 < 0$ are larger than 1. If that is the case, the Jacobi method will be divergent.

It is instructive to consider the difference between the two cases. In the first case the matrix elements are such that:

$$|a_{ii}| > \sum_{j \neq i} |a_{ij}| \tag{7.49}$$

whereas, in the second case, the inequality may be reversed. We say that, in the first case, the matrix is *diagonally dominant* and in the second case, it is not. We thus see that diagonal dominance of the matrix is required for Jacobi's method is to converge. In the remainder of this chapter, we shall investigate the behavior of the methods for Laplace's equation and say only a few words about application to other equations.

We now go on to some examples.

Example 7.1: Jacobi Solution of Laplace's Equation As our example for elliptic equation solvers, we will use Laplace's equation

$$\nabla^2 \phi = 0$$

with the boundary conditions

$$\phi(0,x) = \phi(0,y) = 0$$

$$\phi(x,1) = 100x$$

$$\phi(1,y) = 100y$$

The exact solution of this problem is $\phi = 100xy$. Since this function is linear in both variables, it is differentiated exactly by a second-order finite difference method for any grid size and the converged solution is identical to the exact solution no matter how few grid points are used. This allows us to follow the convergence error of the method accurately.

The program, JACOBI, used to produce the solution to this problem can be found on the Internet site described in Appendix A; the reader is advised to look at it. In particular, note that, because the old value of a variable is used even after the new one has been computed, the Jacobi method requires the use of two arrays (called F and FF in the program) for the dependent variable. This increases not only the memory requirement but also the complexity of the program so that some steps—particularly the initialization of the field and the calculation of the new values of the dependent variable—need to be written twice; some of this could be avoided with a bit more cleverness, but the fact remains that the method is not as easy to program as it could be.

Results were obtained using $\Delta x = \Delta y = h$. The root-mean-square convergence error as a function of the number of iterations is shown for three grid sizes in Figure 7.6. The initial solution guess was zero everywhere inside the domain and iteration was continued until the root-mean-square convergence error, determined by comparison with the exact solution, was less than 2×10^{-4}, the smallest value that could be used without being affected by round-off error on the finest grid. The exponential convergence predicted by the theory presented above is found except for the first few iterations. Specifically, when the mesh size reduced to half its previous value, four times

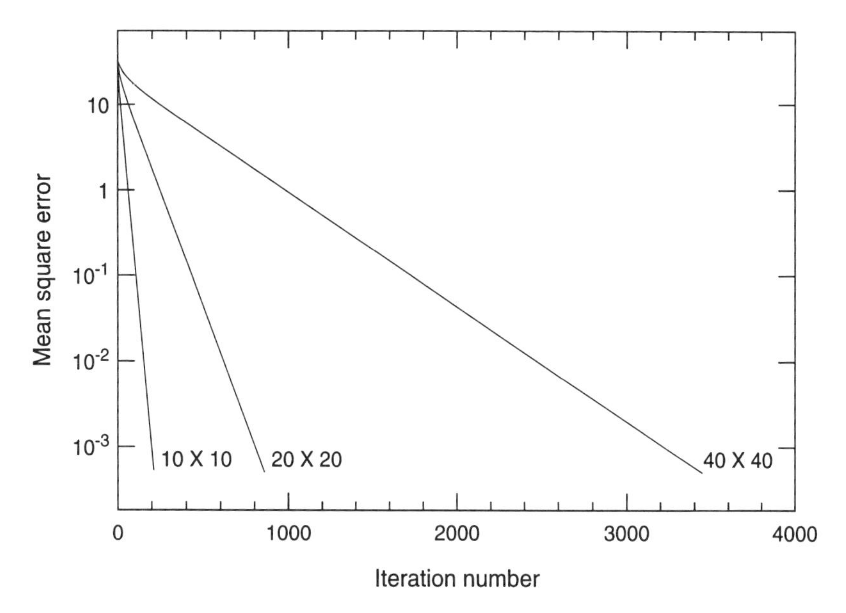

FIGURE 7.6. Convergence error in the solution of Laplace's equation obtained with the Jacobi method on various grids as a function of the iteration number.

as many iterations are required to achieve the same error reduction. The poor rate of convergence of the method predicted by the analysis above is confirmed.

Since the finite difference method is second-order accurate, in order to increase the accuracy by a factor of 4, we need to cut the spatial interval in half. This requires four times as many spatial points and four times as much computation per iteration. To make matters worse, it also requires four times as many iterations to achieve the same accuracy. In a sensible calculation, the method is iterated only to the point at which the convergence error is some fraction of the finite difference truncation error. Thus more than 4 times as many iterations are needed on the finer grid and the total cost of the calculation is increased to more than 16 times the cost of the coarse grid calculation. Clearly, this is unacceptable scaling and we will need to look for other methods.

One method of reducing the cost is suggested by these results. We can begin by doing the calculation on a very coarse mesh since it is relatively cheap. As suggested above, one should continue interating until the convergence error is at least as small as the truncation error. We can then repeat the calculation with the mesh size reduced by half using the converged results on the coarse mesh to provide the initial guess. The initial guesses for the grid points that were not part of the coarse mesh can be obtained by interpolation; bilinear interpolation, which is second-order accurate, is a good and simple method of doing this. Because the error on the coarse mesh is approximately four times the error in the fine mesh result, we only need to iterate long enough to reduce the error by about a factor of 4. This calculation costs much less than a fine mesh calculation started with a poor initial guess, and it can be repeated on finer and finer meshes until the desired accuracy is achieved. This is called the *refinement method.* A more sophisticated version of this idea—one based on methods better than Jacobi and both increases and decreases the mesh size—is known as the multigrid method and is an excellent approach to obtaining accurate solutions to a wide variety of boundary value problems and will be presented in Section 7.10.

One final note. In the preceding chapter, we found that, in solving parabolic problems, expansion of the solution in eigenvectors of the matrix is useful in analyzing the behavior of the method. The high index eigenvectors, which change sign many times in the region of interest, cause the discretized equations to be stiff and are responsible for the severe time step restrictions that are required for the stability of many methods. As noted above, the rate of convergence of the Jacobi method is controlled by two equal and opposite signed eigenvalues. For Laplace's equation, the largest positive eigenvalue corresponds to an eigenvector whose components all have the same sign; it represents a smooth function of the coordinates. The largest negative eigenvalue corresponds to an eigenvector whose components oscillate in sign (this is the one that usually causes instability in the parabolic case). Hence the errors that remain after many iterations are composed of a combination of smoothly varying components and rapidly fluctuating components. In most cases, the initial conditions are such that the smooth component of the error is larger and it usually dominates in the late iterations. Although it was not shown, in Example 7.1 the error has the same sign at all points, which verifies this observation for that case.

7.4. GAUSS–SEIDEL METHOD

Improved iterative methods are needed to reduce the amount of computation required to obtain a converged solution. Improvement is unlikely to be obtained by reducing the amount of calculation per iteration; it is much more likely to come by reducing the number of iterations. Methods that accomplish this are introduced in this section and in the remainder of this chapter.

As noted, each iteration of the Jacobi method for solving Laplace's equation is equivalent to a time step of the Euler method for the heat equation with the largest time step allowed by the stability criterion. Many iterative methods for elliptic equations can be interpreted as relaxing a parabolic or hyperbolic problem to a steady state. We want to get to the steady-state solution as quickly as possible; accurate time behavior is irrelevant. A method that allows a bigger time step would be useful. One possibility is an implicit method. In principle, the Crank–Nicolson method allows an infinite time step and could get the solution in one "iteration" but, since the problem to be solved at the new time step is very similar to the original elliptic problem, this is not practical.

The alternating direction implicit (ADI) method is a possibility that will be considered later, but it is not totally free of difficulties. We begin by looking for ways to speed up the Jacobi method; some of these methods will prove useful as the bases for methods introduced later.

In the Jacobi method, as Eq. (7.38) shows, each new value of the function is computed entirely from old values. The new values can be computed in any order but, for ease of programming, it makes sense to compute them in an orderly manner. As a result, in the Jacobi program used in Example 7.1, two arrays representing the dependent variable and duplicate programming were required. But, when computing a new value of $\phi_{i,j}$, we already have new values of $\phi_{i-1,j}$ and $\phi_{i,j-1}$. Since the new values should be better approximations than the old ones, it is sensible to use the updated values. This leads to a method in which Eq. (7.38) is replaced by

$$\phi_{i,j}^{(p+1)} = \frac{1}{4}(\phi_{i+1,j}^{(p)} + \phi_{i-1,j}^{(p+1)} + \phi_{i,j+1}^{(p)} + \phi_{i,j-1}^{(p+1)} - b_{i,j}) \tag{7.50}$$

This is called the *Gauss–Seidel* method or the method of *successive relaxation* because it computes the new data in a sequential manner.

To find the matrix equivalent of Eq. (7.50), note that it is necessary to separate the parts of the matrix that represent the contributions of points $i-1, j$ and $i, j-1$ to the equation at the point i, j from the other parts of the matrix. This is easily done. The elements of the matrix A of the preceding section that produce these terms are the ones that lie above the main diagonal. We therefore decompose the matrix in the following way:

$$A = -D + C = -D + L + U \tag{7.51}$$

where $-D$ again represents the diagonal part of A, L is the (strictly) lower triangular portion, and U is the (strictly) upper triangular part. Thus the Gauss–Seidel iterative equation (7.50) is equivalent to

$$(D - U)\phi^{(p+1)} = L\phi^{(p)} + \mathbf{b} \tag{7.52}$$

$D - U$ is an upper triangular matrix and is easily inverted [which is another way of saying that Eqs. (7.50) are easily solved] and we can write formally

$$\phi^{(p+1)} = (D - U)^{-1}L\phi^{(p)} + (D - U)^{-1}\mathbf{b} = B_G\phi^{(p)} + \mathbf{d}_G \tag{7.53}$$

Formal analysis of the rate of convergence of the Gauss–Seidel method will not be presented here. We merely state the result that, under fairly broad assumptions, it can be shown that the eigenvalues of the Gauss–Seidel iterative matrix B_G are precisely the squares of the corresponding eigenvalues of the Jacobi matrix B_J. This means that the Gauss–Seidel method converges whenever the Jacobi method does and, in terms of error reduction, one Gauss–Seidel iteration is equivalent to two Jacobi iterations, so the Gauss–Seidel method converges in half as many iterations as the Jacobi method. Since this improvement

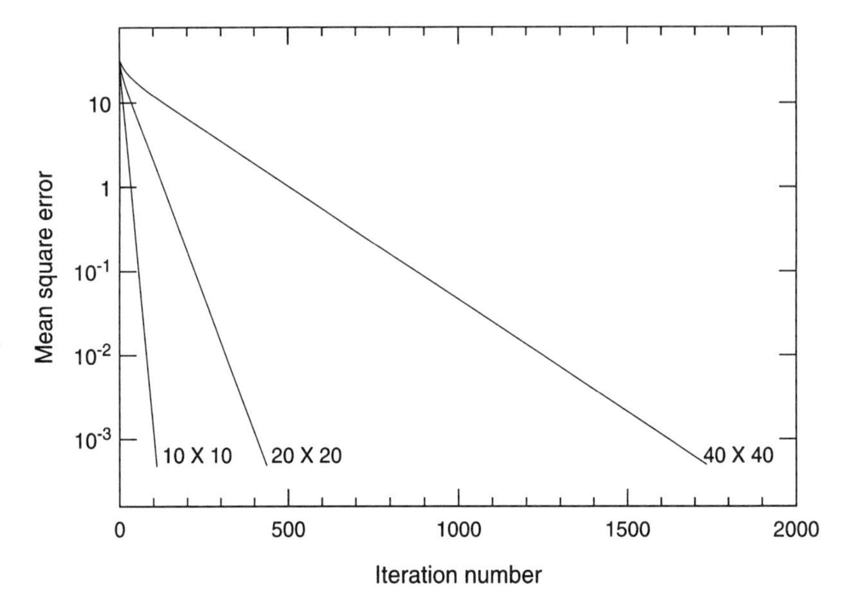

FIGURE 7.7. Convergence error in the solution of Laplace's equation obtained with the Gauss–Seidel method on various grids as a function of the iteration number.

is achieved at no cost in computer time and with a simpler program, there is no reason to use the Jacobi method.

Finally, we note that the Gauss–Seidel method can be used for solving parabolic equations. For the heat equation, with $h = \Delta x = \Delta y$ and $\beta = \alpha\,\Delta t/h^2$ as before, we have

$$(1+\beta)\phi_{i,j}^{(p+1)} = (1-\beta)\phi_{i,j}^{(p)} + \beta(\phi_{i+1,j}^{(p)} + \phi_{i-1,j}^{(p+1)} + \phi_{i,j+1}^{(p)} + \phi_{i,j-1}^{(p+1)} - b_{ij}) \tag{7.54}$$

This method is first-order accurate in time and second-order accurate in space (just as the Euler method is) and is stable for values of β up to 1; for $\beta = 1$ it becomes Eq. (7.50) as might be expected.

Example 7.2: Gauss–Seidel Solution of Laplace's Equation Solving the problem of Example 7.1 with the Gauss–Seidel method is easily done. To construct the program, we need only take the program used for the Jacobi method and eliminate parts of it! Specifically, only one array is required for the dependent variable and the main calculation can be reduced to a single loop.

Results similar to those given in Figure 7.6 for the Jacobi method are given in Figure 7.7 for three values of h. There are no surprises; the Gauss–Seidel method converges almost exactly twice as fast as the Jacobi method for this problem. Despite this, the method still has the principal disadvantage of the

Jacobi method—increased accuracy comes at a high price. The Gauss–Seidel method is a special case of the successive overrelaxation (SOR) method discussed below, and the results shown in Figure 7.7 were obtained with the SOR code. For this reason, a code specifically for the Gauss–Seidel method is not found on the Internet site described in Appendix A.

7.5. LINE RELAXATION METHOD

The Gauss–Seidel method obtains its advantage relative to the Jacobi method by using new values of the solution whenever they are available. In the Gauss–Seidel method, half the data used to compute a new value belongs to the current iteration while the other half is derived from the preceding iteration, and half as many iterations are required. It is natural to ask whether further improvement can be obtained by using more current data to calculate the updated value of the function. If new data were used at all four nearest neighboring points, we would have a fully implicit method and the matrix would be very difficult to invert so that is not practical.

This leaves just one further possibility—using new data at three neighboring points. Such a method is readily constructed. In Eq. (7.50) we can either replace $\phi_{i+1,j}^{(p)}$ by $\phi_{i+1,j}^{(p+1)}$ or $\phi_{i,j+1}^{(p)}$ by $\phi_{i,j+1}^{(p+1)}$. Reasons for choosing one or the other are discussed below. For the moment let us adopt the first choice. We then have

$$\phi_{i,j}^{(p+1)} = \tfrac{1}{4}(\phi_{i-1,j}^{(p+1)} + \phi_{i+1,j}^{(p+1)} + \phi_{i,j-1}^{(p+1)} + \phi_{i,j+1}^{(p)} - b_{i,j}) \tag{7.55}$$

which can be written

$$-\tfrac{1}{4}\phi_{i-1,j}^{(p+1)} + \phi_{i,j}^{(p+1)} - \tfrac{1}{4}\phi_{i+1,j}^{(p+1)} = \tfrac{1}{4}(\phi_{i,j-1}^{(p+1)} + \phi_{i,j+1}^{(p)} - b_{i,j}) \tag{7.56}$$

In this equation all of the terms on the right-hand side are known when we come to the jth line; the new solution on the $(j-1)$st line has already been computed. Equations (7.56) are a tridiagonal system of equations for the solution on the jth line and must be solved by the algorithm for such systems described in Chapter 1.

This method is known as the *line Gauss–Seidel method*, sometimes as *successive line relaxation*, or the *method of lines*, and is equivalent to treating one direction implicitly in solving the heat equation, that is, it is similar (but not identical) to one half of the ADI method. It is possible to construct a line Jacobi method but there is no reason to do so.

We shall not put this method into matrix form or write the equivalent time method; it is neither difficult nor important to do these. It is important, however, to compare this method with the (point) Gauss–Seidel method. Line relaxation requires approximately half as many iterations as the Gauss–Seidel method, or only one fourth as many as the Jacobi method; however, it also

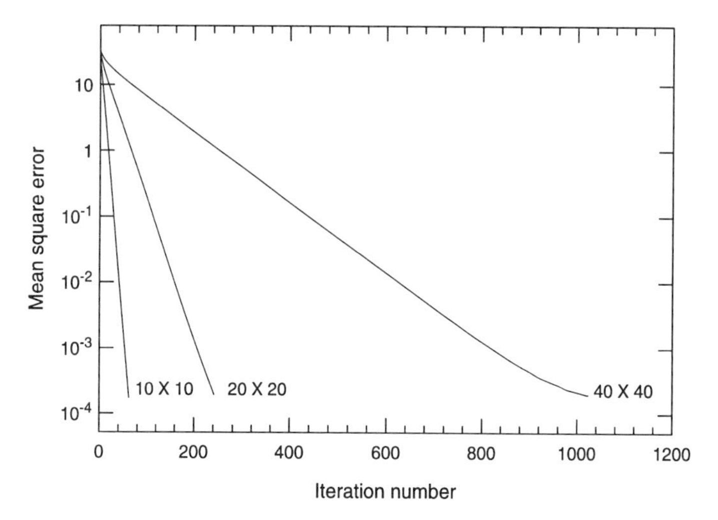

FIGURE 7.8. Convergence error in the solution of Laplace's equation obtained with the line Gauss–Seidel method on various grids as a function of the iteration number.

requires more computation per iteration. The net result is a slight advantage to the line method when the number of points in the two directions are equal. There is another factor to consider. In some problems, the solution varies much more rapidly in one direction than the other, and more points are used in the rapidly varying direction. Here there is considerable advantage to using the line relaxation method, provided the lines on which tridiagonal equations are solved lie in the direction of the more severe variation, which is usually the direction with the larger number of points. (It is better to treat the difficult direction implicitly.)

Example 7.3: Line Gauss–Seidel Solution of Laplace's Equation The problem of Example 7.1 is now done using line relaxation. The results are shown in Figure 7.8 and are in accord with expectations. The method converges in slightly more than half as many iterations as the point Gauss–Seidel method.

7.6. SUCCESSIVE OVERRELAXATION

None of the methods discussed so far yields the solution at a lower cost than the direct solution method. We need further ideas for increasing the rate of convergence of iterative methods. In particular, we would like methods that avoid the scaling in which the number of iterations is proportional to the square of the number of points in each direction, N^2. To see how improved methods might be found, let us look more closely at the convergence behavior of the methods.

7.6.1. Extrapolation

Since many iterative solution methods for elliptic PDEs are akin to solving parabolic PDEs, we can use our knowledge of the latter to understand the former. For the heat equation, in the early stages, the solution varies rapidly in both time and space. After a short time, the solution becomes smooth and relaxes slowly to the steady-state solution. At this stage, the approach to the steady state is monotonic. When a method behaves in this way, we can almost guess the result of the next iteration without computing it. In other words, the past history of the solution contains information that can help the convergence. In practical terms, this means that, instead of using $\phi^{(p)}$ as the starting guess for the next iteration, we can extrapolate the results to find a better guess. This leads to the procedure known as *extrapolation, acceleration*, or *overrelaxation*.

To see how this might work, suppose we use one of the methods previously discussed to compute $\tilde{\phi}^{(p)}$:

$$\tilde{\phi}^{(p)} = B\phi^{(p-1)} + \mathbf{d} \tag{7.57}$$

where the tilde (~) is used to differentiate this result from the extrapolated one that will provide the input for the following iteration. In the methods discussed so far, $\tilde{\phi}^{(p)}$ is the input vector for the next iteration, that is, $\tilde{\phi}^{(p)} = \phi^{(p)}$. We now introduce extrapolation. In particular, we will use

$$\phi^{(p)} = \tilde{\phi}^{(p)} + \alpha(\tilde{\phi}^{(p)} - \phi^{(p-1)}) = (1+\alpha)\tilde{\phi}^{(p)} - \alpha\phi^{(p)} = \omega\tilde{\phi}^{(p)} + (1-\omega)\phi^{(p)} \tag{7.58}$$

in place of $\tilde{\phi}^{(p)}$ as the input to the next iteration. For this to be an extrapolation, we must have $\alpha > 0$. For safety, we expect that $\alpha < 1$ will be required. Thus $\omega = 1 + \alpha$, which is called the *overrelaxation factor*, should lie in the range $1 < \omega < 2$.

The two steps Eqs. (7.57) and (7.58) can be combined into one to give

$$\phi^{(p)} = [\omega B + (1-\omega)I]\phi^{(p-1)} + \omega\mathbf{d} \tag{7.59}$$

and the equation that describes the reduction of the error for this method is

$$\epsilon^{(p)} = [\omega B + (1-\omega)I]\epsilon^{(p-1)} = B_\omega\epsilon^{(p-1)} \tag{7.60}$$

Thus using acceleration is equivalent to replacing the iteration matrix B by $B_\omega = \omega B + (1-\omega)I$. The simple relationship between the iteration matrices of the accelerated and original methods makes it relatively easy to compute the eigenvalues of the new matrix. In fact any eigenvector of B with eigenvalue λ

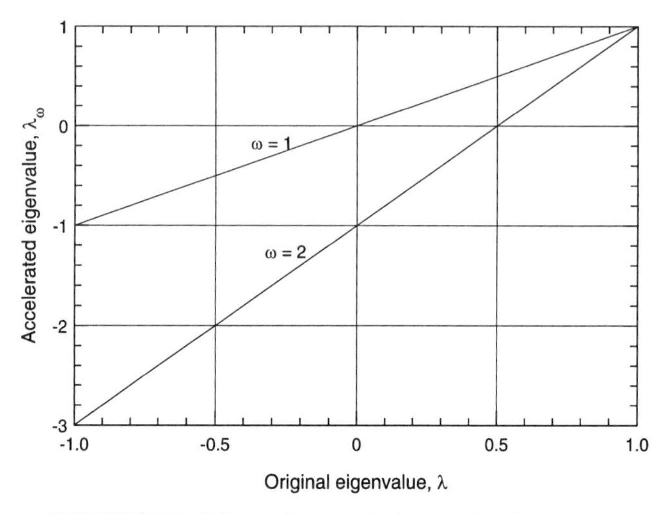

FIGURE 7.9. Effect of extrapolation on the eigenvalue.

is also an eigenvector of B_ω with eigenvalue

$$\lambda_\omega = \omega\lambda + (1 - \omega) \tag{7.61}$$

This relationship is illustrated in Figure 7.9 for $\omega = 1$ and $\omega = 2$, the extreme values of interest.

The eigenvalues of the Jacobi method lie between $-\lambda_1$ and $+\lambda_1$ where λ_1 is the largest eigenvalue of the matrix B_J. From Figure 7.9 we see that if ω is larger than 1, the magnitude of the largest negative eigenvalue is increased by extrapolation, and the extrapolation method actually converges more slowly than the Jacobi method! This may not be in accord with one's intuition, but the Jacobi method cannot be improved by acceleration.

The eigenvalues of the Gauss–Seidel matrix, on the other hand, are the squares of the Jacobi eigenvalues and are therefore all real and positive. For simplicity and safety, we assume that the smallest Gauss–Seidel eigenvalue is zero. The optimum choice of ω is the one that makes the magnitudes of the largest eigenvalues λ_ω equal in magnitude and of opposite sign. Since the largest negative λ_ω of the extrapolated method corresponds to the Gauss–Seidel eigenvalue $\lambda = 0$ and the largest positive λ_ω of the extrapolated method corresponds to $\lambda = \lambda_1^2$ of the Gauss–Seidel method (λ_1 being the largest eigenvalue of the Jacobi matrix), the optimum value of ω satisfies

$$1 - \omega = -\omega\lambda_1^2 - (1 - \omega) \tag{7.62}$$

and is

$$\omega = \frac{2}{2 - \lambda_1^2} \tag{7.63}$$

For λ_1 near 1, say $\lambda_1 \approx 1 - \delta$, since $\lambda_1^2 \approx 1 - 2\delta$, the maximum λ_ω is $(1 - 2\delta)/(1 + 2\delta) \approx 1 - 4\delta$, the accelerated method converges twice as fast as the Gauss–Seidel method.

This method of accelerating the solution can also be applied to the line relaxation method with a similar improvement. However, we are after bigger game and now proceed to the *successive overrelaxation* method.

7.6.2. Point Successive Overrelaxation

Further improvement in the method can be obtained by accelerating or extrapolating the solution at each point as it is calculated rather than waiting for the entire iteration to be completed. For the Jacobi method, which uses only old data, there is no difference between the two approaches and there is nothing to be gained. For the Gauss–Seidel method, however, a surprising improvement is possible.

The method is quite simple. At each point, a new value is computed using Eq. (7.57); this result is immediately extrapolated using the point version of Eq. (7.58). These equations can be combined to give

$$\phi_{i,j}^{(p+1)} = \frac{\omega}{4}(\phi_{i-1,j}^{(p+1)} + \phi_{i+1,j}^{(p)} + \phi_{i,j-1}^{(p+1)} + \phi_{i,j+1}^{(p)} - b_{i,j}) + (1 - \omega)\phi_{i,j}^{(p)} \tag{7.64}$$

which can be written in matrix form

$$\begin{aligned} \phi^{(p+1)} &= (D - \omega U)^{-1}[((1 - \omega)D + \omega L)\phi^{(p)} - \omega \mathbf{b}] \\ &= B_{\mathrm{SOR}}\phi^{(p)} + \mathbf{d}_{\mathrm{SOR}} \end{aligned} \tag{7.65}$$

The problem is now to find the value of the acceleration parameter ω that makes the largest (in magnitude) eigenvalue of B_{SOR} as small as possible. This requires a rather difficult calculation and useful results are known only for particular cases. In the most important such case, the matrix of the Jacobi method possesses Young's property *A*, a property that has to do with the location of the zero and nonzero elements in the matrix. Roughly, a matrix has property *A* if the grid points can be colored red and black in checkerboard or chess fashion such that each equation at a red point contains data only from the point itself and from neighboring black points; the reverse holds for the black points. A more detailed definition of this property and derivation of the results are beyond the scope of this book; for details, see Golub and Van Loan (1989). We merely state that the second-order five-point Laplace operator matrix has property *A*, but the fourth-order approximations to the Laplacian operator do not.

The importance of property A is that, for matrices possessing it, it is possible to determine the eigenvalues of the SOR matrix explicitly. They turn out

TABLE 7.1. Eigenvalues and Iteration Counts for Various Methods

Method	λ_{max}	Iterations
Ten Intervals ($\Delta x = 0.1$)		
Jacobi	0.9595	212
Gauss–Seidel	0.9206	110
SOR	0.5604	24
Twenty Intervals ($\Delta x = 0.05$)		
Jacobi	0.9888	860
Gauss–Seidel	0.9778	437
SOR	0.7406	47
Forty Intervals ($\Delta x = 0.025$)		
Jacobi	0.9971	3446
Gauss–Seidel	0.9941	1736
SOR	0.8578	92

to be

$$\lambda_{SOR}^{1/2} = \tfrac{1}{2}[\omega\lambda \pm (\omega^2\lambda^2 - 4(\omega - 1))^{1/2}] \tag{7.66}$$

where λ is an eigenvalue of the Jacobi iteration matrix. For $\omega = 1$ (for which SOR reduces to the Gauss–Seidel method) $\lambda_{SOR} = \lambda^2$, which demonstrates, for matrices possessing property A, the result that the Gauss–Seidel eigenvalues are the squares of the Jacobi eigenvalues. Not surprisingly, the largest eigenvalue of the SOR matrix corresponds to the largest eigenvalue of the Jacobi matrix. One can then show without much difficulty that the optimum acceleration parameter is given by

$$\omega_{opt} = \frac{2}{1 + (1 - \lambda_{max}^2)^{1/2}} \tag{7.67}$$

and the maximum eigenvalue of the SOR method when the optimum factor is applied is

$$\lambda_{SOR,max} = \omega_{opt} - 1 \tag{7.68}$$

These are the essential results and what they mean is rather surprising. To see what can be achieved by the SOR method, consider a problem for which the maximum Jacobi eigenvalue is $1 - \delta$ where δ is small; for such a problem, the methods considered so far converge rather slowly. The maximum Gauss–Seidel eigenvalue is approximately $1 - 2\delta$, but the maximum SOR eigenvalue turns out to be $1 - 4\sqrt{\delta}$. The number of iterations required to produce accuracy ϵ is approximately $(N/2\sqrt{2}\pi)\ln|\epsilon|$ and is proportional to N rather than N^2. To carry the example further, the largest eigenvalues of the three methods and the number of iterations required for 2×10^{-4} accuracy are given in Table 7.1 for three grid sizes.

The results are extremely impressive. The scaling of the number of iterations with the number of grid points for the Jacobi and Gauss–Seidel methods is again confirmed; the Gauss–Seidel method requires half the number of iterations of the Jacobi method and, for both methods, the number scales as the square of the number of grid points. The number of iterations for the SOR method is linear in the number of grid points, and the method therefore yields the maximum benefit for problems that are most difficult for the other methods. This is further illustrated further in the examples below.

Example 7.4: SOR Solution of Laplace's Equation Since the behavior of the SOR method depends on the overrelaxation parameter ω, its results are not as easily displayed as those of the methods presented earlier. For one thing, the convergence is not always monotonic; indeed, for $\omega > \omega_{\text{opt}}$ the convergence is irregular.

To illustrate the method, we solve the problem treated in the preceding examples:

$$\nabla^2\phi = 0 \qquad \phi(x,0) = \phi(0,y) = 0 \qquad \phi(1,y) = 100y \quad \text{and} \quad \phi(x,1) = 100x$$

with 10, 20, and 40 intervals in each direction with various values of the relaxation factor ω. The number of iterations required to converge to the level used in the preceding examples on each of these three grids is shown in Figure 7.10. For each grid, the minimum is almost exactly at the optimum value of ω given by Eq. (7.64). Agreement between the theory and the minimum found by computation is remarkably good. The program used to make these calculations (SOR) is found on the Internet site described in Appendix A. The SOR method is no more difficult to program or more expensive than the Gauss–Seidel method. It is by far the best method so far and the first one that is considerably more efficient than direct solution of the system of equations. Some of the methods presented below are still more efficient so use of SOR has been declining in recent years.

A comparison between the SOR and Gauss–Seidel methods is easily made by remembering that the latter can be regarded as SOR with $\omega = 1$. We see from Figure 7.11 and Table 7.1 that using the optimum relaxation factor reduces the number of iterations required for convergence considerably. Furthermore, the relative reduction is larger in cases in which the Gauss–Seidel method requires more iterations. This is a very desirable behavior. The results are in good accord with what we anticipated from the discussion of the eigenvalues.

The figures also show that the convergence is monotonic when $\omega < \omega_{\text{opt}}$ because the largest eigenvalue is real in this case. For $\omega > \omega_{\text{opt}}$, the princi-

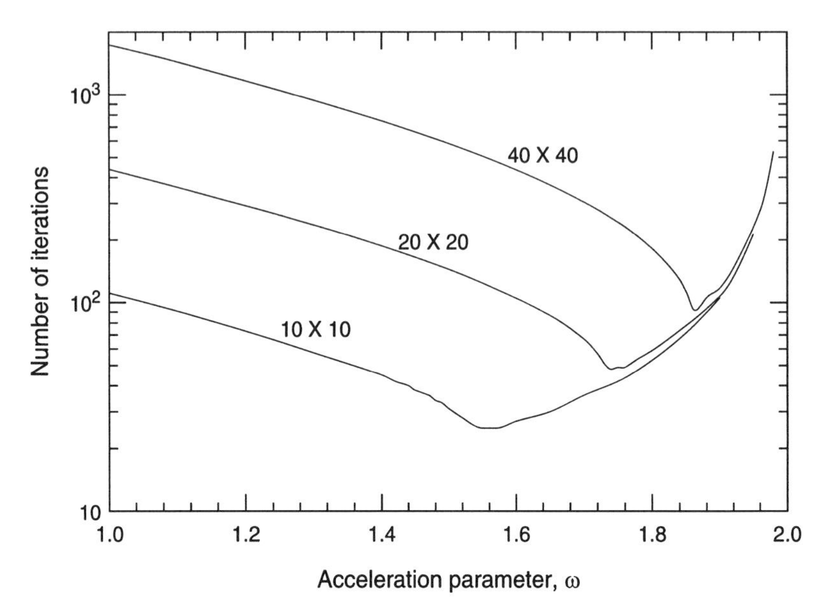

FIGURE 7.10. Number of iterations required for the successive overrelaxation (SOR) method to converge as a function of the overrelaxation factor ω for three different numbers of grids.

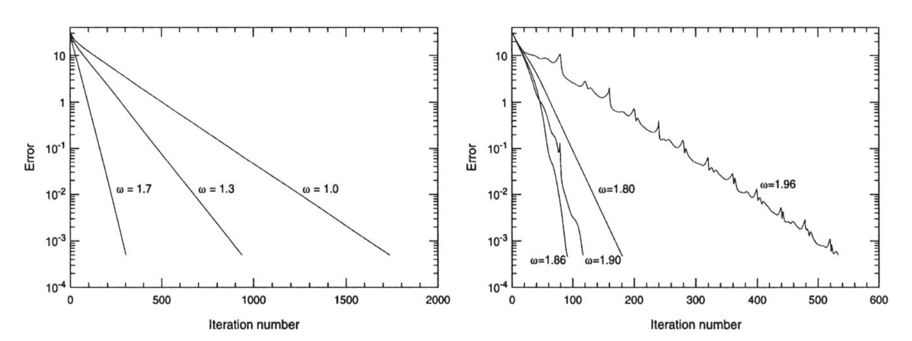

FIGURE 7.11. Convergence error in the solution of Laplace's equation obtained with the SOR method on a 40×40 grid as a function of the iteration number. The left figure shows the results for three overrelaxation factors smaller than the optimum value. The right figure gives the results for four larger overrelaxation factors. (The optimum relaxation factor is 1.8578.)

pal eigenvalue is complex, and this leads to the complicated, nonmonotonic behavior found for higher values of ω.

In practical applications involving more complicated equations or more complex geometries, the optimum relaxation factor cannot be predicted. If a problem is to be solved only once, it may not be important to have the exact optimum; if the procedure is converging slowly, one can stop the calculation, increase the acceleration factor, and continue. On the other hand, if many similar problems with the same mesh and same geometry are to be solved, it may pay to spend some time to find an overrelaxation factor that is nearly optimum. This can be done by increasing ω until the convergence becomes oscillatory. Usually, the first value of ω for which oscillatory behavior is found is a sufficiently accurate estimate of the optimum relaxation factor. A more accurate value can be found by a trial-and-error search, a case of which is shown (for line SOR) in Example 7.5.

Successive overrelaxation can also be applied to nonlinear elliptic problems. In these cases, the nonlinearity is treated by an iterative method such as Newton–Raphson. At each iteration, a linear problem must be solved and SOR can be applied to this part of the job. In an application of this kind, it is best not to iterate the SOR method to complete convergence on the early Newton–Raphson iterations as this wastes computation. A rule of thumb is that the SOR calculation should be continued until changes in the function on successive SOR iterations are about an order of magnitude smaller than the difference between the results at the ends of the two preceding Newton–Raphson iterations. In problems of this kind one speaks of "inner" (SOR) and "outer" (Newton–Raphson) iterations.

7.6.3. Successive Line Overrelaxation

There is no reason that the concepts used in developing SOR cannot be applied to the line relaxation method of the preceding section. All this method involves is using Eq. (7.55) to compute new estimates of the variables on the jth horizontal line and then using acceleration of (7.58) to give the final value. To be more definite, this method solves the tridiagonal system

$$-\tfrac{1}{4}\tilde{\phi}_{i-1,j}^{(p+1)} + \tilde{\phi}_{i,j}^{(p+1)} + \tfrac{1}{4}\tilde{\phi}_{i+1,j}^{(p+1)} = \tfrac{1}{4}(\phi_{i,j+1}^{(p)} + \phi_{i,j-1}^{(p)} - b_{i,j}) \tag{7.69}$$

and then computes

$$\phi_{i,j}^{(p+1)} = \omega\tilde{\phi}_{i,j}^{(p+1)} + (1-\omega)\phi_{i,j}^{(p)} \tag{7.70}$$

This method is no more difficult to program than the line relaxation method and is called *successive line overrelaxation* (SLOR). The equations could be combined as they were in the point SOR method, but there is no important reason for doing so.

We do not have a formula for exact eigenvalues of the line relaxation method, so the theory used to compute the optimum relaxation factor of the SOR method cannot be used here. Despite this, a few estimates can be given. Line relaxation converges approximately twice as fast as Gauss–Seidel, which

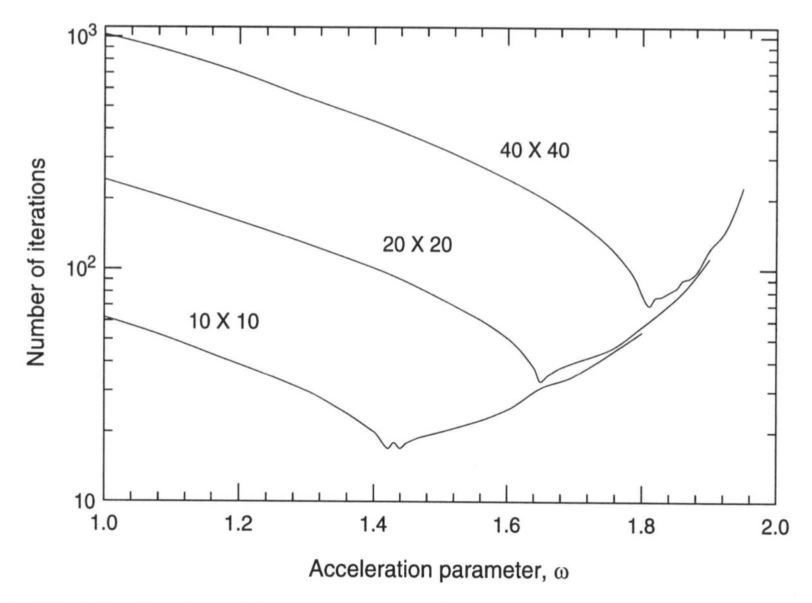

FIGURE 7.12. Number of iterations required for the successive line overrelaxation (SLOR) method to converge as a function of the overrelaxation factor ω for three different numbers of grids.

indicates that its eigenvalues (the largest ones at least) are approximately the squares of the Gauss–Seidel eigenvalues or the fourth powers of the Jacobi eigenvalues. Since line relaxation eigenvalues are smaller than the corresponding Gauss–Seidel eigenvalues, we expect that the optimum relaxation factor for SLOR to be smaller than its SOR counterpart. A first estimate can be obtained by using the SOR formula (7.65) with λ^2 replaced by λ^4. Alternatively, we can use the search procedure suggested above for the SOR method. This is illustrated by the following example.

Example 7.5: SLOR Solution of Laplace's Equation The problem of the preceding examples is solved using the SLOR method. As we have stated, the optimum overrelaxation factor is not known. Using the procedure recommended above, that is, guessing the largest eigenvalue of the line relaxation method to be the square of the largest Gauss–Seidel eigenvalue (or the fourth power of the largest Jacobi eigenvalue) and applying Eq. (7.67), we estimate that the optimum overrelaxation factor on 10×10, 20×20, and 40×40 grids, to be 1.438, 1.653, and 1.805, respectively. These values are very close to the minima shown in Figure 7.12.

Suppose we did not know the optimum value of ω. We might try a calculation on a 10×10 grid with, say, $\omega = 1.30$. By having the program write the result of each iteration at one grid point on the screen, we can see that the convergence is monotonic, that is, the error at that point does not change sign during the iteration process. With $\omega = 1.40$, the convergence is again monotonic. At $\omega = 1.50$, the error oscillates in sign so we assume that the optimum value lies between the last two values of ω. By further narrowing the limits, we find that $\omega = 1.41$ gives monotonic convergence and $\omega = 1.42$ yields oscillatory behavior. This is a narrow enough range and produces convergence in a number of iterations just slightly greater than the minimum. It is also not far from the estimate obtained from the approximation mentioned above. This procedure can also be used in more complex problems.

7.7. ALTERNATING DIRECTION IMPLICIT METHODS

The SOR method is very effective in many problems, but there are cases for which the optimum acceleration parameter is difficult to find and/or a large number of iterations are needed; this provides incentive to look further. To see what might be a good method, recall once more that iterative methods for elliptic problems are analogous to methods for advancing the solutions of parabolic equations in time. The difference is that, in the elliptic case, we want the solution to relax to the steady-state solution in the minimum number of iterations or as few time steps as possible. This suggests that we give some consideration to methods for solving parabolic problems that allow large time steps, in other words, methods for parabolic equations that are unconditionally stable. Of the methods discussed in Chapter 6, the most likely candidate is the ADI or splitting method.

The discretized system of equations we wish to solve is again Eq. (7.13) with the matrix A given by Eq. (7.12); this matrix has five nonzero diagonals. The diagonals of 1's immediately above and below the main diagonal come from the finite difference approximation to the operator $\partial^2/\partial x_1^2$, while the diagonals of 1's that are displaced M rows or columns from the main diagonal arise from finite differencing the operator $\partial^2/\partial x_2^2$. Parts of the main diagonal terms come from both operators. If the mesh spacings were unequal, the structure of the matrix (i.e., the placement of the nonzero elements) would be the same, but the numerical values would differ. If the equation contained terms involving $\partial/\partial x_1$ and/or $\partial/\partial x_2$, the matrix would not be symmetric. Finally, the presence of a term proportional to ϕ in the partial differential equation (as in the Helmholtz equation) would modify the elements on the main diagonal. None of these changes alters the applicability of any of the methods described above, but they may make finding the optimum parameter of the SOR method difficult.

Let us consider adapting the ADI method to the solution of elliptic problems. The original splitting of the matrix A that leads to the ADI method is, in matrix form,

$$A = H + V \tag{7.71}$$

where H (for horizontal) contains the terms arising from the $\partial^2/\partial x_1^2$ operator and V (for vertical) contains those coming from the $\partial^2/\partial x_2^2$ operator. For uniform grids that are the same in both directions, these matrices are defined by

$$(H\phi)_{i,j} = \phi_{i+1,j} - 2\phi_{i,j} + \phi_{i-1,j} \tag{7.72}$$

$$(V\phi)_{i,j} = \phi_{i,j+1} - 2\phi_{i,j} + \phi_{i,j-1} \tag{7.73}$$

Each matrix contributes -2 to the main diagonal; H has two diagonals of $+1$'s immediately adjacent to the main diagonal, while V has two diagonals of $+1$'s displaced by M positions from the main diagonal. In the matrix representing the two-dimensional Laplace operator, we could reverse the roles of the two directions so the structure of these matrices is the matter of choice. Both H and V are essentially tridiagonal matrices.

In terms of these matrices the ADI equations for the parabolic case can be written (see Section 6.5):

$$(I - \tfrac{1}{2}\beta_1 H)\phi^{p+1/2} = (I + \tfrac{1}{2}\beta_2 V)\phi^p + \mathbf{d}_h \tag{7.74}$$

$$(I - \tfrac{1}{2}\beta_2 V)\phi^{p+1} = (I + \tfrac{1}{2}\beta_1 H)\phi^{p+1/2} + \mathbf{d}_v \tag{7.75}$$

where, as earlier, $\beta_i = D\,\Delta t/h_i^2$ and $\mathbf{d}_h$ and $\mathbf{d}_v$ contain the terms due to the boundary conditions in the horizontal and vertical directions, respectively. When applied to the solution of elliptic problems, many authors write the ADI method with $\rho = 1/\beta$ rather than β; the choice is a matter of taste.

Since we want the iteration procedure to reach the steady solution as quickly as possible, it would seem that we should use a very large value of β (since the method is unconditionally stable, we can make it as large as we like), this is equivalent to using a large time step for the heat equation. Unfortunately, the matter is not that simple. To understand the issue better, we again expand the solution (or the error) in terms of eigenvectors of the iteration matrix. The problem we want to solve is posed by Eqs. (7.74) and (7.75); the convergence error satisfies the homogeneous version of these equations, that is, the same equations with the right-hand sides set to zero. The error at iteration p can be expanded in terms of the eigenvectors of A, which as we have seen earlier, are the products of sine functions of the two coordinates. Formally,

$$\boldsymbol{\epsilon}^{(p)} = \sum_{m,n} \epsilon_{mn}^{(p)} \psi^{(m,n)} \tag{7.76}$$

By extending the analysis of Section 7.5, we can show that

$$\epsilon_{mn}^{(p+1)} = \rho_{mn}\epsilon_{mn}^{(p)} \tag{7.77}$$

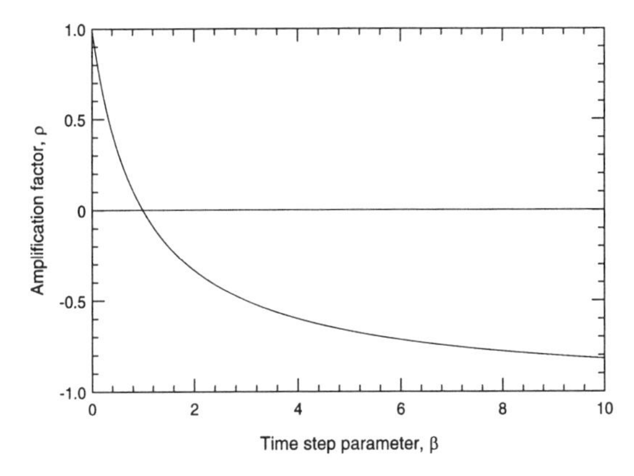

FIGURE 7.13. Amplification factor for the ADI method.

where

$$\rho_{mn} = \frac{1 + \beta_{1,m}}{1 - \beta_{1,m}} \frac{1 + \beta_{2,n}}{1 - \beta_{2,n}} \tag{7.78}$$

and the $\beta_{i,l}$ are given by Eq (6.65).

The behavior of ρ_{mn} as a function of $\beta_{1,m}$ for $\beta_{2,n} = 0$ is shown in Figure 7.13; the behavior for other values of $\beta_{2,n}$ is similar. The important thing to note is that the curve is asymptotic to -1 at large values of $\beta_{1,m}$. So, for very large time steps, $\rho_{mn} \to -1$ and the error oscillates in sign with very little damping, meaning that the method converges very slowly. Since small time steps (or small β) also produce slow convergence, there must be an optimum time step or β for this method. From Figure 7.13, it appears that $\beta = 1$, for which the amplification factor is zero, is ideal. However, each eigenvector because it has its own eigenvalue also has its own β. We can choose the time step to make any one of the βs zero; this will eliminate the component of the error corresponding to that eigenvector in a single iteration.

It is impossible to eliminate all of the components of the error in a single iteration. If the same time step (same β) is used for every iteration, there should be a value of β that yields the most rapid convergence. As we shall see in the example below, the number of iterations required for convergence does not change rapidly near the minimum, so there is a fairly broad range of β that produces good convergence. For a uniform grid with the same number of points, N, in both directions, the optimum β is $\approx N/2$.

It is possible to improve on this. There is no reason why the same value of β must be used each time; the time step can be different on each iteration. A good procedure is to use a fixed but small set of iteration parameters (β) in a cyclic way. On the first iteration, we use $\beta = \beta_1$, on the second iteration, $\beta = \beta_2$, and so on, until on the nth iteration, $\beta = \beta_n$. Then, on the $(n + 1)$st iteration, we again use β_1 and start the cycle over. The trick is to pick the set of iteration parameters such that the product of the amplification factors for n iterations is as small as possible for as many of the eigenvectors as possible. This is not an easy task and is a major reason why the ADI method is not used very often. For geometries other than simple ones like the rectangle, it is difficult to find a good set of parameters.

For a rectangular domain it is possible to find a good set of parameters. If a is a lower bound on the eigenvalues of H and V (i.e., the smallest eigenvalue of either) and b is an upper bound, then Wachspress showed that a good set of parameters is

$$\rho_k = \frac{1}{\beta_k} = b\left(\frac{a}{b}\right)^{(k-1)/(p-1)} \qquad k = 1,2,\ldots,n \tag{7.79}$$

Using these parameters with a cycle of length n, one can show that the method converges in approximately $(n/8)(2N)^{1/n-1}$ iterations, which is much faster than the SOR method. The ADI method is thus extremely efficient when the proper parameters can be found. In cases in which the same elliptic equation must be solved many times (for different sources or boundary conditions), it may pay to take the trouble to find a good set of iteration parameters for the ADI method by experimenting. Otherwise, the ADI method may not be a good choice.

Example 7.6: Fixed ADI Solution of Laplace's Equation In this example, Laplace's equation with the boundary conditions used in the preceding examples is solved with the ADI method with a fixed time step. We shall study the behavior of the method as β is varied. The number of iterations required for convergence as a function of β for several grid sizes is shown in Figure 7.14.

For small values of β the method converges slowly but the rate of improvement as β is increased is quite large. The error decreases monotonically with iteration number for small values of β, indicating that the principal sources of error are the small index eigenvalues and eigenvectors.

The optimum value of β is approximately $N/2$ for this problem. The convergence behavior of the ADI method as a function of β is similar to the behavior of the SOR method with respect to its parameter, ω. The major difference is that the rate of change of the error near the optimum value of β is a little less rapid than that of the SOR method near the optimum value of ω. In other words, the ADI method is a bit less sensitive to its parameter than the SOR method is to its parameter.

For large values of β, the convergence is oscillatory, as might be anticipated from Eq. (7.78). For the 10×10 grid, the number of iterations required to reduce the error to 2×10^{-4} is 15. This is less than the SOR method but slightly more than SLOR. Since the computations involved in ADI are similar to those in SLOR, but two sweeps are required per iteration, the choice between the methods is a close one but favors SLOR.

TABLE 7.2. Convergence of the Cyclic ADI Method

Number of Points	Number of Iterations
10	12
20	15
40	21
60	27

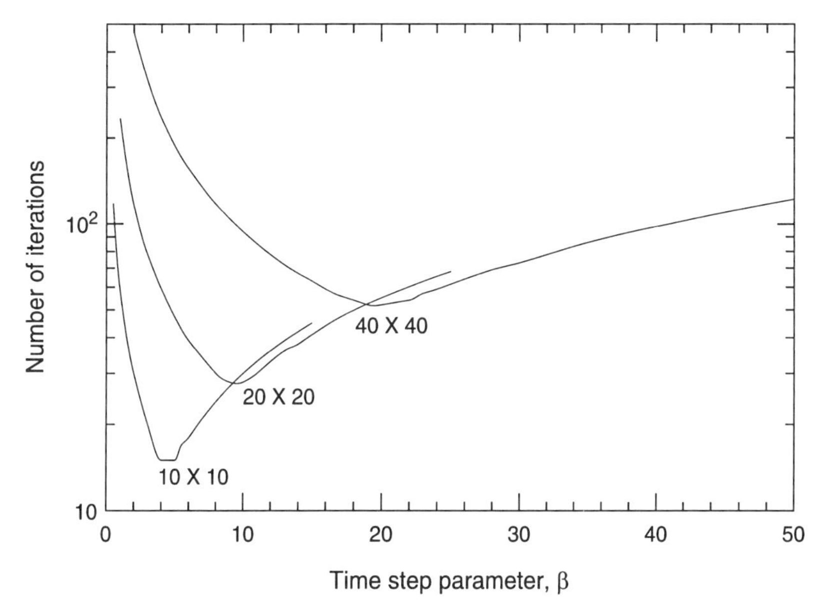

FIGURE 7.14. Number of iterations required by the ADI method as a function of the time step parameter β.

Example 7.7: Variable Parameter ADI Solution of Laplace's Equation Let us repeat the problem of the previous examples with the cyclic ADI method suggested above. Various numbers of points were used in each direction and the effective time steps, β, used in the cycle were determined from Eq. (7.79). We chose a cycle length of three iterations as a compromise. If too small a cycle is used, little benefit is obtained from the cyclic method; with too large a cycle, the number of iterations may be increased because convergence is tested only after a cycle is completed. With three values of the parameter, one acts to remove the high eigenvalue components of the error, the second removes the midrange eigenvalue components of the error and the last one removes the error belonging to the small eigenvalues. The results are shown in Table 7.2 and Figure 7.15.

From the table we see that the number of iterations required is by far the fewest of any method that we have considered so far. Furthermore, the increase

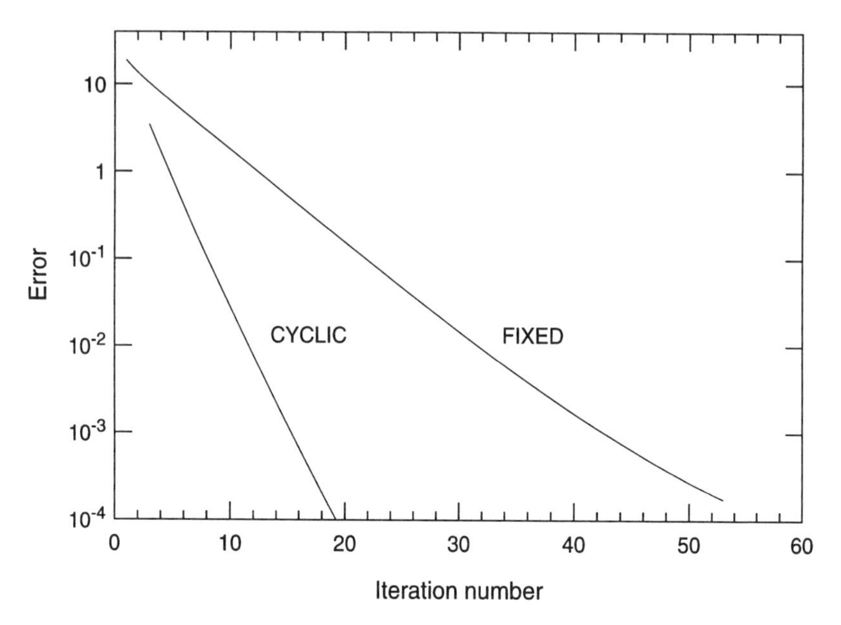

FIGURE 7.15. Convergence error in the solution of Laplace's equation by the cyclic and fixed versions of the ADI method.

in the number of iterations required with the number of points is much slower than for any of the earlier methods, showing that the cyclic ADI method is especially useful when large numbers of points are used.

From Figure 7.15, we see that the cyclic method converges more rapidly than the fixed β method. This shows that ADI is capable of extremely good performance when a good set of parameters is used. For the square region of this example, it was not difficult to find the parameters. The same problem was run on grids that had unequal numbers of points in the two directions and obtained equally good results using the parameters that were optimum for equal numbers of points in each direction.

The advantage of the cyclic method is even greater in difficult problems, in particular, ones with more mesh intervals.

7.8. INCOMPLETE *LU* DECOMPOSITION: STONE'S METHOD

We begin this section with a couple of observations. First, we have seen that there is a wide variation in the effectiveness of iterative methods. The basis for many of these methods is an additive decomposition of the matrix:

$$A = M - N \tag{7.80}$$

An iterative method for the system of equations:

$$A\phi = \mathbf{b} \tag{7.81}$$

is

$$M\phi^{(p+1)} = N\phi^{(p)} + \mathbf{d} \tag{7.82}$$

For some of the methods described above, the decomposition is not obvious. Generally, the closer M is to the original matrix A, the "smaller" N is, and the faster the convergence of the method.

The second observation is to note that, in Chapter 1, LU decomposition was found to be an excellent general-purpose linear systems solver because it takes advantage of the ease of inverting triangular matrices. Unfortunately, the LU method cannot take advantage of the sparseness of a matrix because the factors are not as sparse as the original matrix; this is one reason why LU decomposition is not used to solve the systems of algebraic equations arising from elliptic problems.

These observations suggest use of an *approximate* LU factorization of A as the matrix M. The concept is similar to the approximate factorization used in the ADI method. The idea is that we would like to choose an M that can be expressed as an LU decomposition

$$M = LU = A + N \tag{7.83}$$

in which L and U are as sparse as A and N is small in some sense.

One well-known method of this kind for symmetric matrices is the *incomplete Cholesky* method; it is a special case of the method developed in this section and will not be presented separately. The asymmetric version of this method, called *incomplete LU* factorization, or ILU, has not found widespread use because they converge rather slowly. They do find use as components of other methods.

A more specialized incomplete lower–upper decomposition method was proposed by Stone (1968). This method, called the *strongly implicit procedure* (SIP), is specifically designed for algebraic equations that are discretizations of partial differential equations. We shall describe the method for the five-point difference scheme, but it can be applied to other discretizations. A disadvantage of this method is that it needs to be rederived for each application.

The key to Stone's method is that L and U are allowed to have nonzero elements only on the nonzero diagonals of A. The problem is that a product of matrices with these structures is banded but it has more nonzero diagonals than the original matrix A. For the matrix of the five-point discretization, the product LU has two more nonzero diagonals than A. For the ordering of nodes used here, the extra two diagonals correspond to the nodes NW or $(i-1, j-1)$ and SE or $(i+1, j+1)$. To make the matrices L and U unique, the elements of the main diagonal of U are set to unity. Thus five sets of elements (on three diagonals of L and two of U) need to be determined.

For matrices having the form required of L and U, the definition of matrix multiplication gives the elements of the product of L and U, $M = LU$:

$$\begin{aligned}
M_{Wi,j} &= L_{Wi,j} \\
M_{NWi,j} &= L_{Wi,j}U_{Ni,j-1} \\
M_{Si,j} &= L_{Si,j} \\
M_{Pi,j} &= L_{Wi,j}U_{Ei,j-1} + L_{Si,j}U_{Ni-1,j} + L_{Pi,j} \\
M_{Ni,j} &= U_{Ni,j}L_{Pi,j} \\
M_{SEi,j} &= L_{Si,j}U_{Ei-1,j} \\
M_{Ei,j} &= U_{Ei,j}L_{Pi,j}
\end{aligned} \tag{7.84}$$

The object is to select L and U so that M is a good approximation to A. By construction, N must have at least two nonzero diagonals—the ones corresponding to the two nonzero diagonals of M that are zero in A. We could let N have nonzero elements on just these two diagonals, and force the other diagonals of M to equal the corresponding diagonals of A. This is the standard ILU method; as stated above, it is converges slowly.

Stone (1968) recognized that convergence can be improved by allowing N to have nonzero elements on all the diagonals corresponding to the nonzero diagonals of LU. The method is most easily derived by considering the vector $M\phi$:

$$(M\phi)_P = M_P\phi_P + M_S\phi_S + M_N\phi_N + M_E\phi_E + M_W\phi_W + M_{NW}\phi_{NW} + M_{SE}\phi_{SE} \tag{7.85}$$

where the elements are given by Eq. (7.84). The last two terms are the "extra" ones. Each term in this equation corresponds to a diagonal of $M = LU$.

The matrix N must contain the two extra diagonals of M, and we want to choose the elements on the remaining diagonals so that the norm of $N\phi$ is as small as possible. In other words,

$$N_P\phi_P + N_N\phi_N + N_S\phi_S + N_E\phi_E + N_W\phi_W + M_{NW}\phi_{NW} + M_{SE}\phi_{SE} \approx 0 \tag{7.86}$$

This requires the contribution of the two extra terms in the above equation to be nearly canceled by the contribution of other diagonals. In other words, we want Eq. (7.86) to reduce to

$$M_{NW}(\phi_{NW} - \phi^*_{NW}) + M_{SE}(\phi_{SE} - \phi^*_{SE}) \approx 0 \tag{7.87}$$

where ϕ^*_{NW} and ϕ^*_{SE} are approximations to ϕ_{NW} and ϕ_{SE}.

It is here that Stone takes advantage of the fact that the equations we are trying to solve are derived from an elliptic partial differential equation. As a result, we know that the elements of the vector ϕ at the nodal values of a smooth function so ϕ^*_{NW} and ϕ^*_{SE} can be approximated in terms of the values of ϕ at nodes corresponding to the diagonals of A. Stone proposed the following approximation (other approximations are possible):

$$\begin{aligned} \phi^*_{NW} &\approx \alpha(\phi_W + \phi_N - \phi_P) \\ \phi^*_{SE} &\approx \alpha(\phi_S + \phi_E - \phi_P) \end{aligned} \tag{7.88}$$

If $\alpha = 1$, these are second-order accurate interpolations, but Stone found that stability requires $\alpha < 1$, as will be verified in the example below.

If these approximations are substituted into Eq. (7.87) and the result is equated to Eq. (7.86), the elements of N can be written as linear combinations of M_{NW} and M_{SE}. The elements of M, Eq. (7.84), can now be set equal to the sum of elements of A and N. The resulting equations are not only sufficient to determine all of the elements of L and U, but they can be solved in sequential order beginning at the southwest corner of the grid:

$$\begin{aligned} L_{Wi,j} &= A_{Wi,j}/(1 + \alpha U_{Ni,j-1}) \\ L_{Si,j} &= A_{Si,j}/(1 + \alpha U_{Ei-1,j}) \\ L_{Pi,j} &= A_{Pi,j} + \alpha(L_{Wi,j}U_{Ni,j-1} + L_{Si,j}U_{Ei-1,j}) \\ &\quad - L_{Wi,j}U_{Ei,j-1} - L_{Si,j}U_{Ni-1,j} \\ U_{Ni,j} &= (A_{Ni,j} - \alpha L_{Wi,j}U_{Ni,j-1})/L_{Pi,j} \\ U_{Ei,j} &= (A_{Ei,j} - \alpha L_{Si,j}U_{Ei-1,j})/L_{Pi,j} \end{aligned} \tag{7.89}$$

The coefficients must be calculated in this order. Any matrix element that carries an index corresponding to a boundary is understood to be zero. Thus, along the west boundary ($i = 1$), elements with index $i = 0$ are zero and so forth.

We now turn to solving the system of equations. We use the update formulation of the method given in Section 7.2. The equation relating the update to the residual is

$$M\delta^{n+1} = LU\delta^{n+1} = \rho^n \tag{7.90}$$

These equations are solved as in generic *LU* decomposition. Multiplication of the above equation by L^{-1} leads to

$$U\delta^{n+1} = L^{-1}\rho^n = R^n \tag{7.91}$$

The elements of R^n are easily computed:

$$R_{i,j} = (\rho_{i,j} - L_{Si,j}\rho_{i-1,j} - L_{Wi,j}\rho_{i,j-1})/L_{Pi,j} \tag{7.92}$$

Equations (7.91) are solved by marching in the same way as in the Gauss–Seidel method. When the computation of R is complete, we need to solve Eq. (7.91):

$$\delta_{i,j} = R_{i,j} - U_{Ni,j} R_{i-1,j} - U_{Ei,j} R_{i,j-1} \tag{7.93}$$

in reversed order.

In the SIP method, the elements of the matrices L and U need to be calculated only once, prior to the first iteration. On subsequent iterations, we need calculate only the residual, then R and finally δ, by solving the two triangular systems.

Stone's method usually converges rapidly. The rate of convergence can be improved by varying α from iteration to iteration, but this requires the factorization to be redone. Since computing L and U is as expensive as an iteration, it is more efficient to keep α fixed.

This method can be generalized to yield an efficient solver for other discretizations.

Example 7.8: Solution of Laplace's Equation with the Strongly Implicit Method We again solve the problem used in the previous examples, this time with the SIP method. The program used, called SIP, may be found at the Internet site described in Appendix A. This method is a more complicated than the earlier ones, so the program contains several subroutines. The main ones are SETA, which creates the matrix elements and source terms, SOLVEB, which performs the factorization of the matrix, SWEEP, which performs an iteration, and CONVCHK, which computes the residuals and checks whether the convergence criterion has been met.

The number of iterations required for convergence as a function of the parameter α is shown in Figure 7.16 for 20×20, 40×40, and 60×60 grids. We see that the parameter α must be less than unity for convergence. As with several other methods we have studied, the rate of convergence is quite sensitive to the parameter. The optimum value of α lies in the range 0.92 to 0.94 independent of the grid size but using a value that is a little too large can lead to divergence. For that reason, it is recommended that $\alpha = 0.9$ be used. Recall that the incomplete Cholesky method corresponds to $\alpha = 0$; it converges much slower than SIP.

The numbers of iterations required to solve Laplace's equation with the strongly implicit method with the optimum parameter is given in Table 7.3. We see that the number of iterations is proportional to the number of grid points in each direction, making the SIP method competitive with the SOR and ADI methods.

We have not shown the error distribution, but it is a smooth function of the independent variables. This, together with the relatively rapid convergence of the SIP method make it an ideal base method for the multigrid and conjugate gradient methods described below.

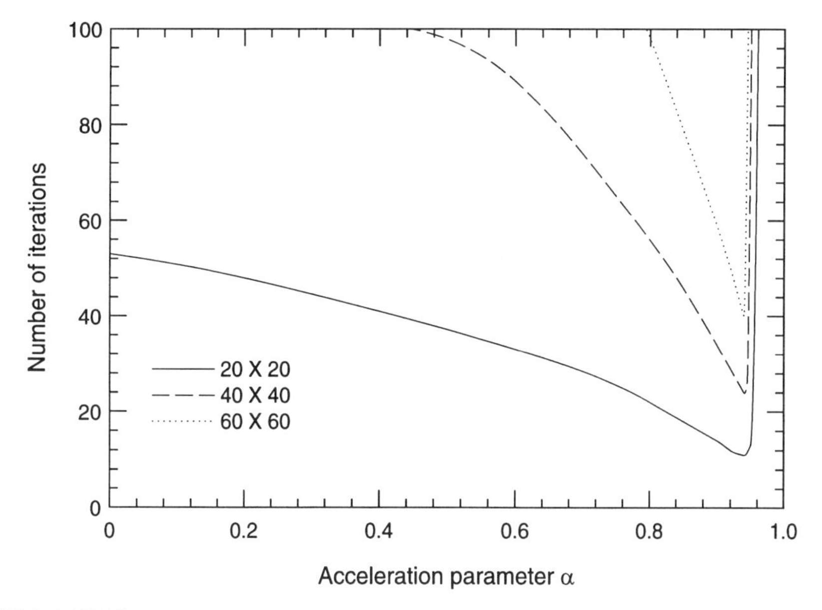

FIGURE 7.16. Number of iterations required for Stone's strongly implicit method to converge as a function of the parameter α.

TABLE 7.3. Convergence of the Strongly Implicit Method

Number of Points	Number of Iterations
10	6
20	12
40	24
60	40

7.9. METHODS FOR PARALLEL COMPUTERS

Over the past 40 years, enormous advances in computer speed and memory capacity were obtained by decreasing the size and increasing the speed of each device, putting more devices on a single chip, and improvements in computer architecture, among others. While these approaches promise further improvement for the future, it has become clear that some limits are being reached and that the rate of improvement in the performance of single processors or serial computers has already decreased.

It appears that the principal way in which increased performance will be obtained in the near future is through the use of massive parallelism. That is, the fastest computers of the future will contain many processors, each with very high individual speed, that will work collaboratively to solve complex problems. Machines of this kind already exist, but designs are still evolving, and, at this point, it is difficult to write programs that can run efficiently on all parallel machines. From the point of view of scientific and engineering computation, a major difficulty is that many of the methods designed to solve problems on serial machines do not work well on parallel computers. This is true of many of the methods described in this book.

A classic example of the difficulty is the method for solving tridiagonal systems of equations (or any other version of Gauss elimination for that matter), which plays an essential role in many of the methods described above. At each stage of the method, the diagonal element of the matrix and the element of the forcing vector in the next row are modified. The modified values are then used to perform the elimination process on the following row. As a result, work cannot begin on row $n + 1$ until all computations for row n have been completed. If this procedure is to be carried out on a parallel machine, the obvious method would assign the calculation at each node or grid point (or a group of neighboring nodes) to a different processor. If this is done, the $(n + 1)$st processor must remain idle until the first n processors have completed their tasks because it must wait for the results produced by the nth processor. With this arrangement, only one processor is active at a time. All of the work could be done in the same time (or less) on a single processor. Clearly, this is very inefficient use of a parallel machine. To use parallel machines effectively requires fresh thinking about the methods used to solve problems. It often turns out that a method that is poor on serial machines is much better suited to parallel machines.

Many of the iterative methods introduced earlier in this chapter do not perform well on parallel machines. The ADI method, which uses Gauss elimination as a building block, is one example. The Gauss–Seidel and SOR methods also do not port well to parallel machines because the result at a given point cannot be computed until work has been completed at the preceding point.

There has naturally been enormous interest in the development of methods for solving problems efficiently on parallel machines in recent years. There is no possibility of covering even a significant part of what has been done here. Rather, we shall give a few methods that illustrate some of the principles.

As we just saw, one of the major problems is the data dependencies found in Gauss elimination and related algorithms. These need to be avoided. There are two major directions that can be taken. One can either choose a method that avoids the data dependencies altogether or retain the method but do many problems in parallel. We shall give an example of each type, but we do not attempt to cover this subject in great detail.

7.9.1. Red–Black Gauss–Seidel Method

As noted, one way around the difficulty is to redesign the method so the data dependencies are removed. Of the iterative methods that have been described above, only the Jacobi method, which computes the new values entirely from old data, does not suffer from the need to wait for the preceding calculation to be completed before updating the solution. However, the Jacobi method has poor convergence properties and we would like a method with faster convergence.

One such method is red–black Gauss–Seidel. The basic equation that defines this method is identical to Eq. (7.50). The difference is that, to remove the data dependency, the sweep through the data is broken into two parts. On the first step, only the values at the points for which $i + j$ is even are updated; these correspond to the red squares of a checkerboard. On the second sweep, only the points with odd values of $i + j$ are updated; these correspond to the black squares on the checkerboard, hence the name of the method. (It is also called the checkerboard method.)

Since the five-point difference scheme for Laplace's equation (or any other elliptic PDE) allows the Jacobi method to compute the new value at a "red" point entirely from old values at "black" points (and, perhaps, its own old value), the data dependency is removed and all of the updates on the red points can be performed simultaneously; this can be followed with a update of the values at the black points. We can then assign each processor of a parallel machine a pair of adjacent points, one red and one black or a number of such pairs. All of the processors can be kept occupied all the time so this method parallelizes well.

Although this method resembles the Jacobi method more than the Gauss–Seidel method, it can be shown to converge at exactly the same rate as the latter and is therefore known as red–black Gauss–Seidel. (The errors produced by the two methods are not identical however.) Red–black Gauss–Seidel is more difficult to program on serial machines than the standard Gauss–Seidel method, so there is no reason to use it except on parallel machines.

Because the method behaves like the Gauss–Seidel method, it can be accelerated using the SOR procedure. The process simply involves using Eq. (7.64) first at the red points and then at the black points. Its behavior is essentially identical to that of the standard SOR method described earlier.

Because the results are identical to what we obtained with the SOR method, no example will be given here. However, a code for the method (for a serial machine) is found on the Internet site described in Appendix A. It is called RBSOR.

The concept behind this method can be generalized to allow more colors, for example, we might label the points with four different colors with each processor being responsible for one point of each color. These methods are required when discretizations other than the five-point scheme are used and are easy to program for parallel machines. They are similar to the two-color method and will not be described in detail here.

7.9.2. Parallelization of Other Methods

An alternative route to successful parallelization is to use an algorithm designed for a serial machine but perform several tasks simultaneously.

As an example, consider what would happen if the ADI method is used to solve an elliptic problem. The heart of this method is the solution of systems of equations on grid lines with the tridiagonal solver, which, as we have seen, is not well suited to parallel treatment. What we might do in the first stage of the method is to assign each horizontal line to a processor. It is possible to solve the equations for each line in a serial manner on a single processor, and the solutions on all of the lines can be done simultaneously on different processors. The problem comes when the second stage, which involves tridiagonal solutions in the vertical direction, is to be carried out. (The ADI method is successful only when the directions are alternated.) The difficulty is that the needed data are distributed across the processors. There are two ways around this difficulty.

In the first method, the tridiagonal solutions are performed with the data in its original location. To get some effectiveness from the parallel architecture, the process begins by doing the first step of the first tridiagonal solver on the first processor. The result is then passed to the second processor. At the same time, the first processor begins work on solving the problem on the second line. By staggering the work in this way, reasonably high efficiency can be obtained.

The other method is to transpose the data. That is, all of the data for each vertical line are transferred to a single processor. The transposition of the data must be carried out before starting the computation. Efficient procedures for transposition exist and are normally supplied with parallel machines; they do, however, take some time, which must be counted as wasted from the point of view of solving the problem.

7.10. MULTIGRID METHODS

One of the best ways of accelerating the solution of linear systems is the multigrid method. A simple version of this method was developed in Russia in the 1960s; it was brought to a state of maturity largely through the efforts of Brandt in Israel in the 1970s. The essential idea is relatively simple and, like many other good ideas, it has many extensions that have made it a useful tool for many types of problems.

The multigrid method is based on some observations about iterative methods. The rate of convergence of any method depends on the eigenvalues of the matrix associated with the method and, in particular, on the eigenvalue with largest magnitude (the *spectral radius*). The eigenvector belonging to this eigenvalue determines the spatial distribution of the convergence error, at least close to convergence; the nature of the error varies considerably from method to method. Let us briefly review the behavior of some of the methods presented above. The properties are given for Laplace's equation; most of them hold for other elliptic partial differential equations as well.

The first observation concerns the spatial distribution of the convergence error. The two largest eigenvalues of the Jacobi method are real and of opposite sign. One eigenvector represents a smooth function of the spatial coordinates, the other, a rapidly oscillating function. The convergence error is thus a mixture of very smooth and very rough components; as we saw, acceleration of the Jacobi method is very difficult. An underrelaxed version of the Jacobi method does yield a smooth error, but it is even more slowly convergent than the standard Jacobi method. On the other hand, the Gauss–Seidel method has a single real positive largest eigenvalue, and the convergence error is a smooth function of the spatial coordinates. The largest eigenvalues of the SOR method with optimum overrelaxation factor lie on a circle in the complex plane, and there are a number of them; consequently, the error behaves in a very complicated manner. In ADI, the nature of the error depends on the value of the parameter but tends to be a rather complicated function of the coordinates. Finally, the strongly implicit method (SIP) produces a smooth error and is more rapidly converging than the Gauss–Seidel method. Thus, we see that the nature of the error depends strongly on the iterative method used. For development of the multigrid method, iterative methods with spatially smooth convergence errors are of most interest; of the methods mentioned above, Gauss–Seidel and SIP are the best in this regard. Underrelaxed Jacobi iteration is another possibility.

Second, it is useful to consider the alternative or update form of iterative methods given in Section 7.2. This method requires solution of

$$M(\mathbf{x}^{n+1} - \mathbf{x}^n) = M\delta^{p+1} = (N - M)\mathbf{x}^n + \mathbf{b} = \mathbf{b} - A\mathbf{x}^n = \rho^n \qquad (7.94)$$

where ρ^n is the *residual*, and δ the update. The new solution ϕ^{n+1} is found by adding the update to the old solution ϕ^n. This is a more complicated way of implementing the iterative method. The major advantage is that, as we saw above, some methods yield an update (which is an approximation to the convergence error) that becomes smooth as the iterative process converges; the solution itself is not necessarily smooth.

Suppose we use a method that, after a few iterations, removes the rapidly varying components of the error. At this point, the update becomes a smooth function of the coordinates. If this is the case, the update could be computed accurately on a coarser grid; doing so yields a huge advantage. On a grid twice as coarse as the original one in two dimensions, each iteration requires only one fourth as much work; in three dimensions, the cost is reduced to one eighth of the fine grid cost. Furthermore, iterative methods converge more quickly on coarser grids. For example, the Gauss–Seidel method converges four times as fast on a grid that is twice as coarse.

This suggests that much of the work can and should be done on a coarser grid. To use this approach, we need to define the relationship between the two grids, the finite difference operator on the coarse grid, the basic iteration or *smoothing* method, a method of passing (*restricting*) a smoothed version of the residual from the fine grid to the coarse one and a method of interpolating (*prolonging*) the update or correction from the coarse grid to the fine one; the words in parentheses are special terms commonly used in the multigrid literature. There are many possibilities for each of these; within reason, the choice is not critical so we shall present just one good choice for each item.

The coarse grid normally consists of every second line of the fine grid, but there is no reason why this must be so, and grid ratios of three have been used successfully.

Although there is no reason to use the multigrid method in one dimension (because the tridiagonal algorithm can solve the problem very effectively), it is easy to illustrate the principles of the method in one dimension and to derive some of the procedures used in the two- and three-dimensional cases. Thus consider the problem:

$$\frac{d^2\phi}{dx^2} = f(x) \tag{7.95}$$

for which the standard finite difference approximation on a grid of size Δx is

$$\frac{1}{\Delta x^2}(\phi_{i-1} - 2\phi_i + \phi_{i+1}) = f_i \tag{7.96}$$

After performing n iterations with the Gauss–Seidel method on this grid, we obtain an approximate solution ϕ^n, and the above equation is satisfied to within the residual ρ^n:

$$f_i - \frac{1}{(\Delta x)^2}(\phi^n_{i-1} - 2\phi^n_i + \phi^n_{i+1}) = \rho^n_i \tag{7.97}$$

Subtracting this equation from Eq. (7.96) gives

$$\frac{1}{(\Delta x)^2}(\delta^{n+1}_{i-1} - 2\delta^{n+1}_i + \delta^{n+1}_{i+1}) = \rho^n_i \tag{7.98}$$

This is the equation we want to iterate on the coarse grid.

To derive the discretized equations on the coarse grid, it is convenient to use the control volume method. Refer to Figure 7.17 for the relationship between the two grids. The control volume around node I of the coarse grid in Figure 7.17 consists of the whole control volume around node i plus half of the control volumes around nodes $i-1$ and $i+1$ of the fine grid. This suggests that we construct the coarse grid equation by adding one half of Eq. (7.98) with indices $i-1$ and $i+1$ to the full equation with index i. The result is (the superscript $n+1$ is omitted)

$$\frac{1}{4(\Delta x)^2}(\delta_{i-2} - 2\delta_i + \delta_{i+2}) = \tfrac{1}{4}(\rho_{i-1} + 2\rho_i + \rho_{i+1}) \tag{7.99}$$

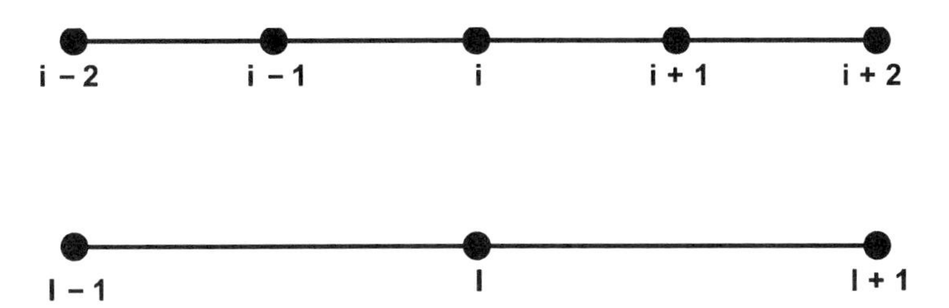

FIGURE 7.17. Illustrative grids for the multigrid technique in one dimension.

Using the relationship between the two grids ($\Delta X = 2\Delta x$, see Figure 7.17), this is equivalent to the following equation on the coarse grid:

$$\frac{1}{\Delta X^2}(\delta_{I-1} - 2\delta_I + \delta_{I+1}) = \overline{\rho}_I \tag{7.100}$$

which serves to define $\overline{\rho}_I$. The left-hand side of this equation is the standard approximation to the second derivative on the coarse grid, indicating that the obvious discretization on the coarse grid is a reasonable one. The right-hand side is a smoothing or filtering of the fine grid forcing term and provides the natural definition of the restriction operation.

The simplest prolongation or interpolation of a quantity from the coarse grid to the fine grid is linear interpolation. At coincident points of the two grids, the value at the coarse grid point is simply *injected* onto the corresponding fine grid point. At fine grid points that lie between the coarse grid points, the injected value is the average of the neighboring coarse grid values (linear interpolation).

A two-grid iterative method is thus:

- On the fine grid, perform a few iterations with a method that gives a smooth error.
- Compute the residual on the fine grid.
- Smooth (restrict) the residual to the coarse grid.
- Perform a few iterations of the correction or update equation on the coarse grid.
- Interpolate (prolong) the correction to the fine grid.
- Update the solution on the fine grid.
- Repeat the entire procedure until the residual is reduced to the desired level.

At this point, it is natural to ask: Why not use still coarser grids to further improve the rate of convergence? This is a good idea. In fact, one should continue the procedure until it becomes impossible to define a still coarser grid; on the coarsest grid, the number of unknowns is so small that the equations can be solved directly at negligible cost.

The multigrid approach is more a strategy than a particular method. Within the framework just described, many parameters can be selected more or less arbitrarily: the coarse grid structure, the smoother or basic iteration method, the number of iterations on each grid, the order in which the various grids are visited, and the restriction and interpolation schemes, among others. The rate of convergence does depend on the choices made, but the range of performance between the worst and the best methods is often less than a factor of 2. The multigrid method tends to be a great leveler.

The most important property of the multigrid method is that the number of iterations on the finest grid required to reach a given level of convergence is essentially independent of the number of grid nodes. This is as good as one can expect to do—the total computational cost of solving a problem is proportional to the number of grid nodes, that is, the cost per grid point is independent of the number of grid points. In two- and three-dimensional problems with about 100 nodes in each direction, the multigrid method may converge in one-tenth to one-hundredth of the time required by the basic method. An example will be presented below.

The iterative method on which the multigrid method is based must be a good smoother; its convergence properties as a stand-alone method are less important. Of the methods presented so far, Gauss–Seidel is a good choice and SIP is an even better one. Some people prefer to use the under-relaxed Jacobi method, which produces a smooth error; in the author's experience, it has no advantage with respect to the Gauss–Seidel method.

In two dimensions, there are many possibilities for the restriction operator. If the method described above is used in each direction, the result is a nine-point scheme. A simpler, but nearly as effective, restriction is the five-point scheme:

$$\overline{\rho}_{I,J} = \tfrac{1}{8}(\rho_{i+1,j} + \rho_{i-1,j} + \rho_{i,j+1} + \rho_{i,j-1} + 4\rho_{i,j}) \tag{7.101}$$

Similarly, an effective prolongator is bilinear interpolation. In two dimensions, there are three kinds of points on the fine grid. Those which correspond to coarse grid points are given the value at the corresponding point (direct injection). Those that lie on lines connecting two coarse grid points receive the average of the two coarse grid values. Finally, the points at the centers of coarse grid volumes take the average of the four neighbor values. Similar schemes can be derived for three-dimensional problems. On nonuniform grids, the averages must be weighted.

The initial guess is usually far from the correct solution (a zero field is often used). It therefore makes sense to solve the equation first on a very coarse grid (which is cheap) and use that solution to provide a better guess for the initial field on the next finer grid, a method we earlier caller refinement. By the time we reach the finest grid, we already have a fairly good starting solution. Multigrid methods based on this idea are called *full multigrid* (FMG) methods. The cost of obtaining the solutions on the coarse grids is usually more than compensated for by the savings in fine grid iterations. In many cases the FMG method reduces the overall cost of computing a solution by about half.

Finally, we remark that it is possible to construct a method in which one solves equations for approximations to the solution rather than for corrections at each grid. This is called the *full approximation scheme* (FAS) and is often used for solving nonlinear problems. It is important to note that the solution on each grid in FAS is *not* the solution that would be obtained if that grid were used by itself but a smoothed version of the fine grid solution; this is achieved by passing a correction from each grid to the next coarser grid.

For a detailed analysis of multigrid methods, see the books by Hackbusch (1984) and Briggs (1987).

Example 7.9: Multigrid Solution of Laplace's Equation Let us now apply the multigrid method to the solution of Laplace's equation with the same boundary conditions that were used in the earlier examples in this chapter. The basic iterative method is the Gauss–Seidel method; the underrelaxed Jacobi method gives slower convergence with no apparent compensating advantage. When the SIP method was used, the number of cycles decreased by about a third, but the extra cost of computing the factorizations reduces that advantage to something much smaller than that.

Before the results are presented, a few programming details should be discussed. As is clear from the description of the method, the multigrid method requires grids and solutions at a number of levels; in obtaining the results presented below, as many as six levels of grids were used. The solutions on the various grids could be kept as separate arrays, but this would require a large number of arrays of varying size. It is simpler (and therefore standard procedure) to store the solution at all levels in a single one-dimensional array. This means that indices or pointers need to be used to locate the location in storage of the solution at a particular point at a particular level.

The code, MG, that was used for the example is found on the Internet site described in Appendix A. In the interests of simplicity, the code uses the straightforward multigrid method, that is, neither the full approximation scheme nor the full multigrid method is used; also, the simplest cycle and prolongation and restriction methods were used; each is found in a separate subroutine. Grid sizes up to 64×64 points may be used; larger grids could be accommodated by changing the dimensioning statements. All of the important operations: iteration, prolongation, and restriction are performed in separate subroutines so that the methods employed can easily be changed. The number of iterations at each level can also be changed.

It is traditional in evaluating multigrid methods to count the number of "work units" used. A work unit is the amount of computation required to perform one iteration on the finest grid. Because an iteration on each succeeding grid costs one fourth as much as an iteration on the next finer grid, the total cost of doing one iteration on each grid is approximately four thirds the cost of one fine grid iteration. This estimate makes the multigrid method look

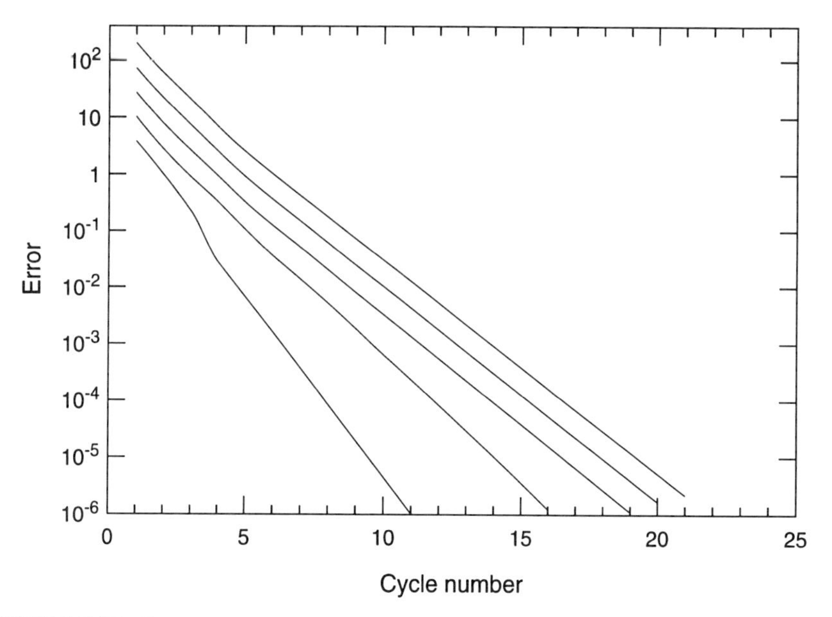

FIGURE 7.18. Convergence error in the multigrid method as a function of the number of cycles for various grids. The lines from left to right represent results for 4×4, 8×8, 16×16, 32×32, and 64×64 grids, respectively. The difference in the initial error is due to the normalization used and helps to separate the curves.

extremely effective. However, it ignores the overhead involved in doing the restriction and prolongation, which may double or triple the total cost. It is best to simply compare computation times.

As noted, the original (correction) form of the multigrid method is used in the code. This means that the finest level is the only one on which the solution is actually computed. At the next coarsest level, the variable is the correction, at the third level, a correction to the correction, and so forth. We experimented with a number of variations of the method and found that the most efficient method was one in which one iteration was performed on each level as the grid was coarsened and no iterations are performed as the grid is refined. The method starts at the fine grid and proceeds through the coarser levels until the coarsest level is reached and then the process is reversed; this is called the V-cycle and is only one of many prescriptions for visiting the grids. One needs to be careful about the updating at each level. All results shown below were obtained with this version of the method.

Figure 7.18 shows the convergence error of the multigrid method at the end of each cycle for various grids. For each grid, the maximum number of levels was used. We see that the multigrid method converges in the same number of cycles for all grids; the number of cycles is about 20 for the case shown here. This is typical of the multigrid method and theoretical arguments demonstrating that this result is to be expected can be constructed. Thus the amount of work per grid point is independent of the problem size, an extremely attractive property. One could hardly hope to do better.

In Figure 7.19, we show, for the 64×64 grid, the convergence error as a function of the number of cycles for the multigrid method with different numbers of levels. With only a single level, the method is Gauss–Seidel and converges very slowly. These results clearly show the advantage of using as many levels as possible.

There is another argument showing why the multigrid method works so well. In elliptic problems, what happens at any grid point affects the solution at all other grid points. For an iterative method to converge, information must travel back and forth through the grid many times. In the simple iterative methods, such as Jacobi and Gauss–Seidel, information travels only one grid point per iteration; that is one reason why the convergence is so slow. By using coarse grids, the multigrid method allows information to propagate all the way across the grid and back in a single cycle. That is what makes it so effective.

7.11. CONJUGATE GRADIENT METHODS

7.11.1. Concept

We now present an unusual class of methods, but they are among the best ones now available. Conjugate gradient methods are based on a class of techniques for solving nonlinear equations called *descent* methods. To solve linear systems of equations, these methods convert the system into a minimization problem. Suppose we wish to solve $A\mathbf{x} = \mathbf{b}$ where the matrix A is symmetric and its eigenvalues are positive; such a matrix is called *positive definite*; we will generalize this method to other types of matrices later. For positive definite matrices, solving the system of equations is equivalent to finding the minimum of

$$F = \tfrac{1}{2}\phi^T A\phi - \phi^T \mathbf{b} = \tfrac{1}{2}\sum_{j=1}^{n}\sum_{i=1}^{n} A_{ij}\phi_i\phi_j - \sum_{i=1}^{n}\phi_i b_i \tag{7.102}$$

with respect to all the ϕ_i as may be verified by taking the derivative of F with respect to each variable and setting it equal to zero.

Other than the method of taking all of the derivatives of the function and setting them to zero (which brings us back to the original equations), the oldest and best known method for seeking the minimum of a function is *steepest descents*. The function F represents a surface in (hyper-)space. The starting guess is a point in that space. We find the steepest downward path on the surface at that point; it lies in the direction opposite to the gradient of the function. We then search for the lowest point on that line. By construction, the new point has a smaller value of F than the starting point; in this sense,

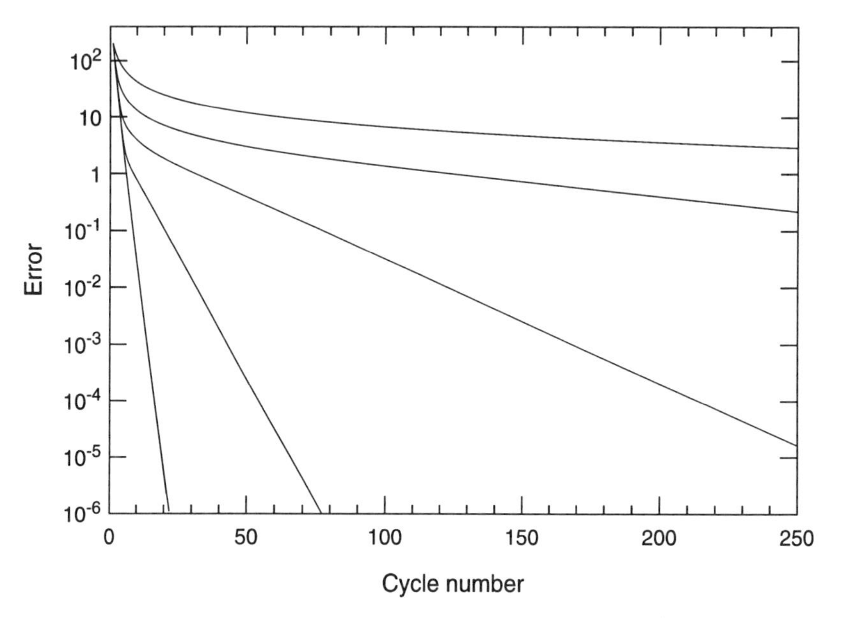

FIGURE 7.19. Convergence error in the multigrid method as a function of the number of cycles for a 64×64 grid for different numbers of coarsening levels. From top to bottom, the curves give the results for 2, 3, 4, 5, and 6 levels, respectively.

it is closer to the solution. The new value is then used as the starting point for another iteration, and the process is continued until convergence. While it is guaranteed to converge, the steepest descent method usually converges very slowly especially if the function F has a narrow valley; when it does, the method generally oscillates back and forth across the valley. In other words, the method tends to use the same search directions over and over again.

Many improvements have been suggested. The best ones require the new search directions to be as different from the old ones as possible. Among these is the *conjugate gradient* method. We shall give only the general idea and a description of the algorithm here; a more complete presentation can be found in Golub and van Loan (1989).

The conjugate gradient method is based on a remarkable discovery: It is possible to minimize a function with respect to several directions simultaneously while searching in one direction at a time. This is made possible by a clever choice of directions. We shall describe this for the case of two directions; suppose we wish to find values of α_1 and α_2 in

$$\phi = \phi^0 + \alpha_1 \mathbf{p}^1 + \alpha_2 \mathbf{p}^2 \tag{7.103}$$

which minimize F; that is, we try to minimize F in the $\mathbf{p}^1 - \mathbf{p}^2$ plane. This problem can be reduced to the problem of minimizing with respect to $\mathbf{p}^1$ and $\mathbf{p}^2$ individually provided that the two directions are conjugate in the following sense:

$$\mathbf{p}^1 \cdot A\mathbf{p}^2 = 0 \tag{7.104}$$

This property is akin to orthogonality; the vectors $\mathbf{p}^1$ and $\mathbf{p}^2$ are said to be conjugate with respect to the matrix A, which gives the method its name.

This property can be extended to any number of directions. In the conjugate gradient method, each new search direction is conjugate to all the preceding ones. If the matrix is nonsingular, the directions are guaranteed to be linearly independent. Consequently, with exact (no round-off error) arithmetic, the conjugate gradient method converges exactly when the number of iterations is the size of the matrix. In practice, exact convergence is prevented by arithmetic errors. For this reason, the conjugate gradient method is treated as an iterative method.

7.11.2. Preconditioning

While the conjugate gradient method guarantees that the error is reduced on each iteration, the size of the reduction depends on the search direction. It is not unusual for this method to reduce the error only slightly for a number of iterations and then find a direction that reduces the error by an order of magnitude or more in one iteration.

It can be shown that the rate of convergence of this method depends on the *condition number* κ of the matrix where

$$\kappa = \frac{\lambda_{\max}}{\lambda_{\min}} \tag{7.105}$$

and $\lambda_{\max}$ and $\lambda_{\min}$ are the largest and smallest eigenvalues of the matrix. The condition numbers of matrices that arise in engineering problems are often fairly large, so the standard conjugate gradient method converges slowly. Although the conjugate gradient method is significantly faster than steepest descents for a given condition number, it is still rather slow.

The basic method can be improved by replacing the problem with one that has the same solution but a smaller condition number. For obvious reasons, this is called *preconditioning*. One way to precondition the problem is to premultiply the equation by another (carefully chosen) matrix. As this would destroy the symmetry of the matrix, the preconditioned system must take the following form:

$$C^{-1}AC^{-1}C\phi = C^{-1}\mathbf{b} \tag{7.106}$$

The conjugate gradient method is applied to the matrix $C^{-1}AC^{-1}$, that is, to the modified problem (7.106). If this is done and the update form is used, the following algorithm results (for a detailed derivation, see Golub and van Loan, 1989). In this description, ρ^k is the residual at the kth iteration, $\mathbf{p}^k$ is the kth search direction, $\mathbf{z}^k$ is an auxiliary vector, and α_k and β_k are parameters used in constructing the new solution, residual, and search direction. The algorithm can be summarized as follows:

- Initialize by setting: $k = 0$, $\phi^0 = \phi_0$, $\rho^0 = \mathbf{b} - A\phi_0$, $\mathbf{p}^0 = 0$, $s_0 = 10^{30}$.
- Advance the counter: $k = k + 1$.
- Solve the system: $M\mathbf{z}^k = \rho^{k-1}$.
- Calculate:

$$
\begin{aligned}
s^k &= \rho^{k-1} \cdot \mathbf{z}^k \\
\beta^k &= s^k / s^{k-1} \\
\mathbf{p}^k &= \mathbf{z}^k + \beta^k \mathbf{p}^{k-1} \\
\alpha^k &= s_k / (\mathbf{p}^k \cdot A\mathbf{p}^k) \\
\phi^k &= \phi^{k-1} + \alpha^k \mathbf{p}^k \\
\rho^k &= \rho^{k-1} - \alpha^k A\mathbf{p}^k
\end{aligned}
$$

- Repeat until convergence.

This algorithm involves solving a system of linear equations at the first step of each iteration. The matrix involved is $M = C^{-1}$ where C is the preconditioning matrix, which is in fact never actually constructed. For the method to be efficient, M must be easy to invert. The choice of M used most often is the incomplete Cholesky factorization of A, but if M is the LU of Stone's SIP method, faster convergence is obtained. Examples will be presented below.

7.11.3. Biconjugate Gradients and CGSTAB

The conjugate gradient method presented above is applicable only to symmetric systems. To apply the method to systems of equations that are not symmetric, we need to convert an asymmetric problem to a symmetric one. There are a couple of ways of doing this of which the following is perhaps the simplest. Consider the system:

$$
\begin{pmatrix} 0 & A \\ A^T & 0 \end{pmatrix} \begin{pmatrix} \psi \\ \phi \end{pmatrix} = \begin{pmatrix} \mathbf{b} \\ \mathbf{0} \end{pmatrix} \tag{7.107}
$$

This system can be decomposed into two subsystems. The first is the original system; the second involves the transpose matrix and is irrelevant. When the preconditioned conjugate gradient method is applied to this system, the following method, called *biconjugate gradients*, results:

- Initialize by setting: $k = 0$, $\phi^0 = \phi_0$, $\boldsymbol{\rho}^0 = \mathbf{b} - A\phi_0$, $\overline{\boldsymbol{\rho}}^0 = \mathbf{b} - A^T\phi_0$, $\mathbf{p}^0 = \overline{\mathbf{p}}^0 = 0$, $s_0 = 10^{30}$.
- Advance the counter: $k = k + 1$.
- Solve the systems: $M\mathbf{z}^k = \boldsymbol{\rho}^{k-1}$, $M^T\overline{\mathbf{z}}^k = \overline{\boldsymbol{\rho}}^{k-1}$.
- Calculate:

$$\begin{aligned}
s^k &= \mathbf{z}^k \cdot \overline{\boldsymbol{\rho}}^{k-1} \\
\beta^k &= s^k / s^{k-1} \\
\mathbf{p}^k &= \mathbf{z}^k + \beta^k \mathbf{p}^{k-1} \\
\overline{\mathbf{p}}^k &= \overline{\mathbf{z}}^k + \beta^k \overline{\mathbf{p}}^{k-1} \\
\alpha^k &= s^k / (\overline{\mathbf{p}}^k A \mathbf{p}^k) \\
\phi^k &= \phi^{k-1} + \alpha^k \mathbf{p}^k \\
\boldsymbol{\rho}^k &= \boldsymbol{\rho}^{k-1} - \alpha^k A \mathbf{p}^k \\
\overline{\boldsymbol{\rho}}^k &= \overline{\boldsymbol{\rho}}^{k-1} - \alpha^k A^T \overline{\mathbf{p}}^k
\end{aligned}$$

- Repeat until convergence.

The above algorithm was published by Fletcher (1976). It requires twice as much effort per iteration as the standard conjugate gradient method but converges in about the same number of iterations. It is quite robust (meaning that it handles a wide range of problems without difficulty).

Other variants of the conjugate gradient method that are stable and robust have been developed. We mention the CGS (conjugate gradient squared) algorithm (Sonneveld, 1989), the CGSTAB (CGS stabilized) method (Van den Vorst and Sonneveld, 1990), and GMRES (generalized minimum residual) (Saad and Schultz, 1986). All of these can be applied to nonsymmetric matrices. We give the CGSTAB algorithm without formal derivation:

- Initialize by setting: $k = 0$, $\phi^0 = \phi_0$, $\boldsymbol{\rho}^0 = \mathbf{b} - A\phi_0$, $\mathbf{u}^0 = \mathbf{p}^0 = 0$.
- Advance the counter $k = k + 1$ and calculate:

$$\begin{aligned}
\beta^k &= \boldsymbol{\rho}^0 \cdot \boldsymbol{\rho}^{k-1} \\
\omega^k &= (\beta^k \gamma^{k-1}) / (\alpha^{k-1} \beta^{k-1}) \\
\mathbf{p}^k &= \boldsymbol{\rho}^{k-1} + \omega^k (\mathbf{p}^{k-1} - \alpha^{k-1} \mathbf{u}^{k-1})
\end{aligned}$$

- Solve the system: $M\mathbf{z} = \mathbf{p}^k$.

- Calculate:

$$\mathbf{u}^k = A\mathbf{z}$$

$$\gamma^k = \beta^k/(\mathbf{u}^k \cdot \boldsymbol{\rho}^0)$$

$$\mathbf{w} = \boldsymbol{\rho}^{k-1} - \gamma^k \mathbf{u}^k$$

- Solve the system: $M\mathbf{y} = \mathbf{w}$.
- Calculate:

$$\mathbf{v} = A\mathbf{y}$$

$$\alpha^k = (\mathbf{v} \cdot \boldsymbol{\rho}^k)/(\mathbf{v} \cdot \mathbf{v})$$

$$\phi^k = \phi^{k-1} + \gamma^k \mathbf{z} + \alpha^k \mathbf{y}$$

$$\boldsymbol{\rho}^k = \mathbf{w} - \alpha^k \mathbf{v}$$

- Repeat until convergence.

Note that **u**, **v**, **w**, **y** and **z** are auxiliary vectors and have nothing to do with the solution. This algorithm can be programmed as given above. Codes for various versions of the conjugate gradient method are available from Netlib. Simple versions used for the example can be found at the Internet site described in Appendix A.

Example 7.10: Solution of Laplace's Equation with the Conjugate Gradient Methods Now let us solve the problem that has been used in the preceding examples with the conjugate gradient methods. We shall use two preconditioners: the incomplete Cholesky method and the strongly implicit procedure of Stone. The codes used, ICCG and CGSIP, respectively, can be found on the Internet site described in Appendix A.

First, in Figure 7.20, we show the error as a function of iteration number for the conjugate gradient method with the SIP preconditioner; detailed parameters are given in the figure caption. We see that the convergence is monotonic (the error decreases at every iteration) but not exponential. As mentioned above, decrease in the error depends on the search direction and is therefore not exponenetial when the conjugate gradient method is used.

The number of iterations required for convergence is given in Table 7.4 for the conjugate gradient method preconditioned with the strongly implicit method with various values of the parameter α and the incomplete Cholesky method. We see first of all that the strongly implicit procedure is the better preconditioner and that its advantage over the incomplete Cholesky method increases as the number of grid points increases; since the cost of the two methods is essentially identical, it is the preferred method. Second, we find that the best value of the parameter in the strongly implicit method is $\alpha = 1$,

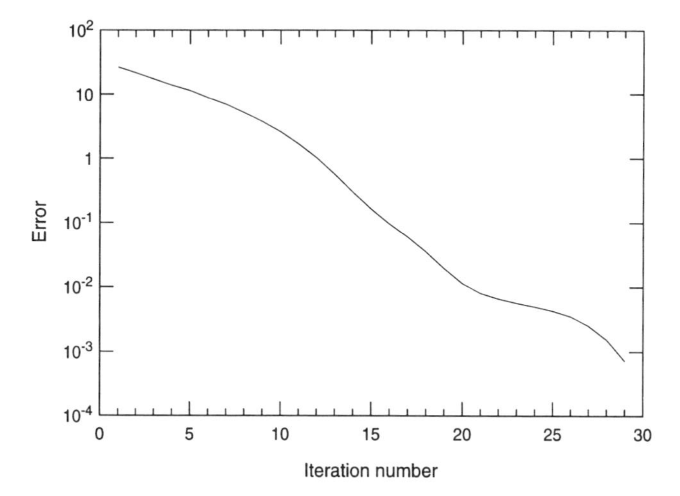

FIGURE 7.20. Convergence error of the conjugate gradient method as a function of iteration number. This case was run on a 60×60 grid with the strongly implicit procedure as a preconditioner; the parameter α of the SIP method was 1.

TABLE 7.4. Number of Iterations for Convergence of Conjugate Gradient Methods

Number Points	Strongly Implicit Procedure				Incomplete Cholesky
	$\alpha = 0.9$	$\alpha = 0.95$	$\alpha = 0.98$	$\alpha = 1$	
10×10	7	6	5	7	10
20×20	12	11	10	10	16
40×40	20	18	16	14	27
60×60	29	25	21	17	38

especially when the number of grid points is large; with a value greater than 1, the method does not converge. With $\alpha = 1$, the number of iterations required for the conjugate gradient method increases approximately as the square root of the number of grid points.

The methods of choice for solving elliptic and, indeed, other types of partial differential equations are the multigrid and conjugate gradient methods. The multigrid method has the advantage that its cost per grid point is independent of the number of points; the number of iterations required by the conjugate gradient method increases as the problem size grows. Thus, the multigrid method is preferred when it can be used but, because it is based on geometric ideas, it can be difficult to apply to complex geometries (defining the coarse meshes may become difficult); it is somewhat difficult to apply on parallel computers but it has been adapted to these machines. The conjugate gradient method, on the other hand, can be applied to any system of equations and can be made to work quite well on parallel machines.

7.12. ADAPTIVE GRIDS

Accuracy is critical in almost every application. Computations intended to test model validity or to be used for design purposes sometimes turn out to be inconclusive because the numerical errors are too large to allow extraction of useful information. It is essential that errors be estimated and reduced. In particular, the contributions of the discretization and convergence errors must be smaller than effects of other parameters.

Solutions to engineering problems often have large regions in which the solution is relatively smooth and a coarse grid can provide sufficient accuracy and other, often much smaller areas, where the rapid variation of the solution requires a much finer grid. The difficulty is that it may be hard, or even impossible, to locate these regions before solving the problem. As we saw in Chapters 3 and 5, the best way to deal with this difficulty is to use an adaptive method. This means that the grid is modified as the solution is computed. An ideal method would produce the solution with the accuracy the user demands throughout the domain automatically at the lowest possible cost. While this is a difficult order, it is clear that such a method must have several essential elements.

The first essential is a method of estimating errors; the Richardson method presented in Chapter 3 is a good choice but other methods can be used.

The objective is to make the error everywhere smaller than a tolerance δ. This could be accomplished by using methods of differing accuracy in different regions of the solution domain, but it is difficult to do this for partial differential equations, especially near boundaries, so this approach is impractical. One could use a highly refined grid everywhere, but this is almost certainly very wasteful and is often impractical as well. A better choice is to locally refine the grid where large errors are found. An experienced user may be able to generate a grid that yields a nearly uniform distribution of discretization error, but this is difficult to do, especially when the geometry and/or the physics of the problem are complex. Local grid refinement is the best choice. A survey of some of the better local grid refinement methods for ordinary differential equations was presented in Chapter 5. The same concepts apply to elliptic partial differential equations.

In the method that was preferred in Chapter 5, one performs new calculations only on the refined portion of the grid, using boundary conditions taken from the coarse grid solution at the refinement interface. If the solution on the unrefined part of the grid is not recomputed, the method is called *passive* (see Berger and Oliger, 1984). This method is not appropriate for elliptic problems because a local change in the solution is likely to affect the solution everywhere. Methods that allow changes in the refined grid to affect the whole domain are called *active* methods.

One active method (Caruso et al., 1985) follows the passive method with the important difference that the procedure is not complete when the solution on the fine grid has been computed. A new coarse grid solution must be computed, but not the one that would be computed on the coarse grid by itself (this solution has already been computed); a smoothed version of the fine grid solution is required. To see how this works, let us suppose that

$$\mathcal{L}_h(\phi_h) = \mathbf{b}_h \tag{7.108}$$

represents the discretized problem on a grid of size h; $\mathcal{L}$ represents the discretized operator matrix. To force the solution on the grid of size $2h$ to be a smoothed version of the fine grid solution on the refinement region, we replace the coarse grid problem by

$$\mathcal{L}_{2h}(\phi_{2h}) = \begin{cases} \mathcal{L}_{2h}(\overline{\phi}_h) & \text{in the refined region} \\ \mathbf{b}_{2h} & \text{in the remainder of the domain} \end{cases} \tag{7.109}$$

where $\overline{\phi}_h$ is the smoothed fine grid solution. The solution to this problem is computed and used to provide modified boundary conditions to the refined grid. Then a new refined grid solution is computed. The solution is iterated between the coarse and fine grids until the convergence error is small enough; four iterations usually suffice. Since the solution on each grid does not need to be iterated to final tolerance each time, this method actually costs only a little more than the passive method.

In another active method (Muzaferija, 1994) all grids are combined into a single global grid. The problem is then solved using a single grid method. This method may produce grid cells of odd shape and/or with variable numbers of neighbors. The computer code thus needs to have a data structure that can handle this situation, and the solver needs to be able to handle the irregular matrix structure that results.

Many levels of grid refinement can be used; use of as many as eight levels has been reported. The advantage of the adaptive grid method is that, because the finest grid occupies only a small part of the domain, the total number of grid points is relatively small, so both the cost of computation and the memory requirements are reduced enormously. Furthermore, with good design of the method, the user need not be an expert at grid construction; the method automatically generates a grid that is well suited to the problem. The only requirement placed on the user is that the initial grid fit the boundaries reasonably well.

Finally, adaptive grid methods combine very well with the multigrid method. The nested grids can be regarded as the ones used in the multigrid procedure; the important difference is that, because the coarse grids provide enough accuracy in some areas, the finest grids do not cover the entire domain. Since the largest cost of the multigrid method is due to iterations on the finest grid, the savings can be very large, especially in three dimensions. For details of some methods, see Thompson and Ferziger (1989).

7.13. FINITE ELEMENT METHODS

Finite element methods for ordinary differential equations were discussed in Section 5.6. Similar ideas can be applied to partial differential equations in two or three dimensions. In recent years finite element methods have become dominant in some engineering fields that require the solution of partial differential equations, particularly in the analysis of stress in solids. As a result, a huge body of literature devoted to these methods has been produced, and courses on finite element methods are offered in many universities.

It is impossible to provide more than a very brief introduction to the subject here. The reader interested in acquiring a deeper knowledge of finite element methods will find a number of well-written books devoted solely to the subject, some of which are given in the bibliography.

As was pointed out in Section 5.6, finite element methods are at their best when the problem to be solved can be stated in variational form. The method can be applied in the absence of a variational statement, but its properties will not be quite as nice. We begin, therefore, with the classic case for which a variational statement is available: Laplace's equation in two dimensions. Solving

$$\nabla^2 \phi = \frac{\partial^2 \phi}{\partial x^2} + \frac{\partial^2 \phi}{\partial y^2} = 0 \tag{7.110}$$

is equivalent to minimizing the integral

$$F = \int (\nabla \phi)^2 \, dA = \int \left[\left(\frac{\partial \phi}{\partial x} \right)^2 + \left(\frac{\partial \phi}{\partial y} \right)^2 \right] dx\,dy \tag{7.111}$$

where the integral is over the region on which the solution is desired.

Approximate solutions to Laplace's equation are generated by creating functions that contain adjustable parameters, computing the integral of Eq. (7.111) as a function of the parameters and minimizing the result with respect to each of the parameters. We give a concrete example below.

Boundary conditions are an important issue. In order to show that a solution of Laplace's equation actually minimizes the integral, we assume that ϕ^* is the function that minimizes the integral and compute the integral using the function

$$\phi = \phi^* + \delta\phi \tag{7.112}$$

where $\delta\phi$ is small compared to ϕ^*. Then substituting Eq. (7.112) into Eq. (7.111) and integrating by parts, we have

$$F = F^* + \delta F = \int_A (\nabla\phi^*)^2 dA - 2\int_A \delta\phi\nabla^2\phi^* dA + \int_S \delta\phi\nabla\phi^* \cdot dS \tag{7.113}$$

where the last integral is over the bounding contour. We have neglected a term quadratic in $\delta\phi$, since, by assumption, it is much smaller than the terms we have kept. The first term on the right is F^*, the exact minimum under the assumptions we have made. Thus the deviation in F due to the small error $\delta\phi$ in the approximating function is

$$\delta F = -2\int_A \delta\phi\nabla^2\phi^* dA + \int_S \delta\phi\nabla\phi^* \cdot dS \tag{7.114}$$

Recall that a function $f(x)$ depends quadratically on the independent variable near a minimum, that is, if $f'(\xi) = 0$, then $f(x) \approx f(\xi) + \text{const.}(x - \xi)^2$ near $x = \xi$. In the same manner, we expect that F should be quadratic in $\delta\phi$ near its minimum. This means that δF defined by Eq. (7.114) must vanish, and, for this to be true for any $\delta\phi$, we must demand that (1) ϕ^* must satisfy Laplace's equation (7.110) and (2) the normal derivative of ϕ^* must be zero on the boundary. The latter is called the natural boundary condition implied by the variational statement (7.111).

Suppose that we wish to solve Laplace's equation subject to some boundary condition other than the natural one implied by the minimization of Eq. (7.111). There are two ways this can be done. The first is to require all test functions to satisfy the desired boundary conditions. The conflict between the desired and natural boundary conditions can cause numerical difficulties, however. A second approach involves adding an integral over the boundary to Eq. (7.111); this integral can be chosen so that the natural boundary condition for the modified variational statement becomes the desired boundary condition. Both things can be done simultaneously, of course, and the result is even better.

In the finite element method, the region for which the equation is to be solved is broken into a number of smaller regions. The name of the method comes from solid mechanics in which the pieces are thought of as elements that together make up the body. The elements can be of essentially any shape, and this flexibility gives the method one of its principal advantages: the ability to handle any geometric domain without much difficulty. A variety of elements have been suggested and used; we discuss just a few to illustrate some of the variety that is possible.

The simplest kind of element (and the first in a historical sense) is the triangular element, an example of their application is shown in Figure 7.21. Using straight-sided triangles, we can fill the region with a set of elements and fit the boundary curve quite accurately. The difficulty that finite difference

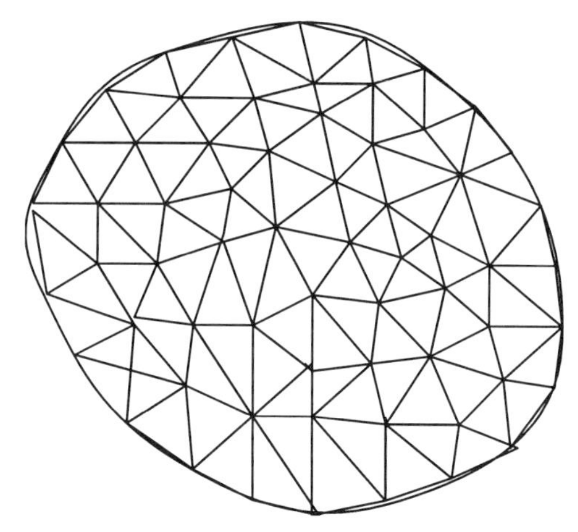

FIGURE 7.21. Typical finite element "tiling" of a domain.

methods have at oddly shaped boundaries is thus eliminated. Furthermore, the elements can be of any size. If the solution varies rapidly in some region, it is possible to subdivide the elements to increase the accuracy in one region without having to do so in other regions. It is also easy to construct error-based adaptive grid methods with finite elements.

Within each region the function can be approximated in a number of ways. Use of constants as trial functions is not a good idea because this makes the solution discontinuous at the boundaries of the elements.

However, the linear approximation

$$\phi(x,y) = c_1 + c_2 x + c_3 y \tag{7.115}$$

is widely used. The parameters in the linear function on a triangular element can be expressed in terms of the values of the function at the corners of the element, and it is better to use the corner values as the unknowns rather than the coefficients of the linear function. The primary reason is that the same values can be used in the adjacent elements. This both reduces the number of unknowns to be computed and ensures that the function is continuous across the element boundaries. This element is illustrated symbolically in Figure 7.22(*a*). Along any edge of the triangle the function is linear and is the same for both triangles containing that edge. Thus continuity of the solution is guaranteed.

Improved accuracy can be obtained with a quadratic element:

$$\phi(x,y) = c_1 + c_2 x + c_3 y + c_4 x^2 + c_5 xy + c_6 y^2 \tag{7.116}$$

The constants can be represented in terms of the values of the solution at the corners and edge midpoints of the triangle symbolized in the second part of Figure 7.22. Still higher accuracy can be obtained by using a cubic (the

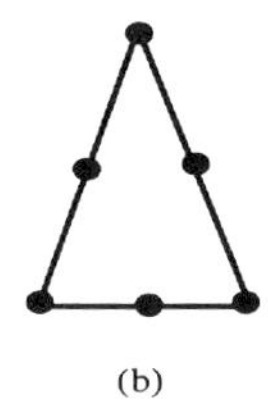

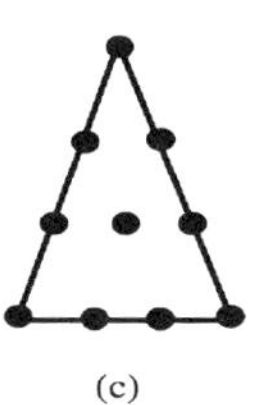

FIGURE 7.22. Some typical triangular finite elements. The dots indicate the locations at which the function values are given: (*a*) a bilinear element; (*b*) a quadratic element; and (*c*) a cubic element; in each case the name designates the type of function used to represent the solution within the element.

highest degree normally used), requiring 10 coefficients in the approximation. This element is shown symbolically in Figure 7.22(*c*).

Another possibility uses a cubic approximation but uses the values of the function and its two first partial derivatives at the corners and the value of the function at the center of the triangle as parameters. This ensures that the first derivatives as well as the solution itself will be continuous across the element boundaries. It is similar to the use of Hermite interpolation in two dimensions and is useful for higher-order equations.

Another type of method uses rectangular elements with curved edges. Curved edges obviously allow one to fit the region boundaries more accurately than is possible with straight-edged triangles. To make this method computationally feasible, irregular quadrilaterals are generated by applying coordinate transformations to a standard rectangular grid on which the calculations are done. When the coordinate transformation is generated from the same kind of polynomials that are used to approximate the function, the elements are said to be *isoparametric*, and the method possesses a number of advantages.

Since we shall not go very deeply into the finite element method here, we adopt the use of straight-edged triangles and linear approximations for the remainder of this section. Formally, one proceeds as described above; the integral (7.111) is evaluated as a sum of integrals over the elements. Within each element, the integral is evaluated by substituting the polynomial approximation for $\phi(x, y)$ and computing the integral analytically. For a linear problem such as the one we are dealing with, the result is bilinear in all of the parameters. Once the integral has been evaluated, the desired equations are obtained by differentiating the result with respect to each of the parameters to be computed and setting the result to zero. In practice there is an easier method, illustrated by the following example.

Example 7.11: Finite Element Approximation of Laplace's Equation Consider the cluster of four elements shown in Figure 7.23. A set of four identical isosceles right triangles has been chosen both for simplicity and ease of comparison with finite difference and volume methods. When the integral described above is computed, terms containing $\phi_{i,j}$ arise from each of the four

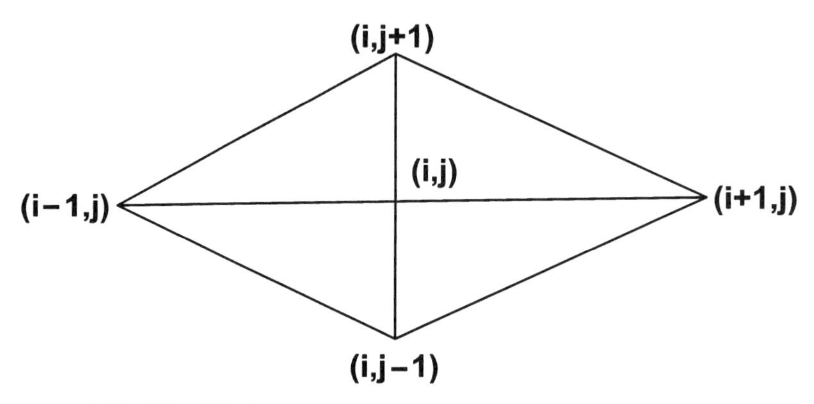

FIGURE 7.23. Cluster of elements within a finite element grid.

adjacent triangles. In fact, it is easy to show that the total contribution of all the terms containing $\phi_{i,j}$ is

$$\phi_{i,j}[8\phi_{i,j} - 4(\phi_{i-1,j} + \phi_{i+1,j} + \phi_{i,j-1} + \phi_{i,j+1})] \tag{7.117}$$

When this is differentiated with respect to $\phi_{i,j}$ and set to zero, we find

$$4\phi_{i,j} - (\phi_{i-1,j} + \phi_{i+1,j} + \phi_{i,j-1} + \phi_{i,j+1}) = 0 \tag{7.118}$$

which is the usual five-point finite difference approximation. This method of collecting the terms containing a particular variable can always be used and is the easiest way of finding the equations, but there are other tricks that are often used in conjunction with finite element methods.

As we saw in Section 5.6, the finite element method normally does not give the same approximations as finite difference methods; it is only the simplicity of Laplace's equation and the choice of elements that caused the two methods to yield the same approximation in this case. If the method used above had been applied to the modified wave or Helmholtz equation

$$\nabla^2\phi + k^2\phi = 0 \tag{7.119}$$

we would get

$$\begin{aligned} &4\phi_{i,j} - (\phi_{i-1,j} + \phi_{i+1,j} + \phi_{i,j-1} + \phi_{i,j+1}) \\ &\qquad + \tfrac{1}{12}k^2(8\phi_{i,j} + \phi_{i-1,j} + \phi_{i+1,j} + \phi_{i,j-1} + \phi_{i,j+1}) = 0 \end{aligned} \tag{7.120}$$

which is different from the usual finite difference approximation. However, Eq. (7.120) can be regarded as a modified finite difference approximation.

Almost any approximation that can be derived by the finite element method using regular arrays of points can be derived by finite differences, and vice versa. The finite element approach however, produces approximations that might not have been thought of in the finite difference context. Which kind of approximation is superior may depend on the problem being solved, but the fact that the finite element method has the conservation laws built into it is a big advantage. Together with the capability of dealing with difficult geometries, this has made the finite element approach the method of choice in a number of applications. Whether its advantages are important in areas with severe nonlinearities for which a variational approach is not available is unclear.

One disadvantage of the finite element approach is that the structure of the matrix in the resulting equations is normally not as simple as the structure of finite difference matrices; this is the price paid for the ability to deal with geometric complexity in such a flexible way. The matrix structure depends on how the points are numbered, so it is important to order the points carefully. Routines for creating good orderings of the points are available. The equations are solved either by a modified Gauss elimination procedure or, more usually, an iterative method. Conjugate gradient methods have been used very commonly in recent years.

7.14. DISCRETE FOURIER TRANSFORMS

We noted earlier that Fourier transform methods can be very useful in solving equations on regular domains. In this section and the next we digress to provide a brief introduction to these methods. Applications to solving PDEs will be given in the following section.

7.14.1. Review of Fourier Series

In this section, it will be assumed that the reader has some familiarity with Fourier series, but we begin with a short review of that subject. Any function defined on $-L < x < L$ can be written as a Fourier series (subject to restrictions that are not usually important in physical applications):

$$f(x) = \frac{a_0}{2} + \sum_{n=1}^{\infty} \left(a_n \cos \frac{2\pi n x}{L} + b_n \sin \frac{2\pi n x}{L} \right) \tag{7.121}$$

The coefficients a_n and b_n can be found from the formulas

$$a_0 = \frac{2}{L} \int_{-L}^{L} f(x)\, dx \qquad a_n = \frac{1}{L} \int_{-L}^{L} f(x) \cos \frac{2\pi n x}{L}\, dx \tag{7.122}$$

$$b_n = \frac{1}{L} \int_{-L}^{L} f(x) \sin \frac{2\pi n x}{L}\, dx \tag{7.123}$$

These results are well known.

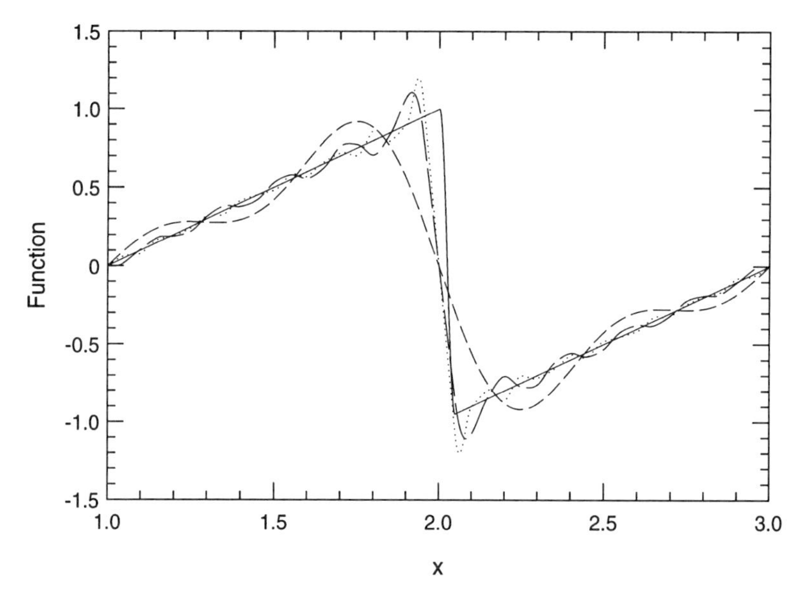

FIGURE 7.24. Gibbs phenomenon resulting from the representation of a function with a discontinuity by a finite number of terms of a Fourier series. The dashed line is the sum of the first 3 terms of the Fourier series, the long dashed line is the sum of the first 10 terms of the series, and the dotted line is the sum of the first 15 terms of the series.

It is also convenient to note that the relationship

$$e^{iy} = \cos y + i \sin y \tag{7.124}$$

allows Eqs. (7.121), (7.122), and (7.123) to be rewritten as

$$f(x) = \sum_{n=-\infty}^{\infty} \alpha_n e^{2\pi i n x/L} \tag{7.125}$$

$$\alpha_n = \frac{1}{L} \int_0^L f(x) e^{-2\pi i n x/L}\, dx \tag{7.126}$$

which is a complex form of the Fourier series; the connection between the two representations is $\alpha_n = (a_n + ib_n)/2$.

It is important to note that although we restricted the range of the independent variables to $-L < x < L$, the series in Eqs. (7.121) or (7.125) can be defined for any x. A representation of a function defined by this extension of the series is depicted in Figure 7.24; it is called the *periodic extension* of $f(x)$. For computational applications we need to deal with finite series. The summation in Eq. (7.24) is truncated after a finite number of terms, and the integral in Eq. (7.24) is replaced by a finite sum. For "smooth" functions this causes no serious problem, but a finite series has difficulty in reproducing a function that has discontinuities. Furthermore, because the Fourier series represents the periodic extension of the function, if $f(-L) \neq f(L)$, the series will "think" that there is a discontinuity in the function at $x = \ldots, -L, L, 3L, \ldots$, and so on. The result is that the Fourier series for the function illustrated in Figure 7.24 produces the result shown. The existence of wiggles near the discontinuities, known as the Gibbs phenomenon, can cause considerable difficulty in applications of Fourier series.

For smooth periodic functions Fourier series converge rapidly, that is, the coefficients fall off rapidly with increasing n. For functions with discontinuities $a_n \sim 1/n$, and truncation removes terms that are not small, resulting in the large errors that are the source of the Gibbs phenomenon illustrated in the figure.

One can differentiate the Fourier series for $f(x)$ term by term to produce a series representation of the derivative $f'(x)$. From Eq. (7.125) we find that the coefficients of the series for the derivative $f'(x)$ are proportional to $n\alpha_n$. It is possible that this series will not converge at all (if f has a discontinuity, $\alpha_n \sim 1/n$ for large n and $n\alpha_n$ does not go to zero as $n \to \infty$). The differentiated series can be used to represent $f'(x)$ only when it converges. Practically, this means that use of Fourier series for calculating derivatives is limited to smooth, continuous functions that are periodic, that is, ones for which $f(-L) = f(L)$.

Everything said about the Fourier series for the function applies to the Fourier series for the derivative. If the derivative is discontinuous at some point (including the endpoint), its series representation will exhibit Gibbs phenomenon. This means that functions possessing discontinuous derivatives (functions with cusps) will be subject to a weaker form of Gibbs phenomenon; the series will produce a wiggly result, but the wiggles will be milder than those shown in Figure 7.24. Still milder forms of the problem will arise for functions with discontinuous higher derivatives. Obviously, Fourier series are at their best for functions that possess continuous derivatives of all orders. This is the case for periodic functions, that is, functions such that $f(x + 2L) = f(x)$, which have continuous derivatives of all orders on $-L < x < L$. This is rather restrictive, but a sufficient number of cases of this kind arise in applications to make the introduction of Fourier methods worthwhile.

This completes our short introduction to Fourier series. A more complete development of Fourier series is found in Bracewell (1967) (on Fourier methods in general) and Brigham (1976) (on discrete transforms and computer applications).

7.14.2. Discrete Fourier Series

For the applications we have in mind, we need a discrete version of Fourier series. Suppose we have a one-dimensional mesh of equally spaced grid points defined by

$$x_j = j\,\Delta x \qquad j = 1,2,\ldots,N \tag{7.127}$$

where $N\,\Delta x = L$.

We assume that the functions we deal with are periodic with a period L, which implies that $x = 0$ and $x = L = N\,\Delta x$ are equivalent points. To write a a Fourier series for a function $f(x)$ whose values are given only at N mesh points requires only N Fourier coefficients. For generality, $f(x)$ is allowed to be complex. The series is

$$f(x_j) = f_j = \sum_{l=1}^{N} \hat{f}_l e^{ik_l x_j} \tag{7.128}$$

[A caret ($\hat{}$) represents the Fourier coefficient of a function.] The set of wavenumbers k_l needs to be chosen carefully. Some constraints on them arise from the requirement that the functions $\exp(ik_l x_j)$ be periodic and, for computational reasons, we want the k_l to be equally spaced. A set of values that accomplishes both of these aims is

$$k_l = \frac{2\pi l}{N\,\Delta x} \qquad l = 1,2,\ldots,N \tag{7.129}$$

and the series (7.128) becomes

$$f_j = \sum_{l=1}^{N} \hat{f}_l e^{2\pi i l j/N} \tag{7.130}$$

The choice of the k_l (7.129) is not unique. To illustrate the nonuniqueness, consider the function $\exp(2\pi i l j/N)$ and replace l by $l+N$. The function then becomes $\exp[2\pi i(l+N)j/N] = \exp(2\pi i l j/n)\exp(2\pi i) = \exp(2\pi i l j/N)$. Thus, in terms of their values at the grid points, the set of functions with the wavenumbers defined by Eq. (7.129) and the set obtained by replacing l by $l+N$ are identical. Between the grid points they are different, of course, but this has no effect on Eq. (7.130).

The existence of Fourier modes with identical values on the grid gives rise to a phenomenon known as *aliasing*. An illustration of how this occurs is given in Figure 7.25, where the functions $\sin \pi x$ and $\sin 9\pi x$ are plotted; at the grid points $x_j = j$, the two functions have the same values. Aliasing has important consequences for computational methods, including ones that do not use Fourier series explicitly.

The inverse of Eq. (7.130) is

$$\hat{f}_l = \frac{1}{N} \sum_{j'=1}^{N} f_{j'} e^{-2\pi i l j'/N} \tag{7.131}$$

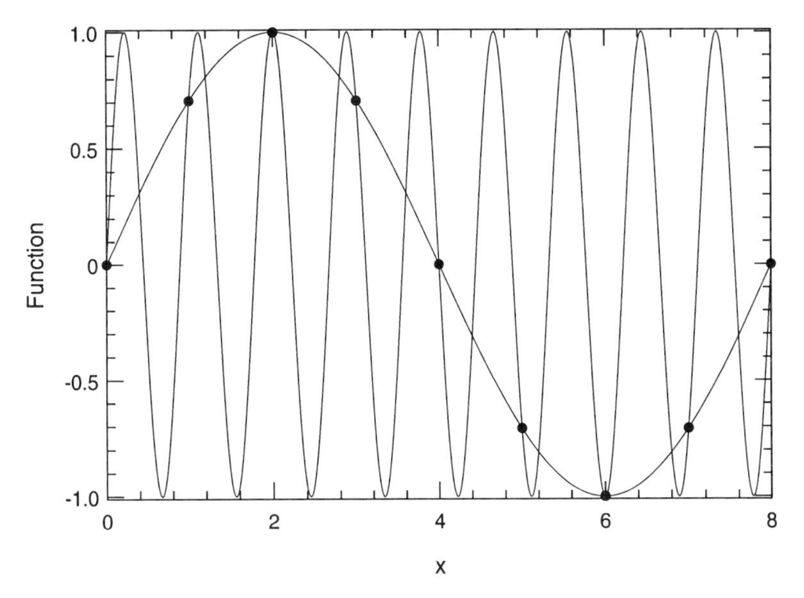

FIGURE 7.25. Illustration of aliasing. The functions $\sin \pi x$ and $\sin 9\pi x$ have the same values at the grid points but not between them.

The proof of this statement is straightforward. If we substitute Eq. (7.131) into Eq. (7.130) and interchange the order of the j' and l summations, we have

$$f_j = \frac{1}{N} \sum_{j'=1}^{N} f_{j'} \sum_{l=1}^{N} e^{2\pi i(j-j')l/N} \tag{7.132}$$

The l summation is a geometric series, which can be summed by the well-known formula for a geometric series to give

$$\sum_{l=1}^{N} e^{2\pi i(j-j')l/N} = e^{2\pi i(j-j')/N} \frac{1 - e^{2\pi i(j-j')N/N}}{1 - e^{2\pi i(j-j')/N}} \tag{7.133}$$

Now, $j - j'$ is an integer that can take the values $-N+1, -N+2, \ldots, N-1$. As a result, the numerator in Eq. (7.132) is always zero; the denominator is not zero unless $j = j'$. So the right-hand side of Eq. (7.132) is zero unless $j = j'$. When $j = j'$, the sum is easily evaluated because each term is exactly unity; the result is simply N, so the expression (7.133) is $N\delta_{jj'}$. Introducing this result into Eq. (7.132), we see that both sides are equal, showing that Eqs. (7.130) and (7.131) are inverses of each other. Some authors use a factor $N^{-1/2}$ in front of the sums in each expression rather than 1 and N^{-1}. The result is more symmetric, but the difference is just a matter of taste. The only other difference between Eqs. (7.130) and (7.131) is the sign in the exponent.

The almost complete symmetry of the two expressions means that f_j and $\hat{f}_l$ may be regarded as discrete Fourier transforms of each other.

7.14.3. Spectral Differentiation

Although the discussion of Fourier series above deals only with the values of the function at the grid points, there is no reason that we cannot use Eq. (7.128) to define the function for any x. That is, the discrete Fourier series can be considered as an interpolation method; we simply replace x_j by the continuous variable x. Having done so, we can use Fourier series to compute integrals and derivatives.

When a Fourier series is to be differentiated, the choice of the set of wavenumbers becomes important. We have seen that there is freedom in the choice of the set of wavenumbers k_l. Any set of integers that spans a range of N values, that is, any set of the type $l = j, j+1, \ldots, j+N$ will reproduce the function at the mesh points. Between the grid points, however, different sets of l give different results. We want the interpolation to be as smooth as possible and must choose the set of l's that accomplishes that. The set that gives the smoothest result is the one that contains the smallest maximum l in the absolute sense. For even N, the most appropriate set is

$$l = -\frac{N}{2}, -\frac{N}{2}+1, \ldots, 0, \ldots, \frac{N}{2}-1 \tag{7.134}$$

and is usually adopted ($-N/2$ could be removed and replaced by $+N/2$).

It is now an easy matter to obtain approximations to the derivatives of the function f. Differentiating Eq. (7.128) (with x_j replaced by the continuous variable x), we have

$$\frac{df}{dx}(x) = \sum_l ik_l \hat{f}_l e^{ik_l x} \tag{7.135}$$

This equation can be interpreted as saying that $ik_l\hat{f}_l$ is the Fourier transform of df/dx. This means that we can compute the derivative of a function by the following sequence of operations:

- Given the values of the function on an evenly spaced grid, calculate its discrete Fourier transform. A fast method of doing this is given below.
- Multiply the Fourier coefficient $\hat{f}_l$ by ik_l.
- Take the inverse Fourier transform of the result.

For smooth periodic functions, the result is a very accurate approximation to the derivative of the original function. For real functions, $\hat{f}_{-l}$ must be the complex conjugate of $\hat{f}_l$; for $l = -N/2$, there is no corresponding $\hat{f}_{-l}$. The resulting derivative may be noisy unless one sets $k_{-N/2} = 0$ in the computation of the derivative. The reason for insisting that N be even is explained below where some examples will be given.

By taking the derivative of Eq. (7.135), we can show that $-k_l^2 \hat{f}_l$ is the Fourier transform of the second derivative $d^2 f/dx^2$, and the second derivative can be computed in a manner very similar to that used for the first derivative. The only difference is that the transform must be multiplied by $-k_l^2$ rather than ik_l. Higher derivatives can be computed in the same manner.

This method is effective only for functions that are periodic and have continuous derivatives of several orders. One can show that the approach outlined above produces derivatives with an order of accuracy (in the usual sense) that is equal to the minimum of (1) one plus the lowest order derivative that possesses a discontinuity or (2) the number of points used in the computation. For functions of the appropriate type, the method is capable of extremely high accuracy. Its practical use as a numerical method requires an efficient means of computing discrete Fourier transforms, which is the subject of the next subsection.

7.14.4. Fast Fourier Transform

The discrete Fourier transform introduced in the preceding section has the potential of providing differential equation solvers of very high accuracy. In order for them to be useful, however, we need a way to compute discrete Fourier transforms efficiently. Direct computation of the series (7.128) requires N multiplications and N additions for each value of j (or l) and, since j has N possible values, the total number of operations is proportional to N^2 (assuming that the complex exponential factors have been precomputed). It is equivalent to multiplying a vector by a matrix. This scaling would make the method too expensive for practical use for large N. A more efficient algorithm is necessary. Such a method was discovered by Cooley and Tukey and is known as the *fast Fourier transform* (FFT) or Cooley–Tukey algorithm.

Since the forward and backward discrete Fourier transforms are identical except for a sign change, we consider only the forward transform. Introducing the abbreviation $\omega = \exp(2\pi i/N)$ and letting the indices run from 0 to $N-1$ rather than from 1 to N, we can write Eq. (7.128) as

$$f_j = \sum_{l=0}^{N-1} \omega^{jl} \hat{f}_l \qquad j = 0, 1, \ldots, N-1 \tag{7.136}$$

The secret to efficient computation of this transform resides in the properties $\omega^{m+n} = \omega^m \omega^n$ and $\omega^N = 1$. As a consequence of these almost trivial facts, we see immediately that many of the exponential factors in Eq. (7.136) are identical. Furthermore, there is a great deal of symmetry.

There are several ways to develop the concepts behind the fast Fourier transform. One of the better ones is to regard ω^{jl} as a matrix in Eq. (7.136) and treat the operation as matrix multiplication. For example, if $N = 4$ we can write

$$\begin{pmatrix} f_0 \\ f_1 \\ f_2 \\ f_3 \end{pmatrix} = \begin{pmatrix} \omega^0 & \omega^0 & \omega^0 & \omega^0 \\ \omega^0 & \omega^1 & \omega^2 & \omega^3 \\ \omega^0 & \omega^2 & \omega^4 & \omega^6 \\ \omega^0 & \omega^3 & \omega^6 & \omega^9 \end{pmatrix} \begin{pmatrix} \hat{f}_0 \\ \hat{f}_1 \\ \hat{f}_2 \\ \hat{f}_3 \end{pmatrix} = \begin{pmatrix} 1 & 1 & 1 & 1 \\ 1 & \omega & -1 & -\omega \\ 1 & -1 & 1 & -1 \\ 1 & -\omega & -1 & \omega \end{pmatrix} \begin{pmatrix} \hat{f}_0 \\ \hat{f}_1 \\ \hat{f}_2 \\ \hat{f}_3 \end{pmatrix} \tag{7.137}$$

In this equation we can change the order in which the f_j appear in the vector if we simultaneously interchange the corresponding rows of the matrix and the result vector:

$$\begin{pmatrix} f_0 \\ f_2 \\ f_1 \\ f_3 \end{pmatrix} = \begin{pmatrix} 1 & 1 & 1 & 1 \\ 1 & -1 & 1 & -1 \\ 1 & \omega & -1 & -\omega \\ 1 & -\omega & -1 & \omega \end{pmatrix} \begin{pmatrix} \hat{f}_0 \\ \hat{f}_1 \\ \hat{f}_2 \\ \hat{f}_3 \end{pmatrix} \tag{7.138}$$

The reordered matrix can be factored into the product of two simpler matrices to give

$$\begin{pmatrix} f_0 \\ f_2 \\ f_1 \\ f_3 \end{pmatrix} = \begin{pmatrix} 1 & 1 & 0 & 0 \\ 1 & -1 & 0 & 0 \\ 0 & 0 & 1 & \omega \\ 0 & 0 & 1 & -\omega \end{pmatrix} \begin{pmatrix} 1 & 0 & 1 & 0 \\ 0 & 1 & 0 & 1 \\ 1 & 0 & -1 & 0 \\ 0 & 1 & 0 & -1 \end{pmatrix} \begin{pmatrix} \hat{f}_0 \\ \hat{f}_1 \\ \hat{f}_2 \\ \hat{f}_3 \end{pmatrix} \tag{7.139}$$

as can be shown by straightforward matrix multiplication.

The structure of the two matrices allows the computation to be simplified. The left-hand matrix is block diagonal; in particular, it is a diagonal 2×2 matrix the elements of which are themselves 2×2 matrices. As a result, the first two elements of the vector produced by multiplying an arbitrary vector **b** by this matrix can be found by multiplying the 2 vector (b_1, b_2) by the upper-left 2×2 submatrix. Similarly, the last two elements of the product vector are obtained by multiplying the vector (b_3, b_4) by the lower-right 2×2 block. In other words, the process of multiplying a 4 vector by a 4×4 matrix of this type can be reduced to two multiplications involving 2×2 matrices.

It is not quite as obvious that the right-hand factor matrix in Eq. (7.139) has a similar structure. The difference is that the first and third elements of the resultant vector depend solely on the first and third elements of the original vector and the second and fourth elements of the resultant vector depend entirely on the second and fourth elements of the original vector. The net result is that the multiplication of a 4 vector by a 4×4 matrix can be replaced by 4 multiplications of 2 vectors by 2×2 matrices.

This result by itself is not of great importance since it produces no savings in computation. The real significance lies in the generalization of this method to larger N. It turns out that as long as $N = 2^n$, in other words, as long as N is an integral power of 2, the matrix that performs the Fourier transform can always be decomposed into the product of $n = \log_2 N$ matrices, each of which is equivalent to a block diagonal matrix whose elements are 2×2 matrices.

To see the advantage in this, recall that the multiplication of a vector by a matrix requires N^2 operations. To compute the Fourier transform by direct application of Eq. (7.137) would require this many operations. Use of the factorization requires $N/2$ multiplications of 2 vectors by 2×2 matrices at each stage—or $(N/2) \times 4 = 2N$ operations. Since the number of factor matrices is $\log_2 N$, the total number of numerical operations is of the order of $2N \log_2 N$. For large N, say $N = 64$, the savings provided by using the factorized form can be an order of magnitude or more. It is the existence of this algorithm that makes Fourier transforms more than a special-purpose tool.

With these basic ideas as a foundation, what remains to be done is to construct a scheme for making the FFT systematic so that one subroutine can be used for any value of N. This means that we need a method of determining which components need to be combined at each stage of the procedure and what the multiplying factors (matrix elements) are. There is some freedom of choice in how this is done; in other words, there is more than one factorization. A simple one, which is the generalization of the method given for $N = 4$ above, combines elements j and $j + N/2$ of the original vector to produce elements j and $j + N/2$ elements of the first intermediate result. At the second stage, elements j and $j + N/4$ are combined, at the third, j and $j + N/8$, and so on. The factors by which each element must be multiplied also form a systematic array. After the first stage, none of the first $N/2$ elements in the array will ever again be combined with any of the last $N/2$ elements. This means that the operations that are performed on each of these sets of $N/2$ elements from then on is precisely an $N/2$-point Fourier transform. In fact, the first stage of the transform can be regarded is the conversion of a single N-point transform into two $N/2$-point transforms. The second stage converts these to four $N/4$-point transforms, and so on, until we have $N/2$ two-point transforms, which are almost trivially done. The multipliers are also easily found.

The results come out in scrambled form; that is, they are not arranged in order of increasing l. It turns out that the relationship between the scrambled ordering and the proper one is quite simple. It is found by writing the binary representation of the number and reversing the order of the bits. Thus in 16-point transform, the seventh element of the array (binary 7 is 0111) is the fourteenth element of the actual transform (binary 14 is 1110).

The algorithm presented above is designed to work when $N = 2^n$. For other values of N one can obtain somewhat smaller advantages. For example, a six-point transform can be reduced to two three-point transforms using the method described above and, in like manner, reductions can be made for other values of N. Newer methods do as well or better for other values of N. If the functions to be transformed are real, it is possible to load the arrays with one function as the real part of the initial data and a second function as the imaginary part. Both transforms then can be obtained simultaneously with a little extra unscrambling.

Fourier sine and cosine transforms are also sometimes useful and N-point transforms of either type can be computed at the cost of one $2N$-point Fourier transform. The algorithm has been applied to a number of other related tasks but space does not permit covering them here.

7.15. FOURIER OR SPECTRAL METHODS

The technique presented in the preceding section can be used to solve some elliptic problems very efficiently. The class of treatable problems is limited, but the technique is so powerful that it is worth presenting. Also, a technique as good as this one results in efforts to extend the range of problems to which it may be applied.

As an illustration of the method, we take one of the simplest possibilities: in two dimensions, the Poisson equation is

$$\frac{\partial^2 \phi}{\partial x^2} + \frac{\partial^2 \phi}{\partial y^2} = \rho(x,y) \tag{7.140}$$

We assume further that the boundary conditions are periodic in one of the dimensions, which, for the sake of definiteness, we choose to be the x direction. Thus

$$\phi(x+L,y) = \phi(x,y) \tag{7.141}$$

The boundary conditions in the y direction can be arbitrary.

Proceeding formally, we write the solution to Eq. (7.140) as a Fourier series of the type (7.128) in the variable x:

$$\phi(x,y) = \sum_{l=1}^{N} \hat{\phi}(k_l,y)e^{ik_l x} \tag{7.142}$$

where the k_l are given by Eq. (7.130) and substitute Eq. (7.142) into Eq. (7.140). An equation for $\phi(k_l,y)$ is then obtained by enforcing Eq. (7.140) only at the grid points x_j, $j = 1,2,\ldots,N$, multiplying the equation by $e^{-ik_l x_j}$, and summing over j—in other words, we take the Fourier transform of Eq. (7.140). We then have

$$\frac{\partial^2 \hat{\phi}}{\partial y^2}(k_l,y) - k_l^2\hat{\phi}(k_l,y) = \hat{\rho}(k_l,y) \tag{7.143}$$

where

$$\hat{\rho}(k_l, y) = \frac{1}{N}\sum_{j=1}^{N} \rho(x_j, y)e^{ik_l x_j} \tag{7.144}$$

Equations (7.143) are a set of N uncoupled ordinary differential equations with respect to y—one for each value of k_l. Each ODE must be solved together with boundary conditions obtained by Fourier transforming the boundary conditions for Eq. (7.140). The ODE boundary value problems may be solved by any of the methods presented in Chapter 5. The usual choice is to use a direct method with second-order differencing and then solving the resulting problems by the standard tridiagonal algorithm. If the boundary conditions in the y direction are periodic, it is possible (in fact, advantageous) to use Fourier transforms to solve the problem in the y direction as well. Having solved the set of equations (7.143) for $\hat{\phi}(k_l, y)$, the desired solution $\phi(x, y)$ is found by another application of the fast Fourier transform algorithm.

Thus we are able to solve the Poisson equation (7.140) by the following sequence of steps:

- Fourier transform the forcing function $\rho(x, y)$ in the x direction for each value of y (the grid should contain $N = 2^n$ evenly spaced x points for maximum efficiency).
- Solve the ODEs (7.143).
- Fourier transform the solutions to the ODEs to produce the desired results.

The most expensive parts of this procedure are the Fourier transforms. Since N transforms must be computed to find $\hat{\rho}$ and another N to compute ϕ, the number of operations required is approximately $4N^2 \log_2 N$ each of additions and multiplications. The number of operations required to solve Eq. (7.143) is approximately $7N^2$ (total operations) and is considerably smaller than the number required by the transforms, but not negligible. The important point is that the solution is obtained by Fourier transforms with a speed that beats even the best iterative methods. Not only that, the Fourier method actually finds the exact solution of the discretized equations; of course, the solution contains errors arising from the finite difference approximations to the ODEs. There is no convergence error. There are a number of areas in which the extra accuracy is important, and the Fourier method yields more accuracy for a given cost than any competing method.

A rather surprising application of the method is to the exact solution of the finite difference version of the Poisson equation. Even though the Fourier method is inherently more accurate than finite differences, there are times when one wants the solution of the finite difference equations to be as accurate as possible. Provided only that the boundary conditions are periodic (certain other simple types of boundary conditions such as the ones that require the function or its derivative to be zero are admissible), this can be achieved by the use of Fourier transforms. The trick is to note that the discrete Fourier transform of the finite difference equations can be written in the form of Eq. (7.143) with k_l^2 replaced by the square of a modified wavenumber, $k_{l,\text{eff}}^2$. Recall that the modified wavenumber for a particular finite difference method can be found by applying the finite difference operator to the function $\exp(ik_l x)$.

There are also a number of other problems to which the method can be applied. If the boundary conditions are not periodic but are homogeneous, the problem can be solved by expansion in an appropriate set of trigonometric functions. For example, if the boundary conditions require the function to be zero, a sine expansion is the proper choice; for zero slope conditions, a cosine expansion is the correct choice. Transforms appropriate to these expansions can be obtained at the same expense as computing a Fourier transform with $2N$ points. The cost of solving these problems is thus about twice that of solving a problem with periodic conditions, but is still quite advantageous. It is also worth noting that a standard way of solving problems with inhomogeneous boundary conditions both analytically and numerically is to use a transformation that introduces an inhomogeneous term into the differential equation and renders the boundary condition homogeneous.

The method has difficulty in dealing with regions that are not rectangular, although methods have recently been developed for cases in which the region is nearly rectangular. The problem is solved on the closest rectangular region and some of the parameters are iteratively adjusted to achieve the desired result. The method is akin to shooting, but it is not reviewed in any detail here. Finally, we note that the method can be applied to nonlinear equations by using iterative techniques.

For a deeper discussion of Fourier methods, the reader is advised to see the text by Hussaini et al. (1989).

7.16. BOUNDARY INTEGRAL METHODS

There are a number of applications in which Laplace's equation needs to be solved on an irregular region, but the solution is needed only on the boundary. For example, one might have a problem in which a potential is given on the boundary, and the desired information is the gradient of the potential on the boundary. No interior information is necessary and, if computation of it can be avoided, so much the better. Although the method we present in this section may seem rather limited in application, it has been used with considerable success in fluid mechanics (especially in aerodynamics) and electromagnetic theory. It is an example of a wider class of methods called *boundary integral methods*. Along with the Fourier method, it shows how analytical methods can be adapted to numerical computation, emphasizing the importance of classical applied mathematics in the computer age.

The method we seek can be derived by the use of Green's formula for Laplace's equation, but in two dimensions it is simpler to use the theory of functions of a complex variable. In three dimensions, Green's formula must be used. The theory of complex variables is not discussed in depth here, but the part of the theory we need is reviewed. If we regard $f(z)$ as a function of the complex variable $z = x + iy$, we can write

$$f(z) = \phi(z) + i\psi(z) \tag{7.145}$$

where ϕ and ψ are two real valued functions of the real variables x and y. If f is to have a uniquely defined derivative as a function of z, it is necessary that ϕ and ψ satisfy the Cauchy–Riemann conditions:

$$\frac{\partial \phi}{\partial x} = \frac{\partial \psi}{\partial y} \tag{7.146}$$

$$\frac{\partial \phi}{\partial y} = -\frac{\partial \psi}{\partial x} \tag{7.147}$$

Eliminating either of the two functions from these equations, we find that both ϕ and ψ satisfy Laplace's equation.

A function of complex variable is called *analytic* in a region if neither it nor any of its derivatives possess singularities in that region. For such a function one can show that Cauchy's theorem applies. This theorem (which is essentially a complex version of Green's formula) states that the integral on any closed contour enclosing a region of analyticity of the function is zero:

$$\int_C f(z)\,dz = 0 \tag{7.148}$$

From this result it is not difficult to prove Cauchy's integral formula:

$$\int_C \frac{f(z)}{z - z_0}\,dz = 2\pi i f(z_0) \tag{7.149}$$

which holds under the same conditions as Eq. (7.148). These formulas are well known and their derivations may be found in standard textbooks on complex variable theory.

The next result is one that is not quite as well known but is the basis of the method discussed in this section. Suppose that in Eq. (7.149) we allow the point z_0 to approach the boundary. Then the singularity due to the factor $(z - z_0)^{-1}$ is on the boundary, and it is not clear how the integral is to be defined. The precise derivation of the result is difficult and can be found in Muskhelishvili (1957). The correct result can be derived in a simple but nonrigorous manner by realizing that what is needed is the limit of Eq. (7.149) as z_0 is allowed to approach the boundary from the interior. This is equivalent to computing the integral in the manner suggested by Figure 7.26. The left side of the figure shows the original contour with z_0 approaching the contour, which is equivalent to deforming the contour slightly as shown in right side

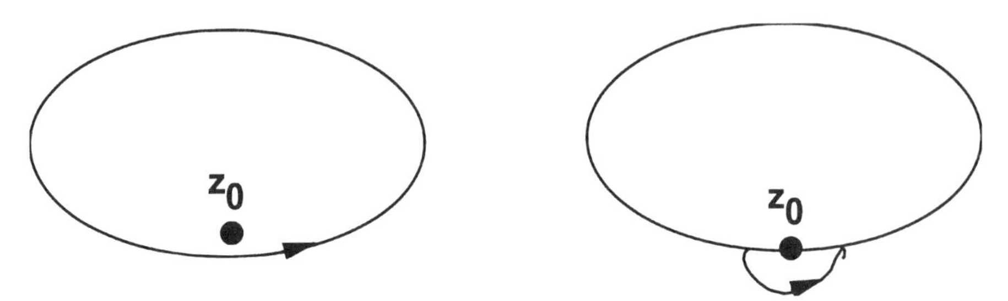

FIGURE 7.26. Contour of integration (left) and an interpretation of the integral as the point z_0 approaches the contour (right).

of the figure. The integral is then composed of two parts—an integral along the contour excluding the point z_0 and an integral over the small semicircle.

For the integral around the contour it is convenient to introduce the concept of a principal value. The principal value is defined as the limit:

$$P\int_A^B \frac{f(z)}{z-z_0}dz = \lim_{\epsilon\to 0}\left(\int_A^{z_0-\epsilon}\frac{f(z)}{z-z_0}dz + \int_{z_0+\epsilon}^B \frac{f(z)}{z-z_0}dz\right) \tag{7.150}$$

The integral around the semicircle is easily evaluated using Cauchy's integral formula and is $\pi i f(z_0)$. Thus, when z_0 is on the boundary, Cauchy's formula becomes

$$P\int_C \frac{f(z)}{z-z_0}dz = \pi i f(z_0) \tag{7.151}$$

which is a version of Plemelj's formula. Another way of viewing this result is that, if $f(z)$ satisfies this equation on the boundary, it must be the limiting value of a function that is analytic within the contour. If we need to deal with the exterior problem, that of solving Laplace's equation outside a contour, and if the function vanishes at infinity, its boundary values satisfy the same equation with a negative sign on the right-hand side. If the function does not vanish at infinity, additional terms appear in the equation.

It is easy to use Eq. (7.151) as the basis for a numerical method. The simplest approach is to enforce the equation at a finite set of points z_i, $i = 1,2,\ldots,N$ on the boundary. The function $f(z)$ can then be approximated by a linear function between each pair of neighboring points and the integrals computed. This requires just a bit of care and reduces Eq. (7.151) to a set of linear algebraic equations:

$$f(z_j) = \sum_{k=1}^{N} A_{jk} f(z_k) \qquad j = 1,2,\ldots,N \tag{7.152}$$

where

$$A_{jk} = \frac{z_{j+1} - z_k}{z_{j+1} - z_j} \ln\left(\frac{z_{j+1} - z_k}{z_j - z_k}\right) + \frac{z_{j-1} - z_k}{z_{j-1} - z_j} \ln\left(\frac{z_j - z_k}{z_{j-1} - z_k}\right) \tag{7.153}$$

For $j = k$:

$$A_{kk} = \ln\left(\frac{z_k - z_{k+1}}{z_k - z_{k-1}}\right) \tag{7.154}$$

To use this method, one needs to specify either ϕ or ψ at each grid point; the variable that is not specified is the unknown. There is one exception; if the data are all values of ϕ, one value of ψ must be given (and vice versa) or the set of equations will not have a unique solution. For computational purposes Eqs. (7.153) are split into $2N$ real equations by equating the real and imaginary parts of both sides of each equation. Since there are only N unknowns, only half of these equations can be used. It has been found by experience that it is usually best to use the imaginary parts of the equations because the resulting matrix is better conditioned than if the other choice were made.

For simple geometries this method turns out to be about equally as fast (for the same accuracy) as the better iterative methods. For more complicated geometries, for which the optimum parameters for the iterative methods cannot be found readily, the boundary integral method is quite advantageous. Versions of this method have long been used in aerodynamics where it is known as the panel method.

7.17. FINITE DIFFERENCES IN COMPLEX GEOMETRY

The method that we deal with here is conceptually the simplest and is reasonably effective for simple equations. We use a rectangular grid despite the irregular boundary. At any point where the neighbors are all interior points, we use the approximation (7.4). At any point that has boundary point(s) as neighbors, we use standard finite difference approximations for uneven mesh spacing. Thus at point (x_j, y_k) of Figure 7.27 we use

$$\left.\frac{\partial^2 \phi}{\partial x^2}\right|_{j,k} \approx \frac{2}{h(h + \Delta x_j)}\phi_{j-1,k} - \frac{2}{h\Delta x_j}\phi_{j,k} + \frac{2}{\Delta x_j(h + \Delta x_j)}\phi_{B,k} \tag{7.155}$$

$$\left.\frac{\partial^2 \phi}{\partial y^2}\right|_{j,k} \approx \frac{2}{h(h + \Delta y_k)}\phi_{j,k-1} - \frac{2}{h\Delta y_k}\phi_{j,k} + \frac{2}{\Delta y_k(h + \Delta y_k)}\phi_{j,B} \tag{7.156}$$

in which the values of $\phi_{B,k}$ and $\phi_{j,B}$ are determined by the boundary conditions. Here Δy_k represents the distance of the boundary point from the last regular grid point on the line $y = y_k$, Δx_j is the distance of the boundary from the last grid point on the line $x = x_j$, and h is the uniform grid size used in the interior

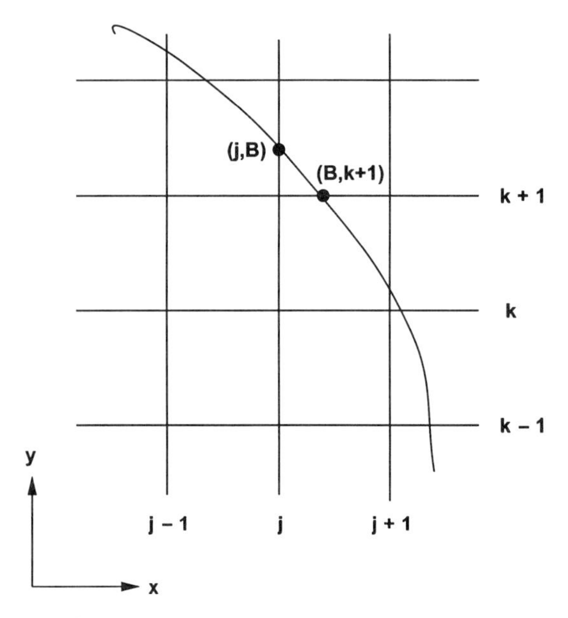

FIGURE 7.27. Grid near an irregular boundary.

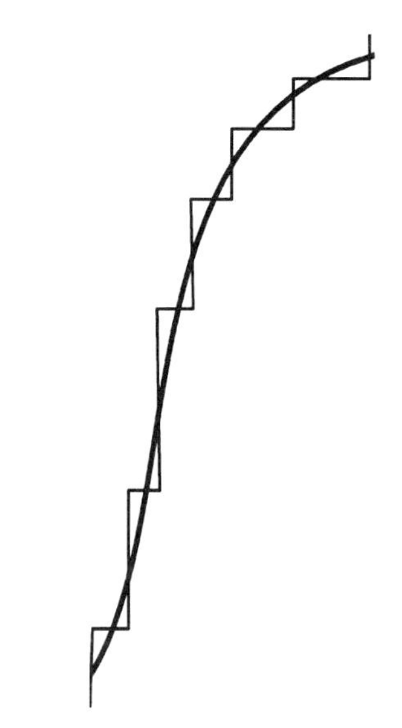

FIGURE 7.28. Stair-step grid at an irregular boundary.

of the domain. The resulting equations are very similar to the finite difference equations obtained earlier. The major complication is in the extra bookkeeping needed to keep track of which points are adjacent to the boundary, since each row has a different number of grid points. Aside from this (which makes programming quite a bit more complicated), the methods of solution used in rectangular regions also work in irregular regions, although sometimes not as effectively.

Another approach, in use a long time ago but which fell from favor, is now making a comeback. In this method, the domain is approximated by a stair-step region such as the one shown in Figure 7.28. Approximation of the equations and boundary conditions is then rather straightforward. The major difficulty is in the data structure, which requires the use of pointers because each line contains a different number of points.

Stair-step grids have also been proposed as refinement regions in adaptive grid procedures. This simplifes the task of fitting irregular regions on which grid refinement is required, allowing the use of fewer refinement regions. For details, see Zhu (1996).

PROBLEMS

1. Solve Laplace's equation on a square using four grid points in each direction (including boundaries, so there are just two interior points in each direction). The boundary conditions are given in Figure 7.29.

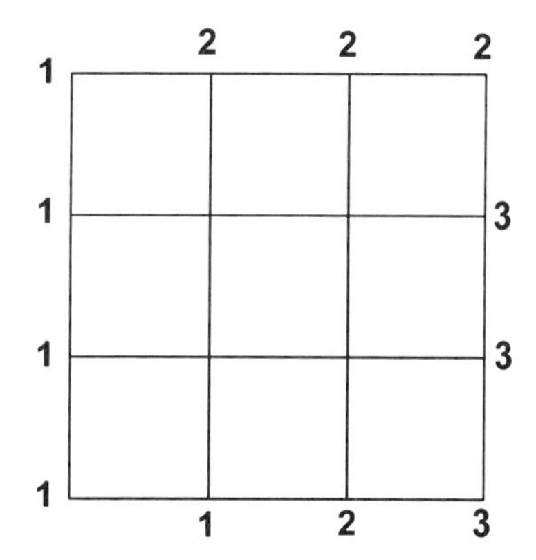

FIGURE 7.29. Grid and boundary conditions for Problem 1.

a. Write the finite difference equations.

b. Start with a guess of $\phi = 1$ at all points at which ϕ is unknown and use Gauss–Seidel iteration to find the solution accurate to two decimal places. At which point is the convergence most rapid?

c. Repeat the preceding part but, on each iteration, compute the new values in reverse order from that of the preceding iteration. Is the convergence any faster or slower? Explain.

2. Compute the optimum SOR relaxation factor for Problem 1 and solve the problem using the SOR method.

3. Solve Problem 1 using the ADI method with a fixed β. Vary β and look at the rate of convergence as a function of β. Compare the results obtained with those for the SOR method obtained in Problem 2.

4. Solve Problem 1 using the multigrid method. Let the coarse grid contain just a single point at the center of the domain so that the problem on this grid can be solved exactly. In this case, the coarse grid point is not a point of the fine grid so you will need to modify the method described in the text a little. Using Gauss–Seidel as the basic method, do two cycles of the multigrid method and compare the rate of convergence with that of the Gauss–Seidel method. Note: this problem is conveniently done using a spreadsheet program.

5. Again solve Problem 1, this time using the conjugate gradient method with Gauss–Seidel as a preconditioner. Discuss what you observe.

6. Higher-order methods (e.g., Adams methods) could be used as the basis of methods for solving elliptic equations.

a. Derive the iterative method equivalent to applying the second-order Adams–Bashforth method to Laplace's equation. For what value of $\beta = \alpha\,\Delta t/\Delta x^2$ will the convergence be most rapid?

b. Does this method have any advantage relative to the Jacobi method?

c. Would still higher-order methods be advantageous? If so, what is the advantage?

7. It is common to use nonuniform grids in elliptic problems. (For Laplace's equation solutions are usually smooth, so this is unnecessary except, perhaps, near corners). The problem is to determine whether use of nonuniform grids slows down the convergence of iterative methods. As a test case, try Problem 1 with $\Delta x_1 = 0.25$, $\Delta x_2 = 0.35$, and $\Delta x_3 = 0.4$. Derive the difference equations and solve them using the Gauss–Seidel method. Compare the rate of convergence with what you found in the uniform grid case.

8. One very common type of elliptic differential equation arises from diffusion problems in inhomogeneous systems. Such systems lead to equations of the type

$$\frac{\partial}{\partial x}\left(\alpha\frac{\partial \phi}{\partial x}\right) + \frac{\partial}{\partial y}\left(\alpha\frac{\partial \phi}{\partial y}\right) = -S(x,y)$$

where α is a function of x and y.

a. Write a finite difference version of this equation.

b. Determine whether the approximation developed in the preceding part satisfies the global conservation statement that the net flux of ϕ out of the region of interest is equal to the total source within the region. If your approximation does not have this property, try to find one that does.

c. Recommend a method for solving the system of equations. Pay particular attention to the effect of the nonuniform diffusion coefficient on the rate of convergence.

9. Solve the Poisson equation

$$\nabla^2 \phi = (1 - x^2)(1 - y^2)$$

subject to the periodic boundary conditions

$$\phi(x + 2, y) = \phi(x, y) \quad \text{and} \quad \phi(x, y + 2) = \phi(x, y)$$

using the Fourier transform method.

10. The Helmholtz equation

$$\nabla^2 \phi - k^2 \phi = 0$$

was considered in the text. Solve it with the boundary conditions:

$$\phi(0, y) = 0 \qquad \phi(1, y) = 1 \qquad \phi(x, 0) = 0 \qquad \phi(x, 1) = 1$$

Use the second-order central difference approximation in the text to find a discrete approximation to this equation and the boundary conditions.

a. Analyze whether the ADI method would converge for this equation and the rate at which it will converge. Note that the coefficient of ϕ in the equation may be of either sign; be sure to consider both possibilities.

b. Carry out several iterations of each method on the grid used in Problem 1 to verify the behavior that you predicted.

11. Apply the conjugate gradient method to the solution of the preceding problem with $k^2 = 1$ and $k^2 = -1$ using 10 intervals in each direction. For this case, it is suggested that you modify the program that is available through the Internet.

12. Do the preceding problem with the multigrid method. Compare the results you obtain.

CHAPTER 8

PARTIAL DIFFERENTIAL EQUATIONS: III. HYPERBOLIC EQUATIONS

The distinguishing feature of hyperbolic partial differential equations (PDEs) is that their characteristics are all real and distinct. This is key to understanding the nature of the solutions of hyperbolic equations and is the basis for an entire class of numerical methods for solving them. In this chapter, some of the salient features of hyperbolic equations and their characteristics will be reviewed; this is followed by a discussion of their numerical treatment.

To begin, we note that all first-order partial differential equations with real coefficients have real characteristics. They can therefore be regarded as hyperbolic equations even though the classification scheme for partial differential equations (hyperbolic, parabolic, elliptic) really applies only to equations of order higher than first. Since first-order equations are simpler to deal with than higher-order equations and their properties are very similar, we shall study them both analytically and numerically in this chapter. In so doing, our purpose is to illustrate features that apply to second-order equations as well. Further, a number of examples will be devoted to the solution of first-order equations.

It is even more important to consider second- and higher-order partial differential equations. In this chapter, we shall consider only partial differential equations that are linear in the highest order derivatives that occur in them. Equations of this kind are called *quasi linear*; essentially all partial differential equations arising in physical problems are of this type.

The most general first-order quasi-linear equation in n independent variables is

$$\sum_{i,k=1}^{n} a_i \frac{\partial u}{\partial x_i} = F(x_1, x_2, \ldots, x_n, u) \tag{8.1}$$

where F, a given function, and the coefficients, a_i, may be functions of the independent variables x_i; $i = 1,2,\ldots,n$ and the dependent variable u. Equation (8.1) usually has to be solved subject to an initial condition that will be given later.

Similarly, the most general second-order quasi-linear partial differential equation in n independent variables is

$$\sum_{i,k=1}^{n} a_{ij} \frac{\partial^2 u}{\partial x_i \partial x_j} = F\left(x_1, x_2, \ldots, x_n, u, \frac{\partial u}{\partial x_1}, \ldots, \frac{\partial u}{\partial x_n}\right) \tag{8.2}$$

where now F and the coefficients a_{ij} may be functions of the independent variables x_i, the dependent variable u, and the first partial derivatives $\partial u/\partial x_i$; $i = 1,2,\ldots,n$. This equation is to be solved subject to initial and/or boundary conditions given later.

8.1. REVIEW OF THEORY

8.1.1. Quasi-Linear First-Order Equations

Let us begin by considering a quasi-linear partial differential equation in two independent variables. It may be written:

$$a\frac{\partial u}{\partial x} + b\frac{\partial u}{\partial y} = c \tag{8.3}$$

where a, b, and c are functions of x, y, and u. We shall not give the complete analysis of this equation here as it can be found in a number of texts on partial differential equations; for example, see Zauderer (1983). It can be shown that the characteristics of this equation are the solutions of the system of ordinary differential equations:

$$\frac{dx}{a(x,y,u)} = \frac{dy}{b(x,y,u)} = \frac{du}{c(x,y,u)} \tag{8.4}$$

The solutions of these equations are a two-parameter family of curves in the three-dimensional x, y, u space with the property that if any point on one of these curves belongs to the solution, so does the entire curve. For most equations, these curves fill all of x, y, u space. A solution to the partial differential equation (8.3) can be represented as a surface made up of a one-parameter subset of these curves. The commonly used nomenclature is a little confusing; sometimes the solutions of Eqs. (8.4), which are curves in the three-dimensional x, y, u space, are called the characteristics; at other times, the term characteristic is used to denote the projections of these curves into the two-dimensional x, y plane. The distinction is important as the three-dimensional curves do not cross but the two-dimensional ones may cross.

To make this more definite, let us consider that case in which a, b, and c are constants. The solutions to the equations (8.4) are then:

$$\begin{aligned} bx - ay &= c_1 \\ cx - au &= c_2 \end{aligned} \tag{8.5}$$

and are straight lines. As stated above, a solution surface is a one-parameter family or subset of these curves. To construct a one-parameter subset of the two-parameter family of curves represented by Eq. (8.5), we can let one parameter become dependent on the other, that is, set $c_2 = f(c_1)$ (or vice versa). This gives

$$cx - au = f(bx - ay)$$

or

$$u = \frac{c}{a}x + f(bx - ay) \tag{8.6}$$

and we see that the general solution to a first-order partial differential equation contains an arbitrary function. That Eq. (8.6) is the solution to the partial differential equation (8.3) may be verified by direct substitution.

8.1.2. Characteristics of Second-Order Equations

Next we present a short analysis of the characteristics of Eq. (8.2). The characteristics are lines or surfaces along which the information about the solution propagates and the equation behaves in a special way on them. It is also useful to introduce them as coordinates. Of course, we do not know the characteristics of Eq. (8.2) as yet. We shall suppose that they are the lines on which the variables:

$$\xi_\alpha = \phi_\alpha(x_1, x_2, \ldots, x_n) \qquad \alpha = 1, 2, \ldots, n \tag{8.7}$$

are constant and change the independent variables from $x_1, \ldots, x_n$ to $\xi_1, \ldots, \xi_n$ and call the dependent variable:

$$w(\xi_1, \xi_2, \ldots, \xi_n) = u(x_1, x_2, \ldots, x_n) \tag{8.8}$$

It is a straightforward and relatively simple calculation to show that under this transformation the PDE (8.2) becomes

$$\sum_{\alpha,\beta=1}^{n} A_{\alpha\beta} \frac{\partial^2 w}{\partial \xi_\alpha \partial \xi_\beta} = G\left(\xi_1, \ldots, \xi_n, w, \frac{\partial w}{\partial \xi_1}, \ldots, \frac{\partial w}{\partial \xi_n}\right) \tag{8.9}$$

where the new coefficient matrix is

$$A_{\alpha\beta} = \sum_{i,j=1}^{n} \frac{\partial \phi_\alpha}{\partial x_i} \frac{\partial \phi_\beta}{\partial x_j} a_{ij} \tag{8.10}$$

Without loss of generality we may suppose that the transformation is chosen in such a way that the initial conditions are given on the surface $\xi_n = 0$, that is,

$$w(\xi_1, \xi_2, \ldots, \xi_{n-1}, 0) = \psi_1(\xi_1, \xi_2, \ldots, \xi_{n-1}) \tag{8.11}$$

$$\frac{\partial w}{\partial \xi_n}(\xi_1, \xi_2, \ldots, \xi_{n-1}, 0) = \psi_2(\xi_1, \xi_2, \ldots, \xi_{n-1}) \tag{8.12}$$

We make the further assumption that the domain for which the solution is desired is infinite in all dimensions other than ξ_n. (It may extend to infinity in ξ_n as well.) This allows us to put the question of boundary conditions aside for now.

This problem can be solved through the use of Taylor series in n dimensions. Given the values of w and $\partial w/\partial \xi_n$ on the surface $\xi_n = 0$ and the PDE, we shall try to find w on a surface defined by $\xi_n = \delta\xi_n$ a short distance away. To do this, we can use the series

$$\begin{aligned} w(\xi_1, \xi_2, \ldots, \xi_{n-1}, \delta\xi_n) &= w(\xi_1^0, \xi_2^0, \ldots, \xi_{n-1}^0, 0) + \sum_{j=1}^{n-1} \left(\frac{\partial w}{\partial \xi_j}\right)_0 (\xi_j - \xi_j^0) + \left(\frac{\partial w}{\partial \xi_n}\right)_0 \delta\xi_n \\ &+ \frac{1}{2} \sum_{j=1}^{n-1} \sum_{k=1}^{n-1} \left(\frac{\partial^2 w}{\partial \xi_j \partial \xi_k}\right) (\xi_j - \xi_j^0)(\xi_k - \xi_k^0) \\ &+ \sum_{j=1}^{n-1} \left(\frac{\partial^2 w}{\partial \xi_j \partial \xi_n}\right)_0 (\xi_j - \xi_j^0)\delta\xi_n + \frac{1}{2}\left(\frac{\partial^2 w}{\partial \xi_n \partial \xi_n}\right)_0 \delta\xi_n^2 + \cdots \end{aligned} \tag{8.13}$$

where $(\xi_1^0, \xi_2^0, \ldots, \xi_{n-1}^0, 0)$ is any point on the surface on which the data are given and the superscript 0 indicates that the function is evaluated at this point.

The initial condition (8.11) provides the value of the function along the initial surface and thus the first term on the right-hand side of this equation. By differentiating this function on the initial surface with respect to $\xi_1, \xi_2, \ldots, \xi_{n-1}$, we can obtain the partial derivatives $\partial w/\partial \xi_1, \partial w/\partial \xi_2, \ldots, \partial w/\partial \xi_{n-1}$ needed to compute the second term. The derivative $\partial w/\partial \xi_n$ needed in the third term is provided by the other initial condition (8.12), so we have all n first partial derivatives. The second partial derivatives that do not involve ξ_n (these occur in the fourth term) can be obtained by differentiating Eq. (8.11) further and the mixed second-order partial derivatives $\partial^2 w/\partial\xi_i\partial\xi_n$ (fifth term) can be obtained by differentiating Eq. (8.12). This leaves only the problem of calculating $\partial^2 w/\partial\xi_n\partial\xi_n$; it is here that the PDE must be used. If $A_{nn} \neq 0$, we can solve Eq. (8.9) for this derivative in terms of data already in hand. We can then proceed to the third partial derivatives, which are readily computed by further differentiation of either the initial conditions or the PDE. Although Taylor series is not recommended as a tool for actually obtaining solutions, it provides the tool needed to show that the problem is well posed as long as $A_{nn} \neq 0$.

When $A_{nn} = 0$, there is trouble. In fact the solutions of the equation

$$A_{nn} = \sum_{i,j=1}^{n} a_{ij} \frac{\partial \phi_n}{\partial x_i} \frac{\partial \phi_n}{\partial x_j} = 0 \tag{8.14}$$

are the *characteristics* of the partial differential equation (8.2). Thus the characteristics are themselves solutions of a first-order PDE. If the coefficients a_{ij} are functions of the dependent variable, that is, if the partial differential equation is nonlinear, the characteristics depend on the solution and can be found only by solving the differential equation itself; we shall see some of the difficulties that can arise in the next subsection. If the initial data are given on a characteristic curve or surface, the problem has no solution. One might think that the difficulty could be cleared up by a clever choice of initial conditions, but this is not possible. An example will illustrate this point.

Example 8.1: Solution of the Wave Equation Let us construct the solution of the wave equation

$$\frac{1}{c^2}\frac{\partial^2 u}{\partial t^2} = \frac{\partial^2 u}{\partial x^2} \tag{8.15}$$

with the initial conditions

$$u(x,0) = \alpha(x) \qquad \frac{\partial u}{\partial t}(x,0) = \beta(x) \tag{8.16}$$

This problem is one of the standards of the PDE repertoire and can be solved in a textbook manner. The general solution of the wave equation is well known:

$$u(x,t) = f(x - ct) + g(x + ct) \tag{8.17}$$

This can be shown by noting that the characteristics of the wave equation are the families of straight lines $\xi = x + ct =$ constant and $\eta = x - ct =$ constant. In this case the coefficients of the PDE are constants, so the characteristics are independent of the solution and can be found a priori. Then the PDE transformed into characteristic coordinates is

$$\frac{\partial^2 u}{\partial\xi\,\partial\eta} = 0 \tag{8.18}$$

whose solution is $u = f(\eta) + g(\xi)$, which is the solution given above.

The particular solution of the wave equation that satisfies the initial conditions is found by substituting the general solution (8.17) into the initial conditions (8.16) to arrive at equations for f and g:

$$f(x) + g(x) = \alpha(x) \tag{8.19}$$

$$-cf'(x) + cg'(x) = \beta(x) \tag{8.20}$$

where a prime denotes ordinary differentiation. If we denote the indefinite integral of $\beta(x)$ by $\gamma(x)$, we find that the solution is given by $f(x) = [c\alpha(x)\gamma(x)]/2c$ and $g(x) = [c\alpha(x) + \gamma(x)]/2c$. If the initial data are given on the line $t = 0$, which is not a characteristic of the equation, a unique solution can be found for any initial data. This is equally true for any other line that is not a characteristic.

Now suppose that the initial data are given on one of the characteristics, for example,

$$u\left(x, \frac{x}{c}\right) = \alpha(x) \qquad \frac{\partial u}{\partial t}\left(x, \frac{x}{c}\right) = \beta(x) \tag{8.21}$$

Instead of Eqs. (8.19) and (8.20), we would have

$$f(0) + g(2x) = \alpha(x) \tag{8.22}$$

$$-cf'(0) + cg'(2x) = \beta(x) \tag{8.23}$$

In general, these equations are inconsistent, that is, there is no solution to the problem unless $\beta(x) = 2c\alpha'(x) + \text{constant}$. If α and β are related in this way, the solutions to these equations are nonunique, that is, if the initial conditions are compatible, the solution is nonunique.

Finally, returning to the original problem, we note that if either α or β has a discontinuity, then so do f and g. Suppose that the discontinuity is at x_0. Then, from Eq. (8.17), we see that the discontinuity will appear at time t at locations $x = x_0 \pm ct$. Thus the characteristics, being the lines along which signals propagate, are also the lines (or in the more general case curves or surfaces) on which the solution may have discontinuities.

This information sets the stage for use of the knowledge of characteristics as a means of solving hyperbolic systems of equations, something we shall do later in this chapter.

8.1.3. Nonlinear Equations and Shocks

Although we shall not go very deeply into methods for solving nonlinear hyperbolic equations in this chapter, they occur so often that something must be said about them. We shall, however, restrict our attention to some simple equations.

In particular, let us consider Burgers' equation:

$$\frac{\partial u}{\partial t} + u\frac{\partial u}{\partial x} = \nu\frac{\partial^2 u}{\partial x^2} \tag{8.24}$$

which is often used as a simple model equation in fluid mechanics; in that application, ν represents the viscosity. This equation is second order and is actually parabolic, but, in many interesting cases, the viscosity is small and it behaves much like a first-order nonlinear equation.

For that reason, the inviscid form of Burgers' equation or the convection equation

$$\frac{\partial u}{\partial t} + u\frac{\partial u}{\partial x} = 0 \tag{8.25}$$

is often studied. Indeed, its solutions are very interesting.

In fact, the solution to this equation can be obtained with the method of characteristics for first-order equations that was presented at the beginning of this section. The solution is very similar to that of the linear first-order equation that was presented earlier. It is

$$u(x,t) = f(x - u(x,t)t) \tag{8.26}$$

This solution is very peculiar. For one thing, the dependent variable occurs on both the left- and right-hand sides, that is, the solution is an implicit one. To see that this is a solution, let us differentiate it with respect to the independent variables. We obtain

$$\frac{\partial u}{\partial t} = -(u - tu_t)f' \qquad \frac{\partial u}{\partial x} = (1 - tu_x)f' \tag{8.27}$$

where f' is the derivative of f with respect to its single argument. From these we find

$$\frac{\partial u}{\partial t} = -\frac{uf'}{1 + tf'} \qquad \frac{\partial u}{\partial x} = \frac{f'}{1 + tf'} \tag{8.28}$$

Substituting these into the inviscid Burgers equation (8.25), we find that the equation is indeed satisfied.

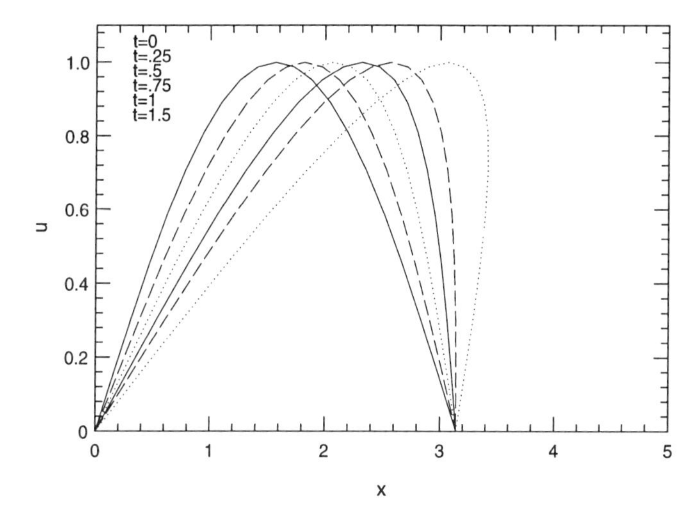

FIGURE 8.1. Solution of the inviscid Burgers equation with a sinusoidal initial condition.

Example 8.2: Solution of the Inviscid Burgers Equation Let us compute the solution to the inviscid Burgers equation with the initial condition:

$$u(x,0) = \sin x \tag{8.29}$$

The solution is shown in Figure 8.1. To generate this solution for purposes of graphing it, one can use a trick to avoid having to solve the implicit equation. We note that the solution is constant on lines defined by $x - ut =$ const. so, for a given time t at which we wish to compute the solution, we can select a value of u and compute the value of x as $\sin^{-1}(u) - ut$.

It is important to note that the solution becomes double valued for $t > 1$. This solution is correct mathematically but does not make any sense from a physical point of view. We need to investigate this more thoroughly. First, let us note that, when the solution first becomes double valued, at minimum, the partial derivative with respect to x (8.28) must become infinite somewhere. This can happen if the quantity in the denominator, $1 + tf'$ becomes zero. This requires, first of all that $f' < 0$ somewhere is the domain, that is, it is the part of the solution in which u is decreasing with x that can cause trouble. Since the minimum (largest negative) value of f' for the initial condition (8.29) is -1, which occurs at $x = 0, \pi, 2\pi, \ldots$, we expect the trouble to begin at these points (and it does). Furthermore, $1 + tf'$ first becomes zero at $t = 1$, as observed in the actual solution.

How do we reconcile the behavior of the solution we just found with the physics of the problem? We begin by giving another explanation of what we have observed. For the Burgers equation, u is both the solution and the speed at which it is convected. Where the velocity is decreasing in the direction

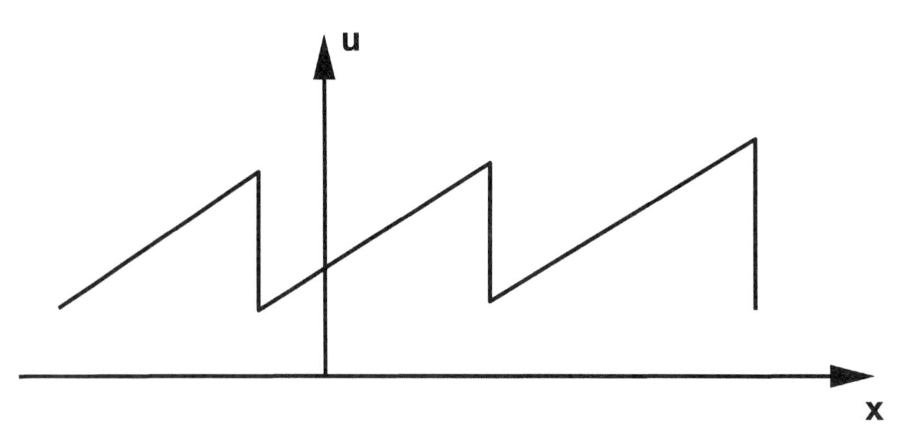

FIGURE 8.2. Solution of Burgers equation and the limit of small viscosity.

of propagation, the faster moving parts of the solution "catch up" with and "overtake" the slower parts, causing the double valuedness. This does not make sense physically; the material at a particular point in space must have a unique speed.

The resolution of this apparent paradox is that the inviscid form of the Burgers equation is only an approximation to physical reality. In the physical world, there is always a little viscosity. The solution to the viscous equation is shown in Figure 8.2 in the limit of the solution as the viscosity becomes zero. The limiting solution has a discontinuity but is not double valued. The limit $\nu \to 0$ is singular; in this limit the order of the equation is reduced, and that is why the solution obtained in the limit as the viscosity goes to zero is different from the solution of the inviscid equation.

In fact, the solutions on the two sides of the discontinuity are solutions of the inviscid equation, that is, the inviscid equation is satisfied everywhere but at one point. We call such a solution a *weak solution* of the differential equation. The discontinuity itself is called a *shock wave* or simply a shock, a term that reflects its aerodynamic origin.

To understand the nature of a shock better, let us consider the simpler problem in which the initial condition is

$$u(x,0) = \begin{cases} u_1 & x < 0 \\ u_2 & x > 0 \end{cases} \tag{8.30}$$

where $u_1 > u_2$. The discontinuity or shock will propagate to the right. One can show (Zauderer, 1983) that the speed of propagation of the shock is

$$u_s = \tfrac{1}{2}(u_1 + u_2) \tag{8.31}$$

There are many problems (especially in aerodynamics) in which it is important to capture the behavior of the shock accurately and doing so presents an enormous challenge to numerical methods.

8.2. METHOD OF CHARACTERISTICS

We begin the discussion of the numerical treatment of hyperbolic PDEs with a method that relies directly on knowledge of the characteristics of the equation. It is thus called the *method of characteristics* and is applicable only to hyperbolic equations. We will begin with first-order equations and then treat second-order equations; in both cases, only the case of two independent variables will be considered.

8.2.1. First-Order Equations

The equation defining the characteristics of first-order equations was given in the preceding section. We repeat it here for completeness. The characteristics of the first-order quasi-linear partial differential equation

$$a\frac{\partial u}{\partial x} + b\frac{\partial u}{\partial y} = c \tag{8.32}$$

are the lines defined by

$$\frac{dx}{a(x,y,u)} = \frac{dy}{b(x,y,u)} = \frac{du}{c(x,y,u)} \tag{8.33}$$

Suppose that we are given initial data, that is, the values of u on a curve. Usually the data are given at $x = 0$ so that they provide $u(0, y)$, but, in principle, any curve in the (x, y) plane (other than a characteristic) could carry the initial data. A typical initial condition is shown in Figure 8.3. The initial data itself can be regarded as a curve in three-dimensional (x, y, u) space. A characteristic (of the three-dimensional kind) runs through each point on this curve. As we saw in the preceding section, the entire characteristic curve is part of the solution surface. If we generate the characteristic curves through a number of the points on the initial condition curve, we will have the elements that make up the solution surface. Thus the task of solving the equation reduces to one of generating these curves, which means that all we need to do is to solve the set of ordinary differential equations (ODEs) (8.33). This is the beauty of the method of characteristics; the solution can be obtained with the use of any of the methods designed for initial value problems for ODEs that were presented in Chapter 4.

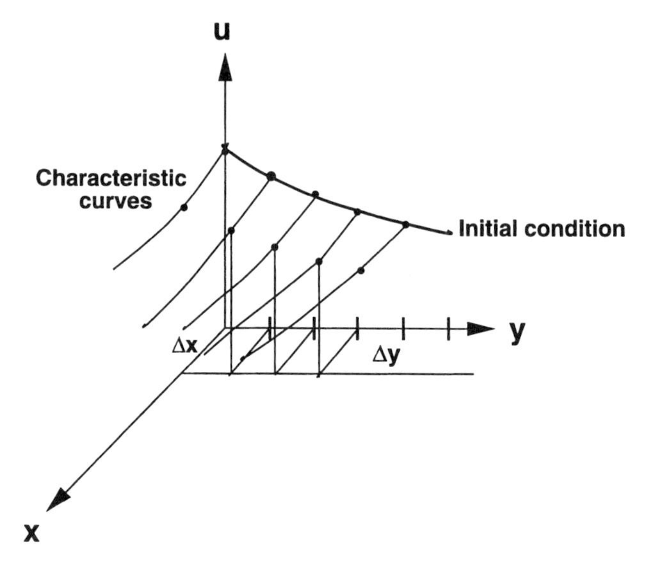

FIGURE 8.3. Illustration of the method of characteristics for a first-order PDE. The calculation is begun at $x = 0$ and continued along the characteristic lines.

Since the purpose is to illustrate how a numerical solution can be obtained, we shall adopt the simplest possible method, the explicit Euler method. We begin by discretizing the initial data. To keep things as simple as possible, we shall use equal spacing in y; the discretization is shown in Figure 8.3; the value of the solution at the point $(0, y_{0,j})$ is $u_{0,j}$; note that the discretized y variable has to carry two indices because y is a function of x along a characteristic. We are now ready to begin the solution process. The step size in x can be different on each characteristic, but, again for simplicity, we shall use the same step size Δx on all of the characteristic curves. If the explicit Euler method is applied to Eqs. (8.33), we have

$$
\begin{aligned}
x_{i+1} &= x_i + \Delta x \\
y_{i+1,j} &= y_{i,j} + \Delta x \frac{a(x_i, y_{i,j}, u_{i,j})}{b(x_i, y_{i,j}, u_{i,j})} \\
u_{i+1,j} &= u_{i,j} + \Delta x \frac{a(x_i, y_{i,j}, u_{i,j})}{c(x_i, y_{i,j}, u_{i,j})}
\end{aligned}
\tag{8.34}
$$

That is all there is to it. The calculation can be continued by the standard marching process for as long as required in the direction of increasing x.

Several things should be noted about this method. If the equation is linear, the coefficients a and b do not depend on u. In this case the characteristic lines in the two-dimensional x, y plane can be found without solving the partial differential equation itself. For the nonlinear case, the solution and the characteristics usually need to be generated at the same time.

In fact, if the method described above is used to solve the linear convection equation (the inviscid Burgers equation in which the advection velocity is constant):

$$\frac{\partial u}{\partial t} + c\frac{\partial u}{\partial x} = 0 \tag{8.35}$$

the method will produce the exact solution for any Δt. This is because, for this equation, the characteristics are straight lines and the dependent variable is constant on them. This is a great advantage of the method; there is no need to illustrate this property with an example.

In the nonlinear case, the characteristics must be generated along with the solution. An exception is the inviscid Burgers equation (8.25) for which the solution is constant on the characteristics, which guarantees that the latter are again straight lines. The difference from the case with constant convection velocity treated above is that, although the characteristics are straight lines, they have different slopes. In fact if $\partial u/\partial x < 0$ in some part of the domain, the characteristics will cross and the solution will become multivalued in that part of the domain as we saw in the preceding section. As we noted there, this is the correct formal mathematical solution to the problem. The method of characteristics described here cannot generate shock waves unless it is modified in a way that allows it to do so. The problem is that the solution on each characteristic line does not "sense" what is happening on the neighboring characteristics and therefore cannot react to it. To allow information to travel across the characteristics, it is necessary to introduce viscosity or an equivalent. We then have the viscous Burgers equation rather than the inviscid one. Technically, the latter equation is parabolic, but, if the viscosity is small, it behaves very much like a hyperbolic equation. We shall deal with this case later.

For other nonlinear equations, the characteristics are usually curved, and their projections into the two dimensional x, y plane will usually cross. (As noted earlier, the three-dimensional curves cannot cross.) As was the case for the inviscid Burgers equation, the solution can be continued along the characteristics, but it becomes multivalued. Again, the physically relevant solution must remain single valued; to do so, it must develop a shock. The numerical method of characteristics cannot create shocks without the introduction of viscosity or special tricks.

8.2.2. Second-Order Equations

Now let us consider the case of a second-order hyperbolic partial differential equation with two independent variables. The most general quasi-linear second-order partial differential equation in two independent variables can be written:

$$a\frac{\partial^2 u}{\partial x^2} + 2b\frac{\partial^2 u}{\partial x \partial y} + c\frac{\partial^2 u}{\partial y^2} = F \tag{8.36}$$

For this case, the results of the preceding section on the characteristics of second-order equations can be simplified somewhat. We found that the characteristics of Eq. (8.36) are the level curves of the function that satisfies the partial differential equation:

$$a\left(\frac{\partial \phi}{\partial x}\right)^2 + 2b\left(\frac{\partial \phi}{\partial x}\right)\left(\frac{\partial \phi}{\partial y}\right) + c\left(\frac{\partial \phi}{\partial y}\right)^2 = 0 \tag{8.37}$$

The equations defining the level curves of a function can be derived as follows. On these curves, $\phi(x, y) =$ constant. By differentiating this equation, we obtain

$$d\phi = \frac{\partial \phi}{\partial x}dx + \frac{\partial \phi}{\partial y}dy = 0 \tag{8.38}$$

which can be rewritten as

$$\frac{dy}{dx} = \frac{\partial \phi / \partial x}{\partial \phi / \partial y} \tag{8.39}$$

Then, dividing Eq. (8.37) by $(\partial \phi / \partial y)^2$, we obtain an ordinary differential equation for the characteristic curves themselves:

$$a\left(\frac{dy}{dx}\right)^2 - 2b\frac{dy}{dx} + c = 0 \tag{8.40}$$

Finally, solving this quadratic equation for dy/dx, we have

$$\frac{dy}{dx} = \frac{b \pm \sqrt{b^2 - ac}}{2a} = \alpha_1, \alpha_2 \tag{8.41}$$

The characteristics are real and distinct if and only if $b^2 - ac > 0$, this is a simplified form of the criterion that determines whether Eq. (8.36) is hyperbolic. The equation for the characteristics is an ODE, which is a consequence of having only two independent variables. For an equation in more than two independent variables, the characteristic equation would be a PDE and the characteristics would be surfaces.

In this section, we will develop a numerical method for solving Eq. (8.36) that makes direct use of the characteristics and the equations that they obey. To set the stage for this, we will specialize the analysis of the preceding section to the two-dimensional case.

Suppose that we have the solution and its normal derivative on some curve C in x, y space. These data might be the initial conditions or they may have been obtained by previous computation. To construct the solution through the use of Taylor series, we will need the second derivatives of u. (The first derivatives are available; the normal derivative is given, and the tangential derivative can be obtained by differentiating the function.) To simplify the analysis we will use a Cartesian (x, y) coordinate system. We thus need to compute the derivatives $\partial^2 u/\partial x^2$, $\partial^2 u/\partial y^2$, and $\partial^2 u/\partial x \partial y$, rather than the derivatives parallel and normal to the curve C; this is a major difference from the preceding section. From the chain rule for differentiation we can write, along any differential line element (dx, dy),

$$d\frac{\partial u}{\partial x} = \frac{\partial^2 u}{\partial x^2}dx + \frac{\partial^2 u}{\partial x \partial y}dy \tag{8.42}$$

$$d\frac{\partial u}{\partial y} = \frac{\partial^2 u}{\partial x \partial y}dx + \frac{\partial^2 u}{\partial y^2}dy \tag{8.43}$$

These two equations, taken together with the PDE (8.36) provide a set of three linear algebraic equations from which the three second-order partial derivatives can be calculated. This system of equations has a unique solution only if its determinant is nonzero, that is,

$$\begin{vmatrix} a & 2b & c \\ dx & dy & 0 \\ 0 & dx & dy \end{vmatrix} \neq 0 \tag{8.44}$$

It should come as no surprise that this condition is equivalent to saying that the line element (dx, dy) does not lie on a characteristic. This also shows that the three second-order partial derivatives of u cannot be computed solely from information given on a single characteristic curve, a result proven in the preceding section. However, there is no reason why we cannot use data on both sets of characteristics to compute these derivatives. We can solve the equations for the second derivatives when the determinant (8.44) is zero provided that the terms on the right-hand sides of these equations obey a solvability condition.

One statement of the solvability condition is that the determinant

$$\begin{vmatrix} a & F & c \\ dx & du_x & 0 \\ 0 & du_y & dy \end{vmatrix} \tag{8.45}$$

must be zero if dx and dy are differential elements of a characteristic curve. In the language of linear algebra, Eq. (8.45) states that the vector (F, du_x, du_y)

must lie in the column space of the matrix (8.44). Here $u_x = \partial u/\partial x$ and $u_y = \partial u/\partial y$. Since, according to Eq. (8.41), the characteristics are the lines on which $dy = \alpha_1\, dx$ or $dy = \alpha_2\, dx$, Eq. (8.45) is equivalent to

$$\alpha_1 a du_x + c du_y - F\, dy = 0 \tag{8.46}$$

on the α_1 characteristics and

$$\alpha_2 a du_x + c du_y - F\, dy = 0 \tag{8.47}$$

on the α_2 characteristics. These equations can be regarded as the PDE in the respective characteristic directions and show that the partial differential equation behaves like an ordinary differential equation on the characteristics.

Equations (8.46) and (8.47) together with the equation obtained by application of the chain rule,

$$du = u_x\, dx + u_y\, dy \tag{8.48}$$

are the three equations that govern the way in which the unknown function u and its two first partial derivatives change with the independent variables x and y along the characteristics. As noted earlier, if the PDE is nonlinear, that is, if a, b, or c are functions of u, these three equations must be solved simultaneously with the equations that define the characteristics. In the linear case the equations for the characteristics can be solved prior to the solution of the equations for the dependent variable.

In summary, what we have done is to use the properties of characteristics to reduce the problem to a set of ordinary differential equations. Equations (8.41) that define the characteristics and are used to locate the mesh points at the next step. Equations (8.46), (8.47), and (8.48) give the values of u, u_x, and u_y at the new points in terms of the values at the old ones. We now introduce numerical approximations that can be used to solve these equations.

As was the case for first-order equations, these equations can be solved by any of a number of methods based on ODE solvers. This time we shall use an implicit method that is essentially the trapezoid rule (see Figure 8.4 for a description of the geometry). Suppose we have the values of the function and its first derivatives at a set of points on curve C and we wish to advance the solution in the direction indicated. The two sets of characteristics have been drawn as if their location were already known; as we have seen, in the nonlinear case they must be computed along with the solution. The problem is to find the coordinates of the point P and the values of the unknown function, u, and its derivatives at point P by using the equations given above.

To do this, we need a set of five discretized equations that can be solved for the solution, its first partial derivatives and the coordinates of the point P. Two of these equations are obtained by applying the finite difference approximation to the ordinary differential equations that define the characteristics

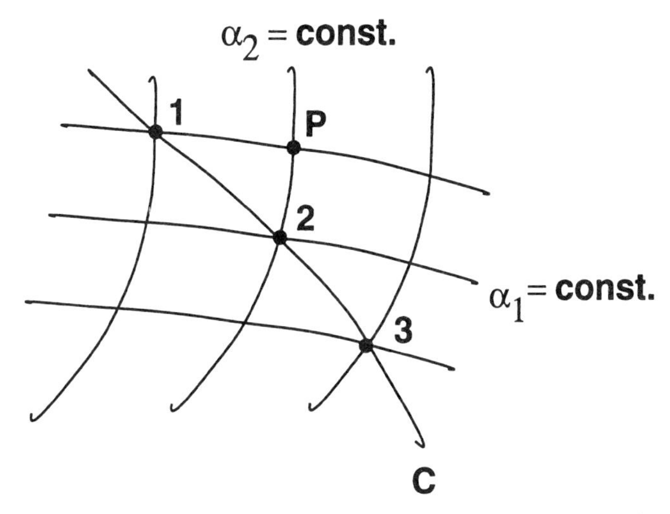

FIGURE 8.4. Illustration of the use of the method of characteristics for solving a second-order hyperbolic equation.

[Eq. (8.41)]:

$$\frac{y(P)-y(1)}{x(P)-x(1)} = \frac{1}{2}[\alpha_1(P)+\alpha_1(1)] \tag{8.49}$$

$$\frac{y(P)-y(2)}{x(P)-x(2)} = \frac{1}{2}[\alpha_2(P)+\alpha_2(2)] \tag{8.50}$$

where α_1 and α_2 are given by Eq. (8.41).

Differencing Eqs. (8.46) and (8.47) in the same manner, we have

$$[a(P)\alpha_1(P)+a(1)\alpha_1(1)][u_x(P)-u_x(1)]+[c(P)+c(1)][u_y(P)-u_y(1)] \\ -[F(P)+F(1)][y(P)-y(1)]=0 \tag{8.51}$$

$$[a(P)\alpha_2(P)+a(2)\alpha_2(2)][u_x(P)-u_x(2)]+[c(P)+c(2)][u_y(P)-u_y(2)] \\ +[F(P)+F(2)][y(P)-y(2)]=0 \tag{8.52}$$

and finally, from Eq. (8.48), we have

$$u(P)-u(1)=[u_x(P)+u_x(1)][x(P)-x(1)]+[u_y(P)+u_y(1)][y(P)-y(1)] \tag{8.53}$$

Equations (8.49) through (8.53) are a set of five equations for the five unknowns $x(P)$, $y(P)$, $u(P)$, $u_x(P)$, and $u_y(P)$ that need to be calculated at the new point P. In general they are nonlinear and so must be solved iteratively. In most cases, use of a predictor–corrector method is a reasonable approach. Since the use of characteristics produces equations that are very much like ordinary differential equations, any other ODE method could be used.

Example 8.3: Method of Characteristics Solution of the Wave Equation: Smooth Initial Condition The equation that we shall use to illustrate the use of methods for second-order hyperbolic partial differential equations described in this chapter is the wave equation:

$$\phi_{tt} = c^2 \phi_{xx} \tag{8.54}$$

We shall use two different initial conditions to illustrate how the methods behave under different circumstances. The first initial condition is smooth, a sine wave:

$$\phi(x,0) = \sin 2\pi x \tag{8.55}$$

The initial time derivative is zero:

$$\phi_t(x,0) = 0 \tag{8.56}$$

The boundary conditions require the function to be zero at both spatial boundaries:

$$\phi(0,t) = \phi(1,t) = 0 \tag{8.57}$$

The exact solution to this problem is easily obtained. Sinusoidal waves, each half the amplitude of the initial condition, propagate to the left and right. When these waves strike the boundary, they are reflected with a change in sign; the sign change allows the solution to satisfy the boundary condition. After enough time has passed for a wave to have propagated the length of the box, that is, at $t = 1/c$, the waves have arrived back at the center of the domain where they reinforce each other, and the solution is identical to the initial condition except for a change in sign. One can study the effectiveness of the numerical method by comparing the negative of the solution obtained at time $t = 1/c$ with the initial condition. The solution can be allowed to continue to propagate further. At time $t = 2/c$, we should recover the initial condition without change in sign. From then on, the process repeats with a period of $2/c$.

In this problem, the characteristics are not difficult to find. In fact $\alpha_1 = c$ and $\alpha_2 = -c$ are the slopes of the two sets of characteristics. Then, it is easy to see that the characteristics are

$$\begin{aligned} x + ct &= \text{constant} \\ x - ct &= \text{constant} \end{aligned} \tag{8.58}$$

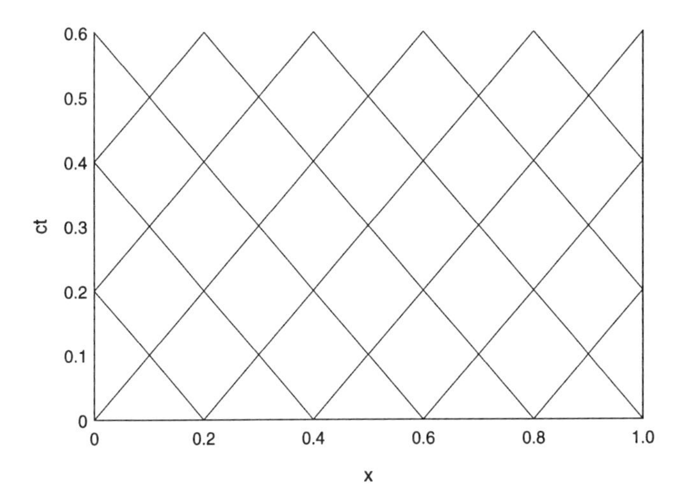

FIGURE 8.5. Method of characteristics solution of the wave equation in one spatial dimension.

The characteristics that pass through the points $x = 0$, 0.2, 0.4, 0.6, 0.8, and 1.0 at $t = 0$ are shown in Figure 8.5. The characteristics "reflected" from the boundary play an important role is solving this equation so they are shown as well. A better description of what happens at the boundary is that information arriving at the boundary on the "outgoing" characteristic is transformed into data on the "incoming" characteristic by the boundary condition. The entire characteristic mesh can be computed in advance, and the equations defining the characteristics need not be treated numerically in this case; since they are straight lines, they would be computed exactly by the numerical method. In this problem, the spatial mesh points at each time step lie midway between the points at the preceding step and on lines of constant time. For this to be the case, the time step must be $\Delta t = \Delta x/2c$.

The equations on the characteristics are also simple in this case. The finite difference versions [Eqs. (8.51) and (8.52)] reduce to

$$c(u_{x,i}^{n+1} - u_{x,i+1/2}^{n}) + c(u_{t,i}^{n+1} - u_{t,i+1/2}^{n}) = 0 \tag{8.59}$$

$$c(u_{x,i}^{n+1} - u_{x,i-1/2}^{n}) + c(u_{t,i}^{n+1} - u_{t,i-1/2}^{n}) = 0 \tag{8.60}$$

These equations are in fact the exact result of integration along the characteristics. Although they have finite difference appearance, they actually contain no approximations. Thus u_x and u_t are propagated as they would be in the exact equations; this is a consequence of the fact that the equation is linear and has constant coefficients.

The initial values of u_x were computed by central differences:

$$u_{x,i}^0 = \frac{u_{j+1}^0 - u_{j-1}^0}{2\Delta x} \tag{8.61}$$

Equations (8.59) and (8.60) can be solved easily for $u_{x,i}^{n+1}$ and $u_{t,i}^{n+1}$. Finally, u_i^{n+1} is obtained from Eq. (8.53) which, for this case, reduces to

$$u_i^{n+1} = u_{i\pm 1/2}^n + \tfrac{1}{2}(u_{x,i}^{n+1} + u_{x,i\pm 1/2})\Delta x + \tfrac{1}{2}(u_{t,i}^{n+1} + u_{t,i\pm 1/2})\Delta t \tag{8.62}$$

Either sign may be used. To keep everything symmetric, we used the average of these two equations in computing the results shown below. Equation (8.62) is, of course, a second-order approximation (trapezoid rule) to the exact result.

This leaves the question of handling the boundary conditions. At the odd-numbered time steps, there is no problem, since all the points at which results are to be computed are interior points. At the even steps, we can apply the equations resulting from solving Eqs. (8.59) and (8.60) at the boundaries. For $i = 0$ the simplest thing we can do is to note that at the boundary at $x = 0$, information can arrive only from inside the domain. Thus only the outgoing characteristics, that is, the members of the set $x + ct =$ constant, can provide data. Thus we must use Eq. (8.59). Since $u_t(0) = 0$, we have

$$u_{x,0}^{n+1} = u_{x,1/2}^n + cu_{t,1/2}^n \tag{8.63}$$

At the other boundary ($i = N$), we must use the $x - ct =$ constant characteristics and we find

$$u_{x,N}^{n+1} = u_{x,N-1/2}^n + cu_{t,N-1/2}^n \tag{8.64}$$

Both u_t and u are zero on the boundary; these conditions have been used in deriving Eqs. (8.63) and (8.64).

The results are given in Figure 8.6. Twenty points were used in this calculation and we see that the method of characteristics (MOC) is very accurate in this case. The errors are quite small and are entirely due to the finite difference approximation of the spatial derivatives at $t = 0$. What happens is that the derivatives u_x and u_t are propagated as the exact wave equation would propagate them and retain their original profiles. However, since $u_x(x,0)$ was obtained from a finite difference approximation, its values are not correct. Then u_t and u "lock onto" the incorrect u_x, and u is essentially the spatial integral of the incorrect u_x. Aside from this error, the profile is propagated without change. When the solution is continued from $t = 1/c$ to $t = 2/c$, no further deterioration of the solution occurs.

The errors due to the finite difference approximations can be reduced by using a finer mesh or, alternatively, using a more accurate approximation for

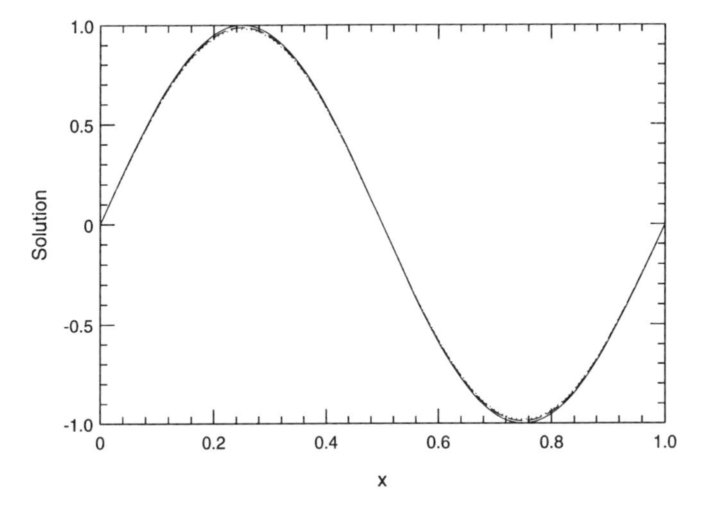

FIGURE 8.6. Method of characteristics solution of the wave equation with a sinusoidal initial condition.

the derivative at $t = 0$. However, the method of characteristics requires that, if the spatial mesh size is reduced by half, the time step must also be halved, which means that the cost of obtaining a solution goes up by a factor of 4.

Other numerical approximations can be applied to the method of characteristics, and it is possible to devise methods that explicitly allow for the possibility of discontinuities. Such methods have been used to fit shock waves in gas dynamic calculations and have been quite successful.

The program used in this example, MOC, is found on the Internet site described in Appendix A.

Example 8.4: Method of Characteristics Solution of the Wave Equation: Discontinuous Slope This example uses the same equation and method; only the initial condition is changed. In the new initial condition the function is a "witch's hat":

$$\phi(x,0) = 5(x - 0.3) \qquad 0.3 < x < 0.5$$

$$\phi(x,0) = 5(0.7 - x) \qquad 0.5 < x < 0.7 \tag{8.65}$$

$$\phi(x,0) = 0 \qquad 0 < x < 0.3 \quad \text{and} \quad 0.7 < x < 1$$

Again the initial time derivative is zero and the boundary conditions are as before. As shown by Figure 8.7, the method has difficulty treating the corner at the top of the hat and also tends to round the corners at the bottom of the profile. The program "thinks" that u_x has the shape shown in Figure 8.8 (it is

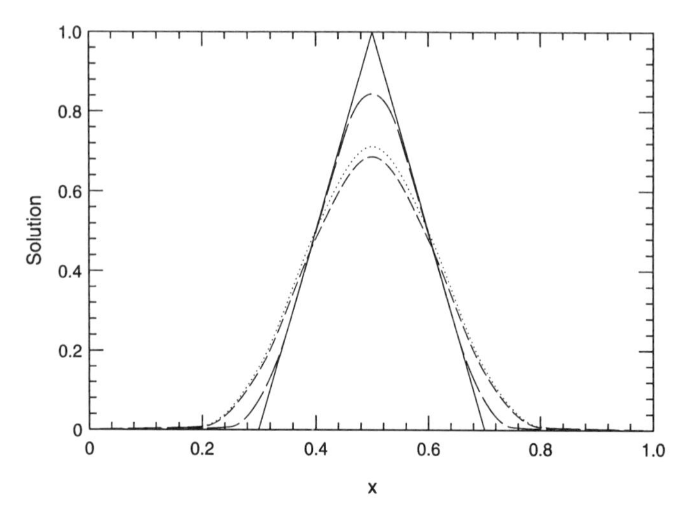

FIGURE 8.7. Method of characteristics solution of the wave equation with a witch's hat initial condition.

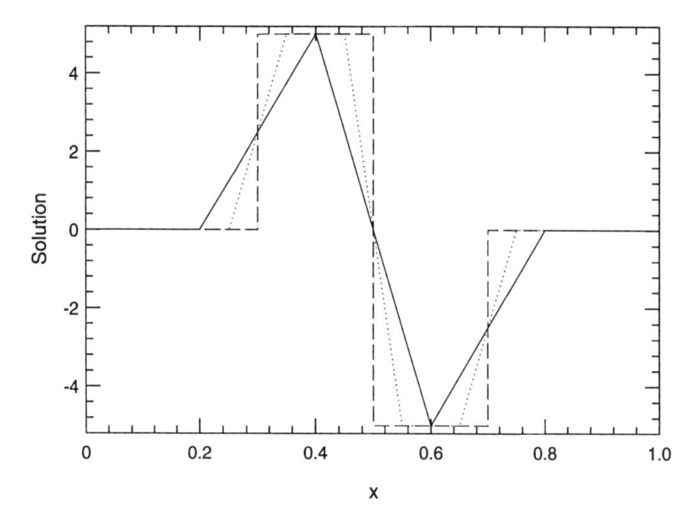

FIGURE 8.8. Derivative $\partial\phi/\partial x$ in the method of characteristics solution of the convection equation with a witch's hat initial condition.

the best interpolation that can be made from the limited data), and then the shape is propagated unchanged in shape through the domain.

For many years the method of characteristics was the dominant method for solving hyperbolic PDEs; because it has some very desirable properties, it is still used. The method of characteristics, by definition, emphasizes the role of characteristics in determining the solution of hyperbolic PDEs. As a result, important properties of the exact solution are preserved in the numerical solution. In particular, because discontinuities propagate along characteristics, variations of the method of characteristics can be applied to the solution of problems that contain discontinuities. The most significant application of this type is found in aerodynamics. In any flow that contains a region of supersonic flow, there is the possibility of the occurrence of shock waves (which are large discontinuities in many of the variables) and other, milder, discontinuities. For this reason, the method of characteristics has been the basis for a large number of methods for treating supersonic flows. A review of methods of this type may be found in some of the works cited in the bibliography. Some of the methods, in simplified form, will be described briefly later in this chapter.

The method of characteristics also has some significant disadvantages. Chief among these is the difficulty of keeping track of the locations of the characteristics and the values of the variables; this becomes especially difficult in complex geometries and in three dimensions and makes writing computer programs based on the method of characteristics very difficult.

The method of characteristics was the first method to find extensive application in the solution of the hyperbolic equations that govern aerodynamic flows. Aerodynamics is probably the principal application area that requires solution of hyperbolic equations; the propagation of electromagnetic waves is likely second. When people became interested in solving for the equations describing flows around bodies of complicated shapes such as aircraft and engine inlets, the method of characteristics was found be almost impossible to program efficiently. Another difficulty arises in these flows. In parts of some flows, the fluid velocity is very close to the speed of sound (transonic flow). In such a region, the characteristics become nearly perpendicular to the direction of the flow. When this happens, it is necessary to take very small steps in the flow direction and the computation time becomes very long. It was these difficulties that motivated the search for methods other than the method of characteristics for use in aerodynamic computation. These will be taken up in the sections that follow. Before doing that, we consider one more issue connected with the method of characteristics—the question of what happens when the grid does not consist of the characteristic lines themselves.

8.2.3. Method of Characteristics on Cartesian Grids

To a large extent, the properties of the method of characteristics exhibited in the examples above are due to the way in which the problem was treated numerically. In particular, the use of the characteristics as the lines along which the equations are finite differenced helped a great deal. To see what difficulties arise if a different coordinate system is used, let us consider an alternative approach. Suppose that, instead of using the characteristics as coordinates as we have done so far, we simply lay down a uniform Cartesian grid similar to the ones that were used for parabolic and elliptic equations. Such a grid is illustrated in Figure 8.9.

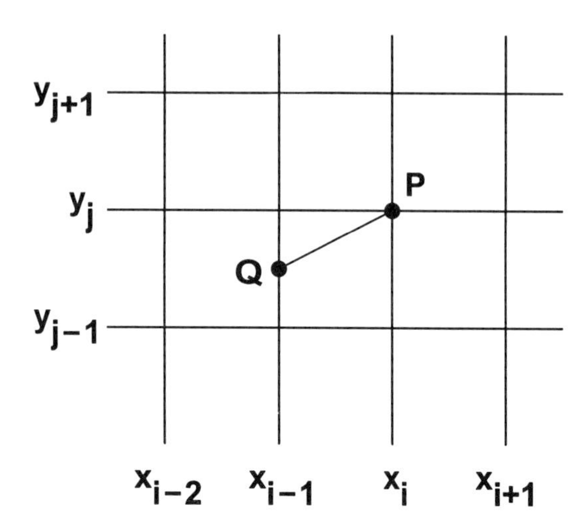

FIGURE 8.9. Uniform Cartesian mesh for the solution of the advection equation in two spatial dimensions.

Suppose that we want to compute the solution of the first-order equation:

$$u\frac{\partial u}{\partial x} + v\frac{\partial u}{\partial y} = 0 \tag{8.66}$$

which is the steady version of the two-dimensional convection equation. (The first-order unsteady version was used in the earlier examples.)

Suppose we want to compute the solution, $u_{ij} = u(x_i, y_j)$, the point labeled P in Figure 8.9. Since the solution is constant on the characteristic that passes through this point, it is natural to set u_{ij} equal to the value of u at the point at which that characteristic crosses the line $x = x_{i-1}$; this is the point labeled Q in Figure 8.9. The problem arises because the point Q lies between two grid points so the only way to compute the value of u at that point is by means of interpolation. However, the use of interpolation introduces an error that can degrade the solution. The problem is illustrated by the following example.

Example 8.5: Grid-Based Method of Characteristics We now solve Eq. (8.66). The characteristics of this equation are the lines:

$$vx - uy = \text{const.} \tag{8.67}$$

which are the lines along which information is convected. (They are usually called streamlines. For this equation, the solution u is constant on each streamline. For the initial condition, we take

$$u(x,0) = \sin 2\pi x$$

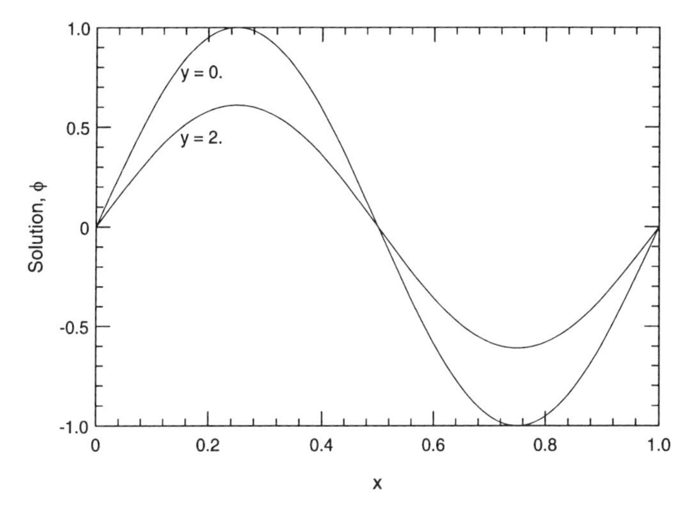

FIGURE 8.10. Method of characteristics solution of the first-order convection equation on a uniform Cartesian mesh.

and pose the problem on an infinite domain. Both the exact and numerical solutions remain periodic, so we can use periodic boundary conditions to simulate the infinite domain. To simplify matters, we shall take $\Delta x = \Delta y$.

If the ratio of the two components of the velocity $a = u/v = 1$, the characteristic line through the point x_i, y_j also passes through the point x_{i-1}, y_{j-1} and no interpolation is needed. The method is then essentially identical to the method of characteristics used in the earlier examples and good results are obtained. Unfortunately, in a more realistic problem, u and v vary with location and we will not be so lucky. Let us see what happens when $a \neq 1$.

In the case in which $a = 0.5$, the characteristic line through x_i, y_j intersects the line $y = y_{j-1}$ midway between x_{i-1} and x_i so interpolation is needed. For purposes of illustration, we shall use linear interpolation. The equation for computing u_{ij} then becomes

$$u_{i,j} = \tfrac{1}{2}(u_{i-1,j-1} + u_{i,j-1})$$

The result is shown in Figure 8.10.

As Figure 8.10 shows, the effect of the interpolation is to reduce the size of the peak. It is as if a diffusive or dissipative term has been added to the equation; indeed, the error created by the interpolation can be interpreted as *numerical* or *artificial* diffusion. The method used in this example is essentially identical to the skew upwind method that will be analyzed in some detail later, so we shall put off detailed discussion of this method. For now, we simply state that the error could be reduced by increasing the accuracy of the interpolation method used or, of course, by reducing the grid size.

As noted above, in a realistic problem the ratio of the velocity components $a = u/v$ that determines the slopes of the characteristics is not constant. The characteristics are then curved and the way in which they are computed becomes important. For example, if the characteristic passing through the point x_i, y_j is approximated by the straight line with slope $a(x_i, y_j)$, then there is an error in the location of the point at which it intersects the line $y = y_{j-1}$ or, if $a > 1$, the point at which it intersects the line $x = x_{i-1}$. This error is proportional to Δx and, no matter how accurate the interpolation method used, the overall method will only be first-order accurate. This error could be reduced by increasing the accuracy of the shape of the characteristic line. The lesson to be learned is that it is important to pay attention to the accuracy of every part of the method.

8.3. EXPLICIT METHODS

The method of characteristics has a number of advantages that were noted and exploited in the preceding section. Among these is the ability to compute the solution to a hyperbolic equation accurately over a long span of one or more of the independent variables. The major disadvantage is that the method becomes very cumbersome in three dimensions. Keeping the coordinates of the characteristic surfaces in good order becomes very difficult, especially for nonlinear equations. Characteristics can merge; that is exactly what happens in the process of generating a shock. The method of characteristics also has difficulty with equations that are hyperbolic in part of the solution domain and elliptic or parabolic elsewhere; other methods have to be used where the equation is not hyperbolic. Finally, for some problems (including slightly supersonic flow), the method is very slow. Consequently, the trend in recent years has been toward techniques based on the direct application of finite difference (or finite volume or finite element) methods to the partial differential equations. These methods are similar to the ones described in the preceding two chapters for the other types of equations and are therefore relatively straightforward to derive. There are, however, a few peculiarities associated with hyperbolic equations that require some attention.

As we have seen, one of the simplest hyperbolic equations is the first-order convection equation with a constant convection velocity. It can be written

$$\frac{\partial u}{\partial t} + c\frac{\partial u}{\partial t} = 0 \tag{8.68}$$

whose characteristics are the straight lines $x - ct =$ constant. This equation is very simple, but it is also important, and knowing how to calculate its solution numerically is essential in many applications areas.

As we have already seen, the well-known second-order wave equation

$$\frac{\partial^2 \phi}{\partial t^2} = c^2 \frac{\partial^2 \phi}{\partial x^2} \tag{8.69}$$

which allows propagation of signals at velocity c in both the positive and negative x directions, is very much like Eq. (8.68) in terms of its behavior and the numerical methods that are applied to it. In fact, it is equivalent to the coupled set of first-order equations:

$$\begin{aligned} \frac{\partial \phi}{\partial t} &= c\frac{\partial \psi}{\partial x} \\ \frac{\partial \psi}{\partial t} &= c\frac{\partial \phi}{\partial x} \end{aligned} \tag{8.70}$$

This factored form demonstrates the similarity to the convection equation (8.68) and shows that one can treat the wave equation with the methods applied to the convection equation. For this reason, we will work mainly with the simpler convection equation and then show how the methods and results apply to the wave equation. Note, however, that this analogy is largely limited to problems in one space dimension. When there are two space dimensions, the situation is quite different as we will see later.

8.3.1. Explicit Central Difference Methods

We begin by looking at the simplest numerical methods, explicit methods. An obvious method of approximating Eq. (8.68) is to use the Euler forward difference method with respect to time and the second-order central difference method for the spatial derivative; this is sometimes called the forward time, centered space, or FTCS method. The difference equation for this method is

$$\frac{u_i^{n+1} - u_i^n}{\Delta t} = -\frac{c}{2\Delta x}(u_{i+1}^n - u_{i-1}^n) \tag{8.71}$$

From our knowledge of the component methods used in constructing this method, we might guess that it is first-order accurate in time and second-order accurate in space; that is correct, but we will not take the space to demonstrate its correctness. Also, based on the properties of the component methods, we might expect it to be conditionally stable, but this turns out not to be true as we shall now show. We investigate the stability of the method with von Neumann stability analysis approach. As we have seen in earlier chapters, the von Neumann method begins by ignoring the boundary conditions and makes the assumption that the solution behaves like e^{ikx} in space. We look for solutions of both the PDE and the finite difference equations that have this form, that is, $\phi(x,t) = e^{ikx}\varphi^n$. It is not difficult to see that the solution of the

PDE of this form is $\phi(x,t) = e^{ikx}e^{-ikct}$. On the other hand, when the assumed form of the solution is substituted into Eq. (8.71), we find

$$\varphi_i^n = \left(1 - \frac{ic\,\Delta t}{\Delta x}\sin k\,\Delta x\right)\varphi_i^{n-1} \tag{8.72}$$

where, as usual, $\varphi_i^n = \varphi(x_i,t_n)$. By induction, the solution is therefore

$$\varphi_i^n = e^{ikx}\left(1 - \frac{ic\,\Delta t}{\Delta x}\sin k\,\Delta x\right)^n \tag{8.73}$$

Thus the FTCS method replaces the factor $e^{-ikc\,\Delta t}$ of the exact solution by $1-(ic\,\Delta t/\Delta x)\sin k\,\Delta x$. [Recall that $(1/\Delta x)\sin k\,\Delta x$ is the effective wavenumber for the second-order central difference approximation; it was introduced in Chapter 4.] More importantly, the exact PDE propagates the solution as a traveling wave whose amplitude is unchanged in time. On the other hand, the amplitude of the finite difference solution (8.73) increases in time no matter what the values of the parameters may be; in other words, the method is unconditionally unstable! This may seem surprising, but we note that when we substitute $\varphi(x,t) = e^{ikx}\varphi(t)$, the PDE is replaced by the ODE $\varphi' = \alpha\varphi$ with α pure imaginary. Euler's method for the equation $y' = \alpha y$ is unstable when α is imaginary, so we have really only reestablished a result that was presented in Chapter 4.

On the other hand, the leapfrog method, which is unstable for $\varphi' = \alpha\varphi$ when α is real and proved to be unsuitable for parabolic equations, turns out to be an excellent choice for the convection equation because it is neutrally stable, that is, it neither amplifies nor dissipates the amplitude of the solution. Because the physics meteorology and oceanography is dominated by convection, the leapfrog method has been widely used in these fields.

It is not our intention to survey all explicit methods. The important result is the one just demonstrated—that good methods for hyperbolic equations are ones that are well behaved for ODEs with purely imaginary coefficients; this is one of the primary reasons why the case of complex α was given so much attention in Chapter 4. To find good methods for hyperbolic equations, we should review the results obtained in Chapter 4 with an eye toward choosing a method that behaves well when the coefficient is imaginary.

Another important idea is embedded in the results derived above. The convection equation moves the solution at constant speed without change in shape and, just as importantly, no change in amplitude. It is a great deal to ask that a numerical method (except, perhaps, a Fourier method) maintain the shape of the solution while moving it about, but at least it should not destroy the amplitude of the function. This issue is closely related to energy conservation; requiring a numerical method to be energy conservative is a difficult constraint. In the earlier chapters we asked only that a numerical solution decay when the solution of the differential equation decays; this is essence of the concept of stability. Now we ask that a method neither increase the amplitude of the solution (which would mean that the method is unstable, as, for example, the Euler method is) nor decrease it (which would eventually drive the solution to zero). That is a very stringent demand.

One consequence of these constraints on methods for solving hyperbolic equations is that it is common to discuss the accuracy of a method in terms of its amplitude and phase errors; this was noted for solutions of ordinary differential equations with imaginary coefficients in Chapter 4. Stability is the requirement that the method not increase the magnitude of the solution with increasing time. A method that produces numerical or artificial damping of the solution is sometimes said to be *overstable*. In time Δt, the solution of the PDE with e^{ikx} spatial behavior changes by a factor $e^{i\theta}$ where $\theta = -kc\,\Delta t$; this means that the solution is simply phase shifted. No numerical method can be expected to produce the exact phase; the difference between the exact and numerically predicted phases is phase error. In general, the phase error is a function of the wavenumber k.

Another way of looking at this is to regard the numerical solution as corresponding to a physical system in which the velocity of propagation is a function of the wavenumber, that is, $c = c(k)$. A physical system with this property is called *dispersive*, and phase error is therefore sometimes known as dispersive error. For a similar reason, *amplitude error* is sometimes known as *dissipative* error. Central difference approximations introduce dispersive error but not dissipative error. Time difference approximations may be a source of error of both types. Note that, although these concepts provide useful tools for comparing numerical methods for linear problems, they may not suffice in the nonlinear case. However, a method that does not behave well for linear equations should be eliminated as a candidate for nonlinear equations, so the investigation of linear equations is a good starting point.

8.3.2. Upwind Methods

In the discussion of characteristics in Section 8.1, we noted the importance of the domains of dependence and influence. These concepts are an expression of the idea of "forbidden signals," that is, not all portions of the solution domain can influence all others. It is reasonable to require that numerical approximations inherit these properties. (Similar considerations played important roles in selecting methods for parabolic and elliptic equations.) For the convection equation with a constant speed of propagation, Eq. (8.68), the domains of influence and dependence reduce to a single straight line. For the wave equation the domains of influence and dependence are defined by a pair of intersecting straight lines in the $x - t$ plane; the regions between the lines are the domains of interest. It would seem important that numerical methods have these domains of influence and dependence.

For the convection equation with positive speed, requiring the method to respect the domain of dependence is equivalent to demanding that the solution at the point x_i at time t_{n+1} not depend on earlier information at points to the right of x_i. This rules out central difference approximations even though they are desirable from an accuracy point of view. A method that uses only information from points to the left of x_i is necessarily one sided; the simplest approximation of this type uses an upwind difference in space and the Euler explicit or forward method in time. Hence it is called the upwind space forward time (USFT) method or, sometimes the backward space forward time, or BSFT, method. It is

$$\frac{\phi_i^{n+1} - \phi_i^n}{\Delta t} = -\frac{c}{\Delta x}(\phi_i^n - \phi_{i-1}^n) \tag{8.74}$$

Analysis of this scheme by the von Neumann method gives

$$\varphi_i^{n+1} = \left[1 + \frac{c\,\Delta t}{\Delta x}(1 - e^{-ik\,\Delta x})\right]\varphi_i^n \tag{8.75}$$

from which it is not difficult to show that this method is conditionally stable. The necessary condition for stability is

$$\frac{c\,\Delta t}{\Delta x} \leq 1 \tag{8.76}$$

This result was first derived by Courant, Friedrichs, and Lewy in 1927 and is therefore known as the *CFL* or, sometimes, the *Courant condition.* We note that on the basis of dimensional analysis, the stability of any conditionally stable method for either the first-order convection equation or the wave equation must depend on the dimensionless parameter $c\,\Delta t/\Delta x$, which is sometimes called the *Courant number.*

The method defined by Eq. (8.74) has interesting properties. For the case in which the Courant number is unity, which is the stability limit of the method, it becomes exact. This seems surprising until we note that when $c\,\Delta t = \Delta x$, both the PDE and the finite difference approximation state that the solution advances a distance $c\,\Delta t$ in time Δt. In essence, the method becomes the method of characteristics discussed earlier in this special case. In general, this method is only first-order accurate in both space and time.

Equation (8.76) states that the numerical method is stable as long as its domain of dependence is contained in the domain of dependence of the PDE. In numerical fluid dynamics the difference scheme (8.74) is known as upwind differencing. It played an important role in the past, but the errors caused by its first-order nature has led to its replacement by more accurate methods.

Note that, if the sign of the velocity were changed, the difference approximation would need to use data from the point x_i and points to the right of it at time t_n. The use of data from the direction from which the information arrives is what gives the method the name "upwind."

It is worth noting that the upwind difference approximation for $c > 0$ can be written:

$$c\frac{\partial u}{\partial x} = c\frac{u_i - u_{i-1}}{\Delta x} = c\frac{u_{i+1} - u_{i-1}}{2\Delta x} + \frac{c\,\Delta x}{2}\frac{u_{i+1} - 2u_i + u_{i-1}}{\Delta x^2} \tag{8.77}$$

which shows that it can be regarded as a combination of the central difference approximation and a diffusive term with an artificial viscosity equal to $c\Delta x/2$.

8.3.3. Lax–Wendroff Method

For the wave equation, which propagates information equally in both directions, the use of one-sided differences does not make sense, so centered difference methods are almost always chosen. The combination of the explicit Euler method in time and central differences in space, which is unstable for the first-order convection equation is also unstable for the wave equation and should not be used. The leapfrog method in time with centered differences in space is stable for Courant numbers less than unity. Many other methods can be used, but we shall not discuss them in detail. Instead, we shall proceed to methods that have more desirable properties and that have received more attention and use.

A method that is important both because it is useful in its own right and because it is the basis for other methods was developed by Lax and Wendroff and is second-order accurate in both space and time. We will apply it first to the convection equation. The basis for this method is the Taylor series approximation introduced in Section 8.2. We write

$$\phi(x,t+\Delta t) = \phi(x,t) + \Delta t\frac{\partial \phi}{\partial t}(x,t) + \frac{\Delta t^2}{2}\frac{\partial^2 \phi}{\partial t^2} \tag{8.78}$$

and note that the first derivative can obtained from the PDE (8.68) itself and the second time derivative can be found by differentiating the PDE to give

$$\frac{\partial^2 \phi}{\partial t^2} = \frac{\partial}{\partial t}\left(-c\frac{\partial \phi}{\partial x}\right) = c^2\frac{\partial^2 \phi}{\partial x^2} \tag{8.79}$$

Thus Eq. (8.78) becomes:

$$\phi(x,t+\Delta t) = \phi(x,t) + c\,\Delta t\frac{\partial \phi}{\partial x}(x,t) + \frac{c^2\,\Delta t^2}{2}\frac{\partial^2 \phi}{\partial x^2} \tag{8.80}$$

When central difference approximations are applied to both spatial derivatives, this method becomes

$$\phi_i^{n+1} = \phi_i^n - c\,\Delta t\left(\frac{\phi_{i+1}^n - \phi_{i-1}^n}{2\,\Delta x}\right) + c^2\left(\frac{\phi_{i+1}^n - 2\phi_i^n + \phi_{i-1}^n}{\Delta x^2}\right) \tag{8.81}$$

and is known as the *Lax–Wendroff* method. It is applied to the wave equation in the example below.

This method is second-order accurate in both space and time, explicit, quite easy to program, and stable for Courant numbers less than unity. For cases in which the coefficients in the PDE are not constant or the equation is nonlinear, the application of this method is not straightforward and a number of variations on it have been developed. Among these are a number of methods which use splitting or approximate factorization. Some of these are discussed later.

Example 8.6: Explicit Solution of the Wave Equation To solve the witch's hat problem for the wave equation of Example 8.4 using the Lax–Wendroff method, the equations are written in the first-order form of Eqs. (8.70). The finite difference equations of the Lax–Wendroff method are then

$$\phi_i^{n+1} = \phi_i^n + \frac{c\,\Delta t}{2\,\Delta x}(\psi_{i+1}^n - \psi_{i-1}^n) + \frac{1}{2}\left(\frac{c\,\Delta t}{\Delta x}\right)^2 (\phi_{i+1}^n - 2\phi_i^n + \phi_{i-1}^n) \tag{8.82}$$

$$\psi_i^{n+1} = \psi_i^n + \frac{c\,\Delta t}{2\,\Delta x}(\phi_{i+1}^n - \phi_{i-1}^n) + \frac{1}{2}\left(\frac{c\,\Delta t}{\Delta x}\right)^2 (\psi_{i+1}^n - 2\psi_i^n + \psi_{i-1}^n) \tag{8.83}$$

which are easily programmed. The boundary condition that requires ϕ to be zero on the boundary is easily satisfied. The first of equations (8.70) shows that $\partial\psi/\partial x = 0$ on the boundary. This condition can be satisfied by introducing an artificial or phantom point at $x_{-1} = -\Delta x$ and letting $\psi_{-1} = \psi_1$ and $u_{-1} = -u_1$. Then Eq. (8.83) with $j = 0$ becomes

$$\psi_0^{n+1} = \psi_0^n + \frac{c\,\Delta t}{\Delta x}\phi_1^n + \left(\frac{c\,\Delta t}{\Delta x}\right)^2 (\psi_1^n - \psi_0^n) \tag{8.84}$$

and at the right boundary

$$\psi_N^{n+1} = \psi_N^n - \frac{c\,\Delta t}{\Delta x}\phi_{N-1}^n - \left(\frac{c\,\Delta t}{\Delta x}\right)^2 (\psi_N^n - \psi_{N-1}^n) \tag{8.85}$$

Results obtained using 20 intervals ($\Delta x = 0.1$) are shown in Figure 8.11. Note that, in contrast to the method of characteristics, the time step in the Lax–Wendroff method can be independent of the size of the spatial mesh. With Courant number $c\,\Delta t/\Delta x = 1$, which is the stability limit of this method, the results are exact, but this is fortuitous. In all other cases, the solutions are much worse than the ones obtained using the method of characteristics. They also continue to deteriorate with increasing time. The results shown in the

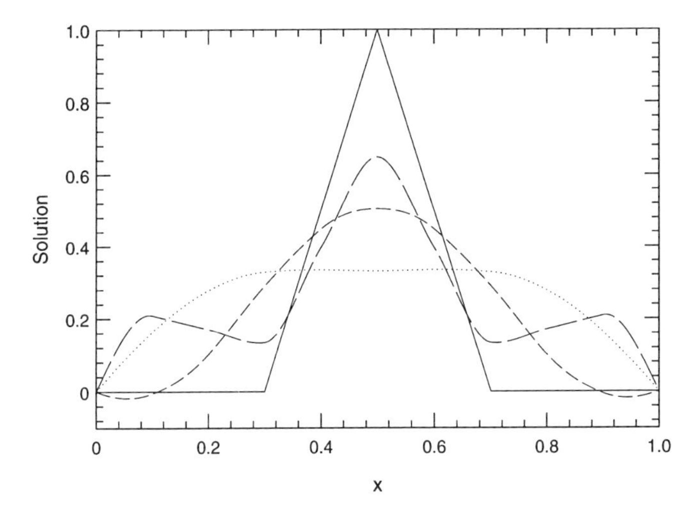

FIGURE 8.11. Lax–Wendroff solution of the wave equation with a witch's hat initial condition with the Lax–Wendroff method; 20 spatial points were used. The solid line is the initial condition, which is also the exact solution at the times shown (except for a sign change at $t = 1$ and 3). The dot–dash curve is the (sign reversed) solution at $t = 1$; the dot–dash curve is the solution at $t = 2$; and the dot curve is the (sign reversed) solution at $t = 3$.

figure were computed with Courant number 0.5, the value for which the errors due to temporal and spatial finite difference approximations are approximately balanced. They are not very good. The peak has been broadened because the method has dissipative or diffusive error. (Diffusion tends to smooth out sharp corners.) The method also contains dispersive error as is shown by the solution becoming negative in part of the domain. Because the solution has been broadened by $t = 1$, the dissipative error is reduced after that, and the decrease of the peak value is not as large between $t = 1$ and $t = 2$ as it was between $t = 0$ and $t = 1$. The dispersion error continues to become more severe however. Reducing the Courant number to 0.01, which eliminates almost all of the time differencing error, improves the solution near the peak, but the solution develops "bumps" or "wiggles" away from the main peak due to the dispersive error.

The witch's hat function is, in a sense, intermediate between a step function that has a discontinuity and a smooth function such as the sine; using the step function as an initial condition would produce even more wiggles while the method would give better results for the sine function.

Using a finer spatial mesh improves the accuracy of the calculation and reduces the dispersion considerably, as shown in Figure 8.11, but the dispersion still increases with increasing time. Reducing the Courant number results in very little change, indicating that the major source of error is the spatial differencing approximation.

The code used to generate these results, LAX, is available from the Internet site described in Appendix A.

8.4. IMPLICIT METHODS

The important properties of hyperbolic equations, namely that information propagates along the characteristics and that there are domains of dependence and influence, are well approximated by explicit methods. Historically, this kept people from using implicit methods, so the latter received relatively little attention. However, we saw earlier that an implicit version of the method of characteristics has excellent properties.

The excellent stability properties of implicit methods strongly suggest that we see what they can do in hyperbolic problems. This is especially true when steady problems are solved by computing the solution to a hyperbolic problem until the solution reaches a steady state because an implicit method will allow large time steps, which should allow the steady state to be reached rather quickly. It turns out that implicit methods do rather well (if one is careful), and they have been increasing in popularity.

There is no reason to survey many implicit methods. Their properties are very nearly what one might expect. As might be anticipated, one of the best methods of this type is the Crank–Nicolson method, which we shall investigate further. It offers an excellent combination of accuracy and stability.

When applied to the convection equation (8.68) with central difference approximations for the spatial derivatives, the Crank–Nicolson method gives

$$\frac{\phi_i^{n+1} - \phi_i^n}{\Delta t} = -\frac{c}{4\,\Delta x}(\phi_{i+1}^{n+1} - \phi_{i-1}^{n+1} + \phi_{i+1}^n - \phi_{i-1}^n) \tag{8.86}$$

Other choices for the spatial differencing approximation are, of course, possible. As was the case with parabolic equations, we are left with a tridiagonal system of equations to solve. This also means that in two or three space dimensions the method will be intractable unless a splitting [alternating direction implicit (ADI)] type of method is used. Splitting methods are covered in the following section.

The stability of this method is readily investigated using the von Neumann approach. Making the substitution $\phi(x,t) = e^{ikx}\varphi^n$, we find without much effort

$$\varphi^{n+1} = \left[\frac{1 - (ic\,\Delta t/2\,\Delta x)\sin k\,\Delta x}{1 + (ic\,\Delta t/2\,\Delta x)\sin k\Delta x}\right]\varphi^n \tag{8.87}$$

The function in brackets represents the approximation to $e^{-ikc\,\Delta t}$ produced by this method. The numerator and denominator of the function in brackets are complex conjugates of each other, independent of the values of the parameters. Consequently, the method is *neutrally stable*, meaning that it does not change the amplitude of the function. In other words, it does not introduce any dissipation and is not unstable. There is, of course, phase or dispersive error, but the second-order nature of the method assures that the error will not be too serious as long as one does not take too large a time step. Thus, despite the fact that implicit methods do not treat the characteristics properly, they (at least the Crank–Nicolson method among them) do have properties that recommend them for application to hyperbolic equations.

We can apply the Crank–Nicolson method to the factored form (8.70) of the wave equation. As might be expected, the resulting method is unconditionally stable. An alternative is to deal directly with the second-order wave equation (8.69) and use a second-order central time difference. For the spatial derivatives, we use the average of the values at time steps $n-1$ and $n+1$:

$$\phi_i^{n+1} - 2\phi_i^n + \phi_i^{n-1} = \frac{c^2\,\Delta t^2}{2\,\Delta x^2}[(\phi_{i+1}^{n+1} - 2\phi_i^{n+1} + \phi_{i-1}^{n+1}) + (\phi_{i+1}^{n-1} - 2\phi_i^{n-1} + \phi_{i-1}^{n-1})] \tag{8.88}$$

This method is also unconditionally stable and is slightly more accurate than the Crank–Nicolson method applied to the factored equation. Note that although Eq. (8.88) represents a multistep method, there is no difficulty in starting the calculation because both the function and its time derivative are provided at the initial time. This information can be used to compute the function at the first time step without the need for a special starting method that multistep methods usually require. Also note that if the spatial derivative in Eq. (8.88) was evaluated at the nth time step, the method would be explicit (similar to the leapfrog method) and the stability would only be conditional—the Courant number would have to be less than unity.

A final word of caution must be injected here. The Crank–Nicolson method has very attractive properties when applied to the simple hyperbolic equations we have chosen as examples. Specifically, the method produces no amplitude error, that is, it is neither unstable nor overstable. When the method is applied to nonlinear equations, it is possible that a small amplitude error will be introduced. There is no simple analytical way to predict how a method will behave in the nonlinear case, so the method may be unstable when applied to a particular nonlinear problem or it may be overstable. A small amount of overstability can probably be tolerated, but instability is difficult to live with. This is probably one reason why the development of implicit methods for hyperbolic problems has proceeded cautiously.

The approach to solving nonlinear hyperbolic problems using implicit methods that has been taken in recent years has been simply to try the method on a number of related problems and see how well it works. As a result, a number of methods have been suggested for nonlinear hyperbolic problems; all of them reduce to the Crank–Nicolson method in the linear case. Some of these are given in the next section.

Example 8.7: Implicit Solution of the Wave Equation We shall treat the problem of the two preceding examples by the Crank–Nicolson method. As has been pointed out, there are at least two versions of the Crank–Nicolson method for the wave equation; it can be applied to either the factored or unfactored forms of the wave equation. We shall use the factored form for this example. The finite difference version of Eq. (8.70) is then

$$\begin{aligned}\phi_i^{n+1} - \phi_i^n &= \frac{c\,\Delta t}{4\,\Delta x}(\psi_{i+1}^n - \psi_{i-1}^n + \psi_{i+1}^{n+1} - \psi_{i-1}^{n+1})\\ \psi_i^{n+1} - \psi_i^n &= \frac{c\,\Delta t}{4\,\Delta x}(\phi_{i+1}^n - \phi_{i-1}^n + \phi_{i+1}^{n+1} - \phi_{i-1}^{n+1})\end{aligned} \tag{8.89}$$

The boundary conditions are the same as those used when this problem was solved with the Lax–Wendroff method. The derivative boundary condition is treated by introducing fictitious or phantom points outside the boundary.

The set of equations, viewed as a whole, is a block tridiagonal system in which the blocks are 2×2. On closer inspection, we find that the equations break into two disjoint sets, one for the variable set

$$(\psi_0, \phi_1, \psi_2, \phi_3, \ldots, \phi_{N-1}, \psi_N)$$

and another for the set

$$(\phi_0, \psi_1, \phi_2, \psi_3, \ldots, \psi_{N-1}, \phi_N)$$

assuming that N, the number of intervals, is even. Each set of equations is tridiagonal and therefore easily solved. The matrix for the second set of variables, which does not contain the boundary values, is $(N-2) \times (N-2)$ and has constant diagonals α, 1, and $-\alpha$, where $\alpha = c\,\Delta t/4\,\Delta x$. The matrix for the first set of variables is $N \times N$ with constant diagonals α, 1, and $-\alpha$, except that the $(1,2)$ element is -2α and the $(N, N-1)$ element is 2α.

Results obtained using 20 intervals are shown in Figure 8.12. It is clear that even though the method conserves the amplitudes of the component waves, they are phase shifted differently; the resultant dispersion is quite evident. The solution is given at both $t = 1.0$ and is fairly similar to the Lax–Wendroff solution at that time. At later times, the dispersion spreads the solution almost uniformly over the domain.

As we noted above, the effect of the numerical approximations is to make each wave travel at a different speed—that is what dispersion is. In particular the wave components that make up the bulk of the peak are not traveling at speed c. We should not be surprised, therefore, if the shape of the peak is somewhat better resolved at some time different than $t = 1$. Figure 8.13 shows the solution at times 1.1, 2.2, and 3.3, the times at which the peak in the function is best resolved; however, the dispersion is still evident. The

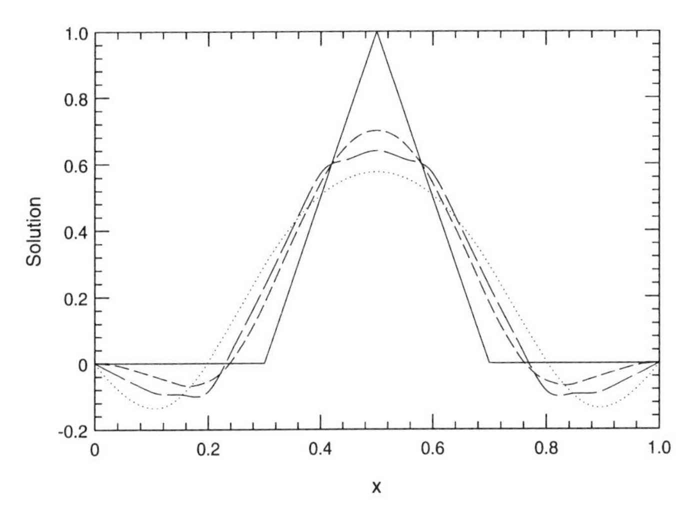

FIGURE 8.12. Lax–Wendroff solution of the wave equation with a witch's hat initial condition with the Crank–Nicolson method; 20 spatial points were used. The solid line is the initial condition, which is also the exact solution at the times shown (except for a sign change at $t = 1$ and 3). The dot–dash curve is the (sign reversed) solution at $t = 1$; the dot curve is the solution at $t = 2$; and the dashed curve is the (sign reversed) solution at $t = 3$.

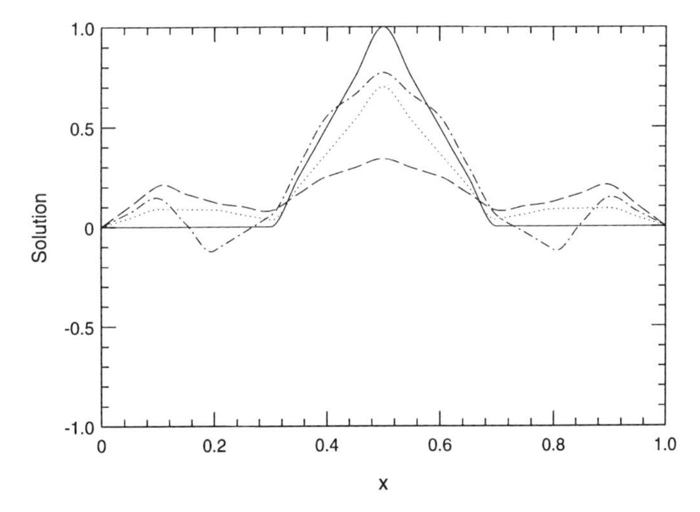

FIGURE 8.13. Same as Figure 8.12 except that the times are $t = 0$, 1.1, 2.2, 3.3.

dispersion is due mainly to the spatial finite difference approximation and cannot be cured solely by changing the time step. Results for smaller Courant numbers are not shown for this reason.

Reducing the mesh size improves the results significantly. Since the results are what one might anticipate, they are not shown.

8.5. SPLITTING METHODS

As one might anticipate from earlier sections, implicit methods are difficult to apply to nonlinear problems. This leads one to consider predictor–corrector methods, a number of which have been proposed for hyperbolic systems. These are useful methods, but since there is considerable overlap between them and splitting methods, only splitting methods are discussed here.

Another problem that poses considerable difficulty is that of solving hyperbolic partial differential equations in two or three space dimensions. If implicit methods applied directly to problems of this type, large computation times are required. The difficulties are very similar to the ones we encountered when we tried to apply implicit methods to parabolic problems in two or three space dimensions. The only methods with generous stability limits that are not too costly are ones of the ADI type that were so successful in the parabolic case. The basis of these methods is the splitting of a multidimensional problem into a set of one-dimensional problems that can be solved successively. The solution of two or three sets of one-dimensional problems requires much less computation than the direct method and retains the favorable stability properties. This suggests the possibility of using splitting methods for hyperbolic equations.

The ADI method, the only splitting method we have investigated in detail, is not exactly equivalent to the Crank–Nicolson method from which it can be derived. To maintain accuracy, a split method needs to match the original method only to the accuracy with which the latter approximates the PDE. Thus, for second-order methods, the split method must be equivalent to the original method to second-order accuracy. The truncation errors for split and unsplit methods are not identical; in fact, a split method may be either more or less accurate than its unsplit ancestor. Because the splitting need not be exact, we noted in Chapter 6 that the split methods are also called approximate factorization methods. Many methods can be split in more than one way that maintain the accuracy of the original method. The issue is further complicated by the fact that splitting methods that produce identical results for linear equations may behave quite differently when applied to nonlinear problems. It appears that the best way to develop a method for nonlinear problems is to choose a method that works well in the linear case (this can usually be tested analytically) and to try several versions of it in the nonlinear case.

Finally, it is difficult to attain high accuracy for nonlinear PDEs in two and three dimensions. In order to achieve more than first-order time accuracy, we need to use one of the following: a multistep method (with problems of starting the calculation and a need for a great deal of storage), an implicit method such as the trapezoid rule (which requires iteration making it expensive), a predictor–corrector method (not a bad choice, but usually requiring increased storage), or a scheme like the Lax–Wendroff method (which requires evaluation of the second time derivative, something that is difficult to do for nonlinear equations). The use of splitting can simplify the computation considerably and thus reduce the difficulty of programming and computation time even for an explicit method. In this section we give a split method that is equivalent to the Lax–Wendroff method and others that are equivalent to the Crank–Nicolson method. By equivalence, we mean that their behavior in the linear case and their stability properties are similar. A great deal of work along these lines was done in Russia and can be found in the works of Marchuk (1982).

8.5.1. Explicit Split Methods

It is not always obvious how an author found the particular splitting. Looking at a split scheme, it is not clear without further analysis that the method is actually equivalent to some unsplit method. As an example, let us take the Richtmyer splitting of the Lax–Wendroff method. Applied to the convection equation it is

$$
\begin{aligned}
\phi_i^{n+1/2} &= \frac{1}{2}(\phi_{i+1}^n + \phi_{i-1}^n) - \frac{c\,\Delta t}{2}\left(\frac{\phi_{i+1}^n - \phi_{i-1}^n}{2\,\Delta x}\right) \\
\phi_i^{n+1} &= \phi_i^n - c\,\Delta t\left(\frac{\phi_{i+1}^{n+1/2} - \phi_{i-1}^{n+1/2}}{2\,\Delta x}\right)
\end{aligned}
\tag{8.90}
$$

The first step of this method is the Euler method with central differences, except that it uses a spatial average in place of ϕ_i^n. This step was proposed as a separate method by Lax. The use of the average stabilizes the Euler method, and in this way it is reminiscent of the Dufort–Frankel method, which followed Lax's method historically. The second equation is the midpoint rule or leapfrog method. Thus Richtmyer's method can be regarded as a predictor–corrector method with a first-order Lax predictor and a second-order leapfrog corrector. As this shows, the relationship between splitting methods and predictor–corrector methods is very close; it is not always clear how a given method ought to be classified. Combining the two equations in the set (8.90) reveals that for linear equations the overall method is in fact the Lax–Wendroff method with spatial differences based on $2\,\Delta x$.

A different splitting, which, because it has some desirable properties, has been very popular in aerodynamic calculations, was proposed by MacCormack. This scheme, which has many variations of its own, is a true split scheme in the sense that both halves of the method are of lower-order accuracy than the complete method. It uses a forward spatial difference in the first half of the method and a backward difference in the second half:

$$\begin{aligned} \phi_i^* &= \phi_i^n - \frac{c\,\Delta t}{\Delta x}(\phi_{i+1}^n - \phi_i^n) \\ \phi_i^{n+1} &= \frac{1}{2}(\phi_i^n + \phi_i^*) - \frac{c\,\Delta t}{2\,\Delta x}(\phi_i^* - \phi_{i-1}^*) \end{aligned} \tag{8.91}$$

This method is equivalent to the Lax–Wendroff method for linear problems. It has the advantage over the Richtmyer method in that it uses only information from points $j-1$, j, and $j+1$ at the preceding time step. In light of what we have already found, it is not surprising that the Richtmyer method is more stable (Courant number up to 2) than the MacCormack method (Courant number up to 1), but the latter is more accurate and easier to program. It also has desirable nonlinear properties, which has made its use popular in fluid mechanics. In the MacCormack method the forward and backward differences can be used in the reverse order with the same accuracy, and it has been found that in the nonlinear case the best results are obtained if the order is reversed at each time step. The method also is readily extended to two and three dimensions.

Example 8.8: Solution of the One-Dimensional Convection Equation with MacCormack's Method First we shall solve the one-dimensional convection equation using the MacCormack method just described. As noted, this method is easy to program and runs very quickly; the program used (MACCORM1) can be found at the Internet site described in Appendix A. Since this problem has been solved earlier, we shall not go into great detail about it here.

The results are presented in Figure 8.14. We see that MacCormack's method is actually less diffusive than the Lax–Wendroff method of which it is a splitting. The peak moves at a speed that is slower than the correct speed by about 5%; this is a smaller phase error than that of the Lax–Wendroff method. While the dispersion produced by the method is apparent, there seems to be less dissipation than is produced by the Lax–Wendroff method. While dispersion continues to grow worse after $t = 1$, there is little additional dissipation.

We shall later solve the two-dimensional convection equation with this method. First we shall introduce the equation and the problems that we will solve.

8.5.2. Convection in Two Dimensions

Most engineering problems are not one dimensional. Moreover, there are major differences between the behavior of numerical methods designed for one-dimensional problems and their counterparts into two and three dimensions, especially for the convection equation. Many methods that have been developed for one-dimensional problems, and work well for them, perform very poorly in two and three dimensions; the difference between two- and three-

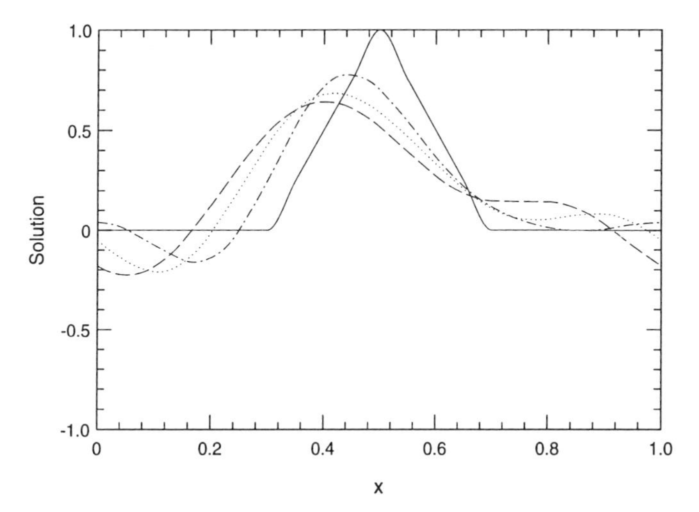

FIGURE 8.14. Solution of the one-dimensional convection equation with a witch's hat initial profile with MacCormack's method. The times are given in Figure 8.11.

dimensional problems is more a matter of increased computational cost than difficulty with the method. In this subsection we shall try to see why this is the case.

The unsteady convection equation in two dimensions is

$$\frac{\partial \phi}{\partial t} + u\frac{\partial \phi}{\partial x} + v\frac{\partial \phi}{\partial y} = 0 \tag{8.92}$$

It describes transport along straight lines in two dimensions. The lines are, of course, the characteristics of the equation and are defined by

$$x - ut = c_1 \qquad y - vt = c_2 \tag{8.93}$$

Thus convection in two dimensions is a lot like one-dimensional convection—information moves on lines. These lines are usually called *streaklines* and are the paths that particles follow in the flow. (The *streamlines* are the paths that a particle would follow if it moved in the instantaneous velocity field. In a flow in which the velocity is time independent, as it is here, the two sets of lines are identical.)

On the other hand, the wave equation in two dimensions is

$$\frac{\partial^2 \phi}{\partial t^2} = c^2\left(\frac{\partial^2 \phi}{\partial x^2} + \frac{\partial^2 \phi}{\partial y^2}\right) \tag{8.94}$$

It describes waves that propagate equally well in all directions, that is, it is isotropic.

Thus there is a major difference between the convection and wave equations in two dimensions. The convection equation has a favored direction; the wave equation does not. This remark holds in three dimensions as well. We therefore expect that at least part of the analogy between the equations that exists in one dimension no longer exists in two and three dimensions. Indeed, the convection equation has some unique problems in two and three dimensions. We saw some of these earlier in Section 8.4.

To study the behavior of methods for two-dimensional convection problems, we shall look at three problems. The first will help us understand the effect of the finite difference approximation on the solution. The second relates to the important issue of artificial viscosity. Finally, the third is an interesting test case that shows some of the points made in the first two examples in more depth.

Example 8.9: One-Dimensional Convection–Diffusion; Upwind versus Central Difference Methods In the first test case, the velocity components are independent of space and time. If one of the velocity components is zero, the problem reduces to the one-dimensional case that we have already studied. We already saw that there is a problem when a grid-based method of characteristics was applied to this problem.

Let us recall some properties of the numerical methods that were discovered in the one-dimensional problem. Central difference methods are more accurate than uncentered ones but have dispersive errors, and they produce solutions that tend to wiggle. On the other hand, the principal error with upwind differences is dissipative; dissipative errors tend to smooth the solution and reduce its amplitude.

As an illustration of this behavior let us consider the ordinary differential equation

$$c\frac{du}{dx} = \alpha\frac{d^2u}{dx^2} \tag{8.95}$$

which represents a one-dimensional steady convection–diffusion problem. As the boundary conditions we shall use

$$u(0) = 0 \qquad u(L) = 1 \tag{8.96}$$

This problem is readily solved with finite difference methods and the tridiagonal algorithm. With central differences, the discrete equations are

$$(\mathrm{Pe}_c - 1)u_{i+1} + 2u_i + (-1 - \mathrm{Pe}_c)u_{i-1} = 0 \tag{8.97}$$

while, with upwind differences, we get

$$u_{i+1} + (2 + \mathrm{Pe}_c)u_i + (-1 - \mathrm{Pe}_c)u_{i-1} = 0 \tag{8.98}$$

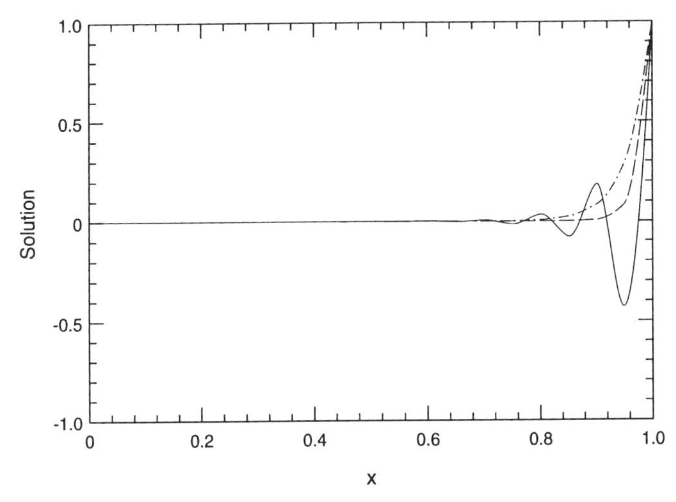

FIGURE 8.15. Solution of the one-dimensional steady convection–diffusion equation for cell Prandtl number $\mathrm{Pr}_c = 2.5$ with two finite difference approximations. The solid line obtained with central differences. The dot–dash curve is the solution obtained with upwind differences. The dashed curve is the exact solution.

where $\mathrm{Pe}_c = c\,\Delta x/\alpha$ is the cell Peclet number, the important parameter in this problem. The boundary conditions are readily applied to these equations and the solution is not difficult to find. It is shown in Figure 8.15 for $\mathrm{Pe}_c = 2.5$, a rather large value.

We see several important things from this figure. First, the central difference solution obviously has severe wiggles or oscillations. However, the upwind difference solution, while it is monotonic like the exact solution, is not very accurate either. The smoothness of the upwind solution may lead one to accept it as correct and that can be dangerous; in fact, neither of the solutions shown should be accepted. The oscillations in the central difference solution can be explained by noting that the elements of the matrix on the main diagonal [the coefficient of u_i in Eq. (8.97)] become negative for $\mathrm{Pe}_c > 2$. This results in the values of the solution at the even-numbered nodes being closely coupled to each other (as are the odd-numbered ones), but the members of each set are weakly coupled to the members of the other set. That produces the oscillations seen in the figure. However, when the grid is refined, reducing Pe_c, the central difference solution will converge to the exact solution more rapidly.

Before we leave this problem, it is worthwhile pointing out that although the differential equation (8.95) is often used as an example against which numerical methods are tested, largely because it is so easy to solve, it actually has only minimal relevance to most problems of interest in engineering or physics. Diffusive effects, when they are important, are usually in a direction normal to the convective velocity. Conclusions reached on the basis of an equation that is very different from the one we really wish to solve can be very misleading and even dangerous.

We study what happens when the conclusion that the upwind method is applicable to convection problems is taken seriously in the next subsection.

Example 8.10: Two-Dimensional Steady Convection Let us now go on to consideration of the steady convection equation in two dimensions. It was considered in Section 8.4.1 and is

$$u\frac{\partial \phi}{\partial x} + v\frac{\partial \phi}{\partial y} = 0 \tag{8.99}$$

If one of the velocity components is zero, the problem becomes one dimensional and easy to solve; it is then not difficult to obtain accurate solutions. Therefore this equation is interesting only if the flow is at an angle to the grid.

There are several ways to difference this equation. One is to specify the values of the variables at the grid points and use a discretization method similar to the Keller box method. That is, we simply integrate the equation over a grid box to obtain

$$\begin{aligned}(u\,\Delta y + v\,\Delta x)\phi_{i,j} &+ (u\,\Delta y - v\,\Delta x)\phi_{i,j-1}\\ &+ (-u\,\Delta y + v\,\Delta x)\phi_{i-1,j} - (u\,\Delta y + v\,\Delta x)\phi_{i-1,j-1} = 0\end{aligned} \tag{8.100}$$

which amounts to a central difference approximation about the center of the box.

On the other hand, the use of backward differences at the point (x_i, y_j) gives the upwind approximation:

$$(u\,\Delta y + v\,\Delta x)\phi_{i,j} - (v\,\Delta x)\phi_{i,j-1} - (u\,\Delta y)\phi_{i-1,j} = 0 \tag{8.101}$$

Another popular way of discretizing Eq. (8.99) is via a control volume approach. In this case, the variables are specified at the centers of the volumes (areas in two dimensions) defined by the grid lines and interpolation must be used to find the values of the variables at the centers of the faces of the control volume. If the face-center variables are computed using an upwind approximation (taking the value at the nearest upwind node as the face-center value), we recover Eq. (8.101). Using a central difference approximation (taking the face-center values as the averages of their nearest neighbors) yields a discrete equation that is a little more complicated than (8.100). For further details on the discretization of the two-dimensional convection equation, see Ferziger and Perić (1997).

Both of these methods have difficulties. The central difference approximation (8.100) may produce oscillations, especially if the upstream boundary condition contains a discontinuity or sharp change; this is similar to its behavior in the one-dimensional problem in the example above.

The upwind approximation (8.101), on the other hand, always produces smooth solutions. This is because it introduces an artificial viscosity. In the one-dimensional example given above, the upwind method introduces diffusion that smoothes rapid changes in the direction of the flow. In many applications, changes in this direction are relatively slow, but those in the normal direction are more rapid. The effect of artificial diffusion in the streamwise direction is normally not very serious. In the two-dimensional case, the upwind approximation also introduces diffusion in the direction normal to the flow, which can have an important effect on the results. This can be demonstrated by transforming the modified equation derived from Eq. (8.101) to a coordinate system in which one coordinate is in the direction of the flow and the other is normal to it. The left-hand side becomes the magnitude of the velocity times the derivative in the streamwise direction (as one would expect). On the right-hand side we find diffusive terms in both directions. The one in the normal direction has an artificial diffusivity or viscosity:

$$\alpha_{\text{art}} = U \sin\theta \cos\theta (\Delta x \cos\theta + \Delta y \sin\theta) \tag{8.102}$$

where $U = \sqrt{u^2 + v^2}$ and θ is the angle that the flow makes with respect to the x axis. As expected, this artificial viscosity becomes zero when the flow is parallel to one of the axis and is largest when the flow is at about a 45° angle to the grid.

Example 8.11: Rotating Convection in Two Dimensions To show the advantages of split methods, we now consider the solution of the two-dimensional convection equation (8.66). A problem of this kind that is commonly used to test methods is one in which the velocity represents solid body rotation:

$$u = -\omega y \qquad v = \omega x \tag{8.103}$$

and the initial condition is a "cone":

$$\phi(x, y, 0) = 1 - 0.5\sqrt{(x - x_0)^2 + y^2} \tag{8.104}$$

where the function is positive and zero elsewhere.

This problem is useful for testing methods because the shape of the solution remains unchanged as it is convected around in a circle. Thus, after a time $t = 2\pi/\omega$ the solution should look exactly as it did at $t = 0$. At time $t = \pi/\omega$, the cone should be rotated 180°, that is, it is on the other side of the domain.

The solution at $t = 0, \pi, 2\pi$ obtained with MacCormack's method is shown in Figure 8.16; the Courant number in this case was $\pi/2$ and the grid size was 40×40. We see that the method behaves very much as it does in one dimension. It is also very easy to program and runs quite quickly.

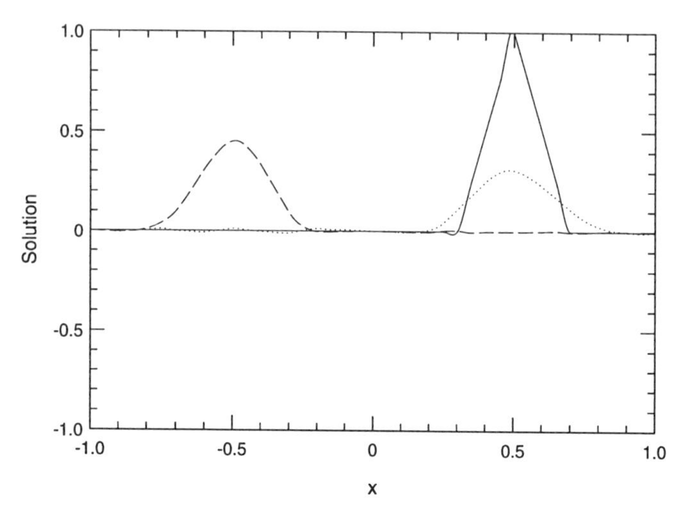

FIGURE 8.16. Solution of the two-dimensional convection–diffusion equation obtained with MacCormack's split method. The solid line obtained is the initial condition. The dashed curve is the solution at $t = \pi$ and the dotted curve is the solution at $t = 2\pi$.

8.5.3. Implicit Split Methods

Now let us consider implicit splitting methods for hyperbolic equations. Beam and Warming (1978) showed the effectiveness of an ADI method in aerodynamic applications. Of course, this method is meaningful only in two- or three-dimensional problems. When applied to the convection or wave equations, the results are very much what one might expect—the method is neutrally stable. There are a great many ways to produce ADI methods in nonlinear problems and the particular choice can have a great effect on the results.

We will satisfy ourselves with a simple example—the application to the convective equation in two dimensions. When the Crank–Nicolson method is applied to Eq. (8.92), we have

$$\left(1 + \frac{u\,\Delta t}{2}\frac{\delta}{\delta x} + \frac{v\,\Delta t}{2}\frac{\delta}{\delta y}\right)\phi^{n+1} = \left(1 - \frac{u\,\Delta t}{2}\frac{\delta}{\delta x} - \frac{v\,\Delta t}{2}\frac{\delta}{\delta y}\right)\phi^{n} \tag{8.105}$$

where $\delta/\delta x$ and $\delta/\delta y$ represent finite difference approximations to $\partial/\partial x$ and $\partial/\partial y$, respectively. These could be either upwind or central difference approximations. This method is second-order accurate in time. We can make an approximate factorization of Eq. (8.105):

$$\left(1 + \frac{u\,\Delta t}{2}\frac{\delta}{\delta x}\right)\left(1 + \frac{v\,\Delta t}{2}\frac{\delta}{\delta y}\right)\phi^{n+1} = \left(1 - \frac{u\,\Delta t}{2}\frac{\delta}{\delta x}\right)\left(1 - \frac{v\,\Delta t}{2}\frac{\delta}{\delta y}\right)\phi^{n} \tag{8.106}$$

which is equivalent to Eq. (8.105) within a term of order Δt^3 as required. Then the two-step method:

$$\left(1 + \frac{u\,\Delta t}{2}\frac{\delta}{\delta x}\right)\phi^* = \left(1 - \frac{u\,\Delta t}{2}\frac{\delta}{\delta x}\right)\phi^n \tag{8.107}$$

$$\left(1 + \frac{v\,\Delta t}{2}\frac{\delta}{\delta y}\right)\phi^{n+1} = \left(1 - \frac{v\,\Delta t}{2}\frac{\delta}{\delta y}\right)\phi^* \tag{8.108}$$

is a method of solving Eq. (8.92), which has all of the desired properties. It requires only a standard tridiagonal system solver, it is second-order accurate, and it is unconditionally stable. Methods that are generalizations of this one have been playing an increasing role in computational fluid dynamics.

PROBLEMS

1. Propagation of waves through nonuniform media are governed by the wave equation:

$$\frac{\partial^2 \phi}{\partial t^2} = \frac{\partial}{\partial x}\left(c^2(x)\frac{\partial \phi}{\partial x}\right)$$

a. What are the characteristics of this equation?

b. Give a numerical method for solving this problem.

c. Suppose we have an infinite medium in which c^2 varies in the manner shown in Figure 8.17(*a*).

d. A pulse of witch's hat shape [see Figure 8.17(*b*)] is introduced into the system at time $t = 0$. When the position of the disturbance that propagates to the right reaches the ramp in $c^2(x)$, part of it will be transmitted and part reflected. Compute the fractions that are reflected and transmitted to an accuracy of 1%.

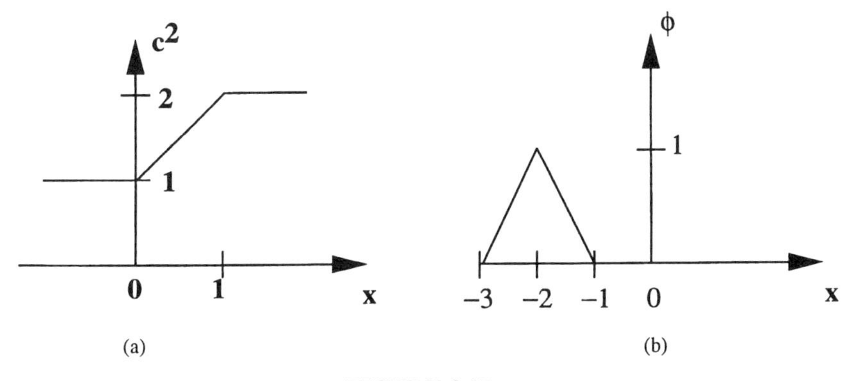

FIGURE 8.17.

2. Show how the leapfrog method could be applied to solving the wave equation with constant propagation speed. Find the amplitude and phase errors of the method in a manner similar to those used in the text.

3. Consider the convection equation.

a. Apply the Euler method to this equation.

b. Write this method in operator form.

c. Show how the operator can be approximately factored into the product of an operator involving only x and another involving only y.

d. Consider a solution of the form $\phi = e^{ik_1x}e^{ik_2y}e^{i\omega t}$. Find the exact (dispersion) relation among k_1, k_2, and ω, by substituting this solution into Eq. (8.92). Then find the analogous relations for the methods developed in the earlier parts of this problem. How does the approximate factorization method compare with the original method in terms of its treatment of waves of this type?

4. Repeat the preceding problem for the Crank–Nicolson method.

5. An equation frequently used as a model for fluid mechanics is Burgers' equation:

$$\frac{\partial u}{\partial t} + u\frac{\partial u}{\partial x} = \nu\frac{\partial^2 u}{\partial x^2}$$

a. Give an explicit method for solving this equation (there are many).

b. Given the initial condition

$$u(x,0) = \sin \pi x \qquad 0 < x < 1$$

and the boundary conditions

$$u(0,t) = u(1,t) = 0$$

solve the problem, plot the solution for various times, and explain the results.

APPENDIX A

LIST OF COMPUTER CODES

The codes mentioned in the text can be found at the Internet site http://ecoule.stanford.edu/ferziger under the listing for the chapter in which they are described. Here we give a list of the programs that are found at the site and a very brief description of what they do.

We again note these programs are intended to show how the methods can be implemented and are not the most efficient programs of their type. Many comments are included in the programs to help you interpret what is being done at each stage.

CHAPTER 1

BLKTRI: Solves block tridiagonal systems of algebraic equations by a variant of Gauss elimination.

GAUSS: Uses Gauss elimination to solve a system of algebraic equations; it is best applied to systems with full matrices.

LU: Solves a full system of algebraic equations using *LU* decomposition, a variant of Gauss elimination.

TRDIAG: Solves a tridiagonal system of equations using a simplified version of Gauss elimination.

CHAPTER 2

DIVDIFF: Finds the interpolated value of a tabulated function using divided differences.

SPLINE: Has two separate subroutines; the first finds the coefficients of a cubic spline; the second evaluates the spline.

TENSPL: Similar to SPLINE but for a tension spline; requires specification of the tension parameter.

CHAPTER 3

ADAPQUAD: Computes the integral of a function using adaptive quadrature.

MCQUAD: Computes the integral of a function using Monte Carlo integration.

ROMBRG: Computes the integral of a function by the Romberg method.

CHAPTER 4

EULER: Solves a system of ordinary differential equations by the explicit Euler method.

PRECOR: Solves a system of ordinary differential equations by the predictor–corrector method.

RUNKUT: Solves a system of ordinary differential equations by the fourth-order Runge–Kutta method.

CHAPTER 5

ADAPGRID: Solves a stiff problem by a direct method on an adaptive grid.

BLASIUS: Solves the Blasius boundary value problem by the shooting method.

NEUDIFF: Solves the diffusion equation by the direct method.

CHAPTER 6

CRANIC: Solves the heat equation using the Crank–Nicolson method.

HEATEUL: Solve the heat equation using the Euler explicit method.

CHAPTER 7

ADI: Solves Laplace's equation using the alternating direction implicit method with a fixed time step parameter β.

CGSIP: Solves Laplace's equation with the conjugate gradient method using the strongly implicit method as a preconditioner.

CYLADI: Solves Laplace's equation using the alternating direction implicit method with the time step parameter β varied in a cyclic manner.

ICCG: Solves Laplace's equation with the conjugate gradient method using the incomplete Cholesky method as a preconditioner.

JACOBI: Solves Laplace's equation using the Jacobi iterative method.

MG: Solves Laplace's equation using the multigrid method with the Gauss–Seidel method as the smoother.

RBSOR: Solves Laplace's equation using the red-black version of the successive overrelaxation iterative method.

SIP: Solves Laplace's equation using Stone's strongly implicit method.

SOR: Solves Laplace's equation using the successive overrelaxation iterative method, which includes the Gauss–Seidel method as a special case.

CHAPTER 8

HYPCN: Solves the wave equation using the Crank–Nicolson method.

LAX: Solves the wave equation using the Lax–Wendroff method.

MACCORM1: Solves the one-dimensional convection equation with a witch's hat initial profile using the MacCormack split method.

MOC: Solves the wave equation using the method of characteristics.

APPENDIX B

ANNOTATED BIBLIOGRAPHY

In this bibliography we have listed not only the works cited in the text but several significant textbooks on individual subject. Some provide a general introduction to the mathematics of a particular subject while others present material similar to that given in the text in greater depth. In a few cases, older books that are regarded as classics have been included.

General

Some general texbooks on numerical analysis are the following. Most are written for an audience that is more mathematically oriented than the typical reader of this book.

Atkinson, K. E., *Elementary Numerical Analysis*, Wiley, New York, 1993.

Buchanan, J. L. and Turner, P. R., *Numerical Methods and Analysis*, McGraw-Hill, New York, 1992.

Eldben, L. and Wittmeyer-Koch, L., *Numerical Analysis: An Introduction*, Academic Press, Boston, 1990.

Forsythe, G. E., Malcolm, M. A. and Moler, C. B., *Computer Methods for Mathematical Computations*, Prentice-Hall, Englewood Cliffs, NJ, 1977.

Gerald, C. F., *Applied Numerical Analysis*, Addison-Wesley, Reading, MA, 1994.

Isaacson, E. and Keller, H. B., *Analysis of Numerical Methods*, Wiley, New York, 1966.

Kincaid, D. R., *Numerical Analysis: Mathematics of Scientific Computing*, Brooks/Cole, Monterey, CA, 1996.

Press, W. H., *Numerical Recipes: The Art of Scientific Computing*, Cambridge Univ. Press, New York, 1988.

Schwarz, H. R., *Numerical Analysis: A Comprehensive Introduction*, Wiley, New York, 1989.

Smith, W. A., *Elementary Numerical Analysis*, Prentice-Hall, Englewood Cliffs, NJ, 1986.

Chapter 1

Linear algebra is a popular subject so a number of excellent textbooks on the subject are widely available. One that I have used as a text in a course is:

Strang, G., *Linear Algebra and Its Applications*, Harcourt Brace Jovanovich, San Diego, 1988.

Other introductory books on the subject are:

Anton, H., *Elementary Linear Algebra*, 7th ed., Wiley, New York, 1994.

Edwards, C. H., and Penney, D. E., *Elementary Linear Algebra*, Prentice-Hall, Englewood Cliffs, NJ, 1988.

Friedberg, S. H., Insel, A. J. and Spence, L. E., *Linear Algebra*, 3rd ed., Prentice-Hall, Englewood Cliffs, NJ, 1997.

Kolman, B., *Introductory Linear Algebra with Applications*, Macmillan, New York, 1988.

Finally some books devoted specifically to the numerical solution of algebraic problems are:

Golub, G. H. and van Loan, J., *Matrix Computation*, Johns Hopkins Univ. Press, Baltimore, 1986.

Jennings, A. and McKeown, J. J., *Matrix Computation*, 2nd ed., Wiley, New York, 1992.

Trefethen, L. N., and Bau, L., *Numerical Linear Algebra*, SIAM, Philadelphia, 1997.

Chapter 2

Lagrange and Hermite interpolation were developed a long time ago, so it is difficult to find a book devoted to them today. On the other hand, splines and computer graphics are important topics and books about them are not difficult to find. Below are some books on interpolation. Most are written for mathematicians and may be hard going for many of the readers of this book. See also the list of general books on numerical analysis given above.

Chui, C. K., Schumaker, L. L. and Utreras, F. I., *Topics in Multivariate Approximation*, Academic Press, Boston, 1987.

de Boor, C., *A Practical Guide to Splines*, Springer, Berlin, 1978.

Farin, G. R., *Methods for Computer Aided Graphic Design*, Academic Press, Boston, 1993.

Lancaster, P. and Salkauskas, K., *Curve and Surface Fitting: An Introduction*, Academic, Boston, 1986.

Schultz, M. H., *Spline Analysis*, Prentice-Hall, Englewood Cliffs, NJ, 1973.

Varga, R. S., *Topics in Polynomial and Rational Interpolation and Approximation*, Presses de l'Universite de Montreal, 1982.

Chapter 3

This is another subject that is rarely the entire focus of a book. A good source for quadrature formulas is the reference book:

Abramowitz, M. and Stegun, I. A., *Handbook of Mathematical Functions*, Dover, New York, 1970.

A text on the subject is:

Engels, H., *Numerical Quadrature and Cubature*, Academic Press, Boston, 1980.

Chapters 4 and 5

There are many textbooks on ordinary differential equations as it is the subject of a popular undergraduate course. There are also advanced books suitable for a second course on the subject and books devoted to numerical solution of ordinary differential equations. This is a sampling of some of what is available; the titles are, for the most part, self-explanatory.

Boyce, W. E. and DiPrima, R. C., *Elementary Differential Equations and Boundary Value Problems*, 6th ed., Wiley, New York, 1997.

Butcher, J. C., *The Numerical Analysis of Ordinary Differential Equations: Runge–Kutta and General Linear Methods*, Wiley, New York, 1987.

Caruso, S. C., Ferziger, J. H. and Oliger, J., *An Adaptive Grid Method for Incompressible Flows*, Rept. TF-23, Dept. Mech. Engr., Stanford, 1985.

Davis, H. T., *Introduction to Nonlinear Differential and Integral Equations*, Dover, NY, 1962.

Edwards, C. H. and Penney, D. E., *Elementary Differential Equations with Boundary Value Problems*, 3rd ed., Prentice-Hall, Englewood Cliffs, NJ, 1992.

Fatunla, S. O., *Numerical Methods for Initial Value Problems in Ordinary Differential Equations*, Academic, Boston, 1988.

Ferziger, J. H. and Perić, M., *Computational Methods for Fluid Mechanics*, Springer, Berlin, 1996.

Gear, C. W., *Numerical Initial Value Problems in Ordinary Differential Equations*, Prentice-Hall, Englewood Cliffs, NJ, 1971.

Gill, P. E., Murray, W. and Wright, M. H., *Practical Optimization*, Academic, New York, 1981.

Hairer, E., Norsett, S. P. and Wanner, G. W., *Solving Ordinary Differential Equations*, Springer, Berlin, 1987.

Hughes, T. J. R., *The Finite Element Method: Linear Static and Dynamic Finite Element Analysis*, Prentice-Hall, Englewood Cliffs, NJ, 1987.

Humi, M. and Miller, W., *Second Course in Ordinary Differential Equations for Scientists and Engineers*, Springer, Berlin, 1988.

Ince, E. L., *Integration of Ordinary Differential Equations*, 7th ed., Interscience, New York, 1956.

Iserles, A., *A First Course in the Numerical Analysis of Differential Equations*, Cambridge Univ. Press, New York, 1996.

Lambert, J. D., *Numerical Methods for Ordinary Differential Systems*, Wiley, New York, 1997.

Mattheij, R. M. M. and Molenaar, J., *Ordinary Differential Equations in Theory and Practice*, Wiley, New York, 1996.

Poinsot, T. J. and Lele, S. K., "Boundary conditions for direct simulations of compressible viscous flows," *J. Comp. Phys.*, **101**, 104–129, 1992.

Schlichting, H., *Boundary Layer Theory*, 7th ed. Translated by J. Kestin, McGraw-Hill, New York, 1987.

Shampine, L. F. and Gordon, M. K., *Computer Solution of Ordinary Differential Equations: The Initial Value Problem*, W. H. Freeman, 1975.

Van Dyke, M. D., *Perturbation Methods in Fluid Mechanics*, Parabolic Press, Stanford, CA, 1975.

Chapters 6, 7, and 8

Ames, William F., *Numerical Methods for Partial Differential Equations*, 3rd ed., Academic Press, Boston, 1992.

Berger, M. J. and Oliger, J., "Adaptive mesh refinement for hyperbolic partial differential equations," *J. Comp. Phys.*, **53**, 484, 1984.

Bleecker, D. and Csordas, G., *Basic Partial Differential Equations*, Van Nostrand Reinhold, New York, 1992.

Bracewell, R., *Fourier Methods*, McGraw-Hill, New York, 1967.

Briggs, W. L., *A Multigrid Tutorial*, Siam, Philadelphia, 1987.

Brigham, R., *The Fast Fourier Transform*, Prentice-Hall, Englewood Cliffs, NJ, 1976.

Carrier, G. F., *Partial Differential Equations*, Academic, Boston, 1988.

Colton, D. L., *Partial Differential Equations: An Introduction*, Random House, New York, 1988.

Courant, R. and Hilbert, D., *Methods of Mathematical Physics*, 2 vols., Interscience, New York, 1953.

DiBenedetto, E., *Partial Differential Equations*, Birkhauser, Boston, 1995.

Fletcher, C. A. J., *Computational Techniques for Fluid Dynamics*, Springer, Berlin, 1976.

Folland, G. B., *Introduction to Partial Differential Equations*, 2nd ed., Princeton Univ. Press, Princeton, NJ, 1995.

Garabedian, P., *Partial Differential Equations*, Wiley, New York, 1964.

Hackbusch, W., *Advances in Multigrid Methods*, Vieweg, Braunschweig, 1985.

Hall, C. A., *Numerical Analysis of Partial Differential Equations*, Prentice-Hall, Englewood Cliffs, NJ, 1990.

Hussaini, M. Y., Quateroni, A., Canuto, and Zang, T., *Spectral Methods for Flow Computation*, Springer, Berlin, 1989.

Johnson, C., *Numerical Solution of Partial Differential Equations by the Finite Element Method*, Cambridge Univ. Press, 1987.

Kevorkian, J., *Partial Differential Equations: Analytical Solution Techniques*, Brooks/Cole, Monterey, CA, 1990.

Marchuk, G. I., *Methods of Numerical Mathematics*, Springer, Berlin, 1982.

Morton, K. W. and Mayers, D. F., *Numerical Solution of Partial Differential Equations: An Introduction*, Cambridge Univ. Press, New York, 1994.

Muzaferija, S., *Adaptive Finite Volume Method for Flow Predictions Using Unstructured Meshes and Multigrid Approach*, dissertation, Univ. of London, 1994.

Pinsky, M. A., *Partial Differential Equations and Boundary-Value Problems with Applications*, McGraw-Hill, New York, 1991.

Quarteroni, A. and Valli, A., *Numerical Approximation of Partial Differential Equations*, Springer, Berlin, 1994.

Renardy, M., *An Introduction to Partial Differential Equations*, Springer, Berlin, 1993.

Rubinstein, I. and Rubinstein, L., *Partial Differential Equations in Classical Mathematical Physics*, Cambridge Univ. Press, New York, 1993.

Saad, Y. and Schultz, M. H., "GMRES: A generalized residual algorithm for solving nonsymmetric linear systems," *SIAM J. Sci. Stat. Comp.*, **7**, 856–869, 1986.

Sonneveld, P., "CGS: A fast Lanczos solver for nonsymmetric linear systems," *SIAM J. Sci. Stat. Comp.*, **10**, 36–52, 1989.

Stone, H. L., "Iterative solution of implicit approximations of multidimensional partial differential equations," *SIAM J. Num. Anal.*, **5**, 530–558, 1968.

Strauss, W. A., *Partial Differential Equations: An Introduction*, Wiley, New York, 1992.

Taylor, M. E., *Partial Differential Equations*, Springer, Berlin, 1996.

Thomas, J. W., *Numerical Partial Differential Equations: Finite Difference Methods*, Springer, Berlin, 1995.

Thompson, M. C. and Feziger, J. H., "A multigrid adaptive method for incompressible flows," *J. Comp. Phys.*, **82**, 94–121, 1989.

Van den Vorst, H. A., Bi-CGSTAB: A fast and smoothly converging variant of BI-CG for the solution of nonsymmetric linear systems," *SIAM J. Sci. Stat. Comp.*, **13**, 631–644, 1992.

Van den Vorst, H. A. and Sonneveld, P., *CGSTAB, A More Smoothly Converging Variant of CGS*, Rept. 90-50, Delft Univ. Tech., 1990.

Zauderer, E., *Partial Differential Equations of Mathematical Physics*, Wiley, New York, 1983.

Zhu, X., *Adaptive Grid Methods for Irregular Domains*, dissertation, Comp. Sci. Dept., Stanford Univ., 1996.

APPENDIX C

A NOTE ON THE NEWTON–RAPHSON METHOD

The Newton–Raphson method is one of the best ways of solving nonlinear algebraic equations. Suppose that we need the root of some generic algebraic equation:

$$f(x) = 0 \tag{C.1}$$

The procedure starts with an initial guess of the root; call it x_0. This can be a pure guess, but it is better if it is close to the solution. One good method is to find two points at which the function has opposite signs (there has to be a root between them) and choose an x_0 somewhere in the interval, for example, the midpoint. We then locally approximate the function by linearizing it about x_0, keeping two terms of the Taylor series:

$$f(x) = f(x_0) + (x - x_0)\frac{df}{dx}(x_0) \tag{C.2}$$

This is equivalent to approximating the curve $y = f(x)$ by its tangent at the point x_0.

The next approximation is obtained by setting the linear approximation (C.2) equal to zero. The result is

$$x_1 = x_0 - f(x_0)/f'(x_0) \tag{C.3}$$

Taking this as a new guess, we compute still another guess and then another, and so forth until convergence is achieved. Each new estimate is obtained from

$$x_k = x_{k-1} - f(x_{k-1})/f'(x_{k-1}) \tag{C.4}$$

The method converges very quickly if the initial guess is "close enough" to the root. One can show that the error ϵ (the difference between the present estimate x_k and the exact root ξ) decreases according to

$$\epsilon_k = \text{const. } \epsilon_{k-1}^2 \tag{C.5}$$

once the estimate is close enough to the root. Thus if $\epsilon_{k-1} \approx 10^{-3}$, then $\epsilon_k \approx 10^{-6}$.

If the initial guess is too far from the root, the method may converge to the wrong root or diverge.

Practice: Use this to find the square root of a number, a. (Solve $x^2 - a = 0$ using this method.) This is how computers do it.

The secant method is a variation that essentially approximates the derivative in Eq. (C.2) by a finite difference approximation. To do this, two previous guesses need to be available. Then, instead of the update equation (C.4), we get

$$x_k = x_{k-1} - f(x_{k-1}) \frac{x_{k-1} - x_{k-2}}{f(x_{k-1}) - f(x_{k-2})} \tag{C.6}$$

INDEX